Reviews of

105 Physiology, Biochemistry and Pharmacology

Editors

P. F. Baker, London · H. Grunicke, Innsbruck
E. Habermann, Gießen · R. Jung, Freiburg
R. J. Linden, Leeds · P. A. Miescher, Genève
H. Neurath, Seattle · S. Numa, Kyoto
D. Pette, Konstanz · B. Sakmann, Göttingen
W. Singer, Frankfurt/M · U. Trendelenburg, Würzburg
K. J. Ullrich, Frankfurt/M

With 74 Figures, Some in Colour

Springer-Verlag
Berlin Heidelberg GmbH

ISBN 978-3-662-31046-5   ISBN 978-3-540-47152-3 (eBook)
DOI 10.1007/978-3-540-47152-3

Library of Congress-Catalog-Card Number 74-3674

© by Springer-Verlag Berlin Heidelberg 1986
Originally published by Springer-Verlag Berlin Heidelberg New York in 1986
Softcover reprint of the hardcover 1st edition 1986

2127/3130-543210

Contents

Reflex Control of the Human Cardiovascular
   System
   By J. T. SHEPHERD, Rochester, Minnesota/USA,
   and G. MANCIA, Milan/Italy
   With 19 Figures   1

Vascular Capacitance: Its Control and Importance
   R. HAINSWORTH, Leeds/Great Britain
   With 13 Figures 101

Electrical Breakdown, Electropermeabilization
   and Electrofusion
   U. ZIMMERMANN, Würzburg/Federal Republic
   of Germany
   With 42 Figures, Some in Colour 175

Subject Index 257

Indexed in Current Contents

Rev. Physiol. Biochem. Pharmacol., Vol. 105
© by Springer-Verlag 1986

Reflex Control of the Human Cardiovascular System

JOHN T. SHEPHERD[1], and GIUSEPPE MANCIA[2]

Contents

1 Measurement of Cardiovascular Events 3
   1.1   Cardiac Output 3
   1.2   Limb Blood Flow 3
   1.3   Efferent Sympathetic and Vagal Activity 5
   1.4   Dynamic Regulation of Adrenoceptors 6
   1.5   Vascular Capacitance 6
   1.6   Limb Volume–Venous Pressure Relationships 7
   1.7   Venous Pressure at Constant Blood Volume 7
   1.8   Vein Diameter 7
   1.9   Circulating Vasoactive Substances 7

2 Regulation of Sympathetic Outflow 10

3 Methods of Studying Human Cardiovascular Reflexes 13
   3.1   Arterial Baroreceptors 13
   3.1.1 Application of Negative and Positive Pressure to the Neck 13
   3.1.2 Occlusion of Common Carotid Arteries and Stimulation of Carotid
           Sinus Nerves 15
   3.1.3 Pressor and Depressor Drugs 15
   3.2   Cardiopulmonary Mechanoreceptors 16
   3.2.1 Negative Pressure to the Lower Body 16
   3.2.2 Congesting Cuffs 17
   3.2.3 Tilting 17
   3.2.4 Elevation of Legs or Legs and Lower Trunk 17

4 Carotid Baroreflexes 17
   4.1   Heart Rate 18
   4.2   Respiratory Sinus Arrhythmia 19
   4.3   Standing 20
   4.4   Atrioventricular Conduction 21
   4.5   Arterial Blood Pressure 21
   4.6   Vascular Beds 23
   4.6.1 Skeletal Muscles 23
   4.6.2 Skin 27
   4.6.3 Splanchnic Region 28
   4.6.4 Kidney 28
   4.6.5 Venous System 28

[1] Mayo Clinic and Foundation, Rochester, Minnesota, USA
[2] Centro di Fisiologia Clinica e Ipertensione, University of Milan, Milan, Italy

 5   Aortic Baroreflexes   29
 6   Sleep   29
 7   Age   29
 8   Exercise and Arterial Baroreflexes   30
 9   Chemoreflexes   34
10   Diving Reflex   34
11   Cardiopulmonary Reflexes   35
12   Axon Reflex   37
13   Renin   37
14   Antidiuretic Hormone, Plasma and Extracellular Fluid Volume   39
15   Salt and Neural Cardiovascular Control   42
16   Static Exercise and Ergoreceptors   43
17   Thermoregulation   47
18   Rhythmic Exercise   49
19   Emotion   51
20   Fainting   54

21   Reflex Control in Cardiovascular Disease   56
     21.1   Autonomic Failure   56
     21.2   Cardiac Failure   58
     21.3   Aortic Stenosis   60
     21.4   Myocardial Ischemia   61
     21.5   Sudden Death from Coronary Artery Disease   64
     21.6   Hypertension   65
     21.7   Autonomic Hyperreflexia   72
     21.8   Carotid Sinus Syndrome   73

References   74

The surge of knowledge about the autonomic regulation of the human cardiovascular system which commenced in the 1940s was a consequence of many events. These included the information gained from animal studies of the various reflexogenic zones which modulate the circulation, the demonstration of the feasibility of cardiac catheterization in man; the urgent need in the Second World War for information on the effects of gravitational forces on the human cardiovascular system which were encountered in military aircraft; the development of instrumentation and techniques applicable to human studies; and the growing number of clinician-scientists able to divise and conduct appropriate studies and of informed volunteers willing to act as subjects for the experiments. To these should be added the demonstration in 1948, in the splenic nerves of the ox, that norepinephrine was the sympathetic neurotransmitter (von Euler 1948) and in the same year the development of the concept of two types of adrenoceptors, alpha and beta (Ahlquist 1948).

1 Measurement of Cardiovascular Events

1.1 Cardiac Output

The development of the cardiac catheter permitted the application of the direct Fick principle for the determination of cardiac output (Snellen et al. 1981) and the strain gauge pressure transducer made possible accurate measurements of intracardiac and intravascular pressures (Lambert and Wood 1947). The later use of the indicator-dilution method (Fox and Wood 1960; Lassen et al. 1983), particularly after the development of indocyanine green dye as the indicator (Fox et al. 1957), facilitated the repeated determinations of cardiac output in individual studies (Wood 1978). Later, the use of thermodilution provided an alternative method (Swan and Ganz 1983). In recent years, the use of two-dimensional Doppler echocardiographic scanners with quantitative spectral analysis of the Doppler signal has provided a noninvasive means of estimating the cardiac output and beat-to-beat changes in stroke volume (Valdes-Cruz et al. 1984).

The Fick principle also had wide application in the calculation of the blood flow to the organs and tissues of the body, for example, the use of inert gases to measure brain (Kety 1960) and coronary blood flow (Bing et al. 1949), of indocyanine green to measure splanchnic or hepatic blood flow (Bradley et al. 1945; Rowell 1974), and of the organic acid para-aminohippurate and of inulin, a beta-glycosidic polymer of fructose, to measure renal plasma flow and glomerular filtration rate respectively (Pitts 1968).

Later, it was possible to measure regional myocardial blood flow by the localized uptake of a diffusible indicator or by determination of the rate of its washout (Dwyer et al. 1973).

1.2 Limb Blood Flow

By the 1940s venous occlusion plethysmography was well established as a reliable method for the measurement of blood flow in human limbs (Greenfield et al. 1963). The relative ease of making these measurements, both for the subject and investigator, as compared to the determination of the blood flow to other parts of the body, and the fact that the plethysmographic method permits moment-to-moment changes in flow to be followed, paved the way for the understanding of the cardiovascular reflexes in man. Using this method, Barcroft et al. (1943) were the first to demonstrate the significance of the sympathetic innervation of the resistance blood vessels in human muscles. They found that the blood flow through the forearm was about doubled after all three deep nerves to the forearm had been blocked with a local anesthetic. Since this paralyzed the forearm muscles, it was possible that the increased flow was due to a decrease in mechanical hindrance. However, similar blocks in sympathectomized limbs did not increase the flow. Thus the increased flow was due to active dilatation of the resistance blood vessels as a result of interruption of sympathetic vasoconstrictor impulses. When the skin circulation to the forearm was suppressed by iontophoresis of epinephrine, the flow still increased after deep nerve block, indicating that the increased flow was in vessels beneath the skin, presumably in muscle. Barcroft and his colleagues (1943) concluded that the muscle vessels in man were supplied by sympathetic vasoconstrictor fibers. This was confirmed later by the finding of increased oxygen saturation of venous blood draining muscle immediately following the nerve block (Roddie et al. 1957).

These studies in 1943, followed a year later by the observation of an increase in forearm blood flow during fainting, attributed to activation of sympathetic vasodilator fibers (Barcroft et al. 1944), set the stage for studies to assess the importance of reflex changes in muscle blood flow in the regulation of the human cardiovascular system in normal and abnormal circumstances.

The muscles of the human body constitute about 45% of the body mass. The blood flow to the resting forearm and calf muscles is approximately 3–5 ml/100 gm/min and with severe exercise it may be more than 10–15 times this amount, due to the action of local vasodilator metabolites. With maximal dynamic exercise of the knee extensors, blood flows of 250 ml/100 gm/min have been reported (Andersen and Saltin 1985). By contrast, blockade of the sympathetic nerves to the blood vessels in the resting limb muscles results in a two- to threefold increase in blood flow; when the sympathetic nerves are activated maximally the resting blood flow is decreased to about 75% or more. Thus, as compared to the local mechanisms regulating muscle blood flow, the sympathetic nerves control a relatively small portion of the flow capacity of the muscles. However, because of the large proportion of the body mass which is muscle, reflex changes in vascular resistance in the muscles can make contributions equivalent to that of the splanchnic and renal circulation to the reflex control of total systemic vascular resistance and thereby of the arterial blood pressure (Shepherd 1983).

Skin blood vessels, which are also richly innervated by sympathetic nerves, have little role in the regulation of arterial blood pressure. In addition to its role as a waterproof and protective layer, the skin has a vital role in maintaining a normal body temperature. Thus, like the kidney, its blood flow usually far exceeds its small nutritional needs. There are arteriovenous anastomoses in certain parts of the skin which are characterized by a thick muscular wall and a rich sympathetic innervation. They are most numerous in the nail bed, but are also present in the palmar surface of the hand and feet, and in the ears. They have not been identified in the skin of the forearm or calf. These anatomoses provide a direct connection between the arterioles and the venules in the dermis. Thus, when they are open, much of the blood can bypass the high-resistance arterioles and capillaries of the papillary plexus and be shunted directly to the capacious subpapillary venous plexus (Roddie 1983).

1.3 Efferent Sympathetic and Vagal Activity

The recording of efferent sympathetic activity, using tungsten microelectrodes, from human skin and muscle nerve fascicles of awake subjects has added new important information to our knowledge of the reflex control of the circulation (Vallbo et al. 1979). Usually, multiunit recordings are obtained, but occasionally single units have been used. Evidence that it is sympathetic efferent activity that is being recorded is afforded by the fact that injection of local anesthetic around the nerve proximal to the recording site eliminates the activity, whereas injections made distally are without effect, and by the facts that ganglionic-blocking agents eliminate the activity and that the conduction velocity is about 1 m/s. In addition, changes in nerve activity are followed within a few seconds by the expected changes in vascular resistance, in electrical resistance of skin as evidence of sweating, and in arterial blood pressure. In skin fascicles, the resting activity consists of irregular burst of impulses, unrelated to arterial pulse pressure, whereas in muscle fascicles, it is characterized by fairly regular pulse-synchronous bursts, with intervening periods of silence. The sympathetic activity in skin nerves is a mixture of vasoconstrictor and sudomotor impulses, whereas that in muscle nerves is due to vasoconstrictor traffic. These studies have again emphasized the selectivity of the sympathetic outflow to different effector organs (Abboud et al. 1975; Shepherd 1982). They have also demonstrated the similarity of the outflow to different muscles. This suggests that there is a homogeneous regulation of the postganglionic sympathetic neurons to skeletal muscle vessels, presumably originating in supraspinal neurons (Delius et al. 1972a−c; Hagbarth and Vallbo 1968; Hagbarth et al. 1972; Janig et al. 1983; Sundlöf and Wallin 1977).

It has not been possible to obtain direct recordings of vagal efferent activity in man. The degree of vagal tone to the sinus node has been estimated indirectly by the increase in heart rate following atropinization.

Another approach has been to use the magnitude of sinus arrhythmia as an estimate of the degree of the tonic vagal influence on the heart (Eckberg 1983; Katona and Felix 1975).

1.4 Dynamic Regulation of Adrenoceptors

The finding that beta-adrenoceptors are present on human leukocytes and alpha-adrenoceptors on human platelets permits the study of adrenoceptor regulation in man. In subjects with normal or elevated arterial blood pressure, aging was found to be associated with a reduced beta-adrenoceptor density (Bertel et al. 1980; Dillon et al. 1980; Fitzgerald et al. 1984) and no evident change in alpha-adrenoceptor density (Elliott et al. 1982). Beta-adrenoceptor density is reduced in patients with pheochromocytoma or otherwise high circulating catecholamines and is increased in subjects with low circulating or urinary catecholamines (Greenacre and Connolly 1978; Fraser et al. 1981a,b). A reduction in human lymphocyte beta-adrenoceptors and platelet alpha-adrenoceptors occurs after prolonged exposure to epinephrine (Cooper et al. 1978). These studies raise the possibility of down- or up-regulation of the numbers of adrenoceptors as a consequence of a prolonged increase or decrease, respectively, in sympathetic nerve activity (Nahirski and Bennett 1982; Tohmen and Cryer 1979). In studies on healthy volunteers at the 30th min of treadmill exercise sufficient to increase their heart rate to 75% of predicted maximum, the beta-adrenoceptor density fell during the exercise and was only 51% of the control value 15 min after cessation of the exercise. On a change from the resting supine to the upright posture, the beta-adrenoceptor density declined to 65% of control after 2 h. When the subject lay supine over this period, receptor density did not change. Thus it seems that changes in sympathetic tone even for minutes rather than weeks may result in alterations in adrenoceptor density (Fitzgerald et al. 1981).

1.5 Vascular Capacitance

Observations on the behavior of the capacitance vessels in the human have been restricted to the limbs (Shepherd and Vanhoutte 1975). Although splanchnic blood volume can be estimated by indicator-dilution techniques, it is not possible to determine whether changes in this volume are induced actively or passively (Rowell 1974). Sections 1.6–1.9 describe the methods used to study the limb veins.

1.6 Limb Volume-Venous Pressure Relationships

Measurements are made at equilibrium of the stepwise changes in volume of the forearm, hand, calf, or foot that accompany changes in venous pressure. A plethysmograph is used to determine the changes in blood volume. Venous pressure is changed by inflation of a pneumatic cuff proximal to the site of measurement. Since paralysis of the limb muscles does not alter these pressure-volume relationships and since capillary filtration makes a minimum contribution over the short-term period of the measurements, shifts in these pressure-volume curves indicate alterations in distensibility of the capacitance vessels due to local, humoral, or neural stimuli.

1.7 Venous Pressure at Constant Blood Volume

The blood flow to a segment of the limb is temporarily arrested by means of a pneumatic cuff inflated to suprasystolic pressure and the pressure within one of the large veins in the limb is measured. Since the limb contains a constant volume of blood, the changes in venous pressure are proportional to changes in tension of the smooth muscle of the vein wall. This is a convenient way of studying rapid changes in venomotor tone mediated by the sympathetic nerves (Samueloff et al. 1966).

1.8 Vein Diameter

Changes in diameter of cutaneous vein at a given venous distending pressure are estimated by a microscope (Nachev et al. 1971).

1.9 Circulating Vasoactive Substances

Radioimmunoassay and radioenzymatic assay have been used to measure the plasma concentrations of dopamine-beta-hydroxylase, epinephrine, norepinephrine, renin, and vasopressin. Plasma levels of norepinephrine are frequently used as an index of sympathetic activity. Such levels are governed by the balance between release of norepinephrine from sympathetic nerve endings, reuptake into the endings, and catabolism of the amine. Only a small fraction of neuronally released norepinephrine reaches the circulation and substantial differences exist in the rate of norepinephrine clearance between individuals under basal conditions (Fitzgerald 1984). Thus local variations in sympathetic activity may occur without altering the plasma norepinephrine concentration measured in peripheral plasma (Brown et al. 1981; Esler et al. 1982). When pressor and depressor responses were obtained in subjects by applying positive and negative pressures to the neck, plasma norepinephrine and epinephrine levels did not reflect these baroreceptor-induced alterations in sympathetic activity (Mancia et al. 1983a). However, in normal subjects in whom nitroprusside hypotension and phenylephrine hypertension were induced, the plasma norepinephrine levels increased and decreased respectively. The amount of change in plasma norepinephrine level associated with a drop in blood pressure was greater than that with a rise in blood pressure. This is consistent with known changes in sympathetic activity (Grossman et al. 1982).

Arteriovenous differences in epinephrine and norepinephrine concentration were measured across several organs in nine resting patients with mild essential hypertension and five with renal artery stenosis. The estimated percentage contribution of various organs to the norepinephrine concentrations in the venous effluent was: heart 21, kidney 47, and legs 68. The arteriovenous difference across the adrenal glands was more than tenfold greater for epinephrine than for norepinephrine. The usual adrenal contribution to circulating norepinephrine in resting subjects is likely to be less than 2% (Brown et al. 1981). The splanchnic bed contributes little to the norepinephrine levels, since the liver clears catecholamines from the blood. Splanchnic fractional uptake of epinephrine in normal subjects was 90% ± 3%, and for norepinephrine 68% ± 4% (Keller et al. 1984). Removal of norepinephrine has also been demonstrated in human lungs (Gillis et al. 1972). Since the plasma norepinephrine levels depend in part on the rate at which norepinephrine is removed from the circulation, the measurement of the spillover rate of norepinephrine to plasma, by use of a small amount of radiolabeled norepinephrine, is a better index of sympathetic nerve activity. The latter measurement, however, does not identify the sources of the released norepinephrine (Esler 1982). Wallin et al. (1981) have compared, in healthy resting recumbent subjects aged 22–58 years, the plasma norepinephrine concentration with the level of multiunit sympathetic activity in muscle branches of the peroneal nerve. The results showed that both the norepinephrine levels and sympathetic activity vary widely between subjects, and both increased with age. There was a significant positive correlation between a given subject's level of sympathetic activity and plasma concentration of norepinephrine. The authors suggest that overflow of transmitter from the sympathetic terminals in skeletal muscles makes an important contribution to the plasma level of norepinephrine at rest. This is supported by the studies of Grassi et al. (1985a) in subjects with mild hypertension. They found that there was a significant positive relationship between the vasoconstriction and vasodilation in the

Fig. 1. Correlation between forearm vasoconstriction and changes in plasma norepinephrine. Relationship between the increase and reduction in forearm vascular resistance (*FVR*) induced by lower-body suction and passive leg elevation respectively and corresponding changes in plasma norepinephrine (*NA*) and epinephrine (*A*) in nine normal subjects. The changes that occurred at minutes 5, 10, and 20 of each maneuver are shown by the *triangles* and *white* and *black circles* respectively. Changes are expressed as percentages of basal values. (From Grassi et al. 1985a)

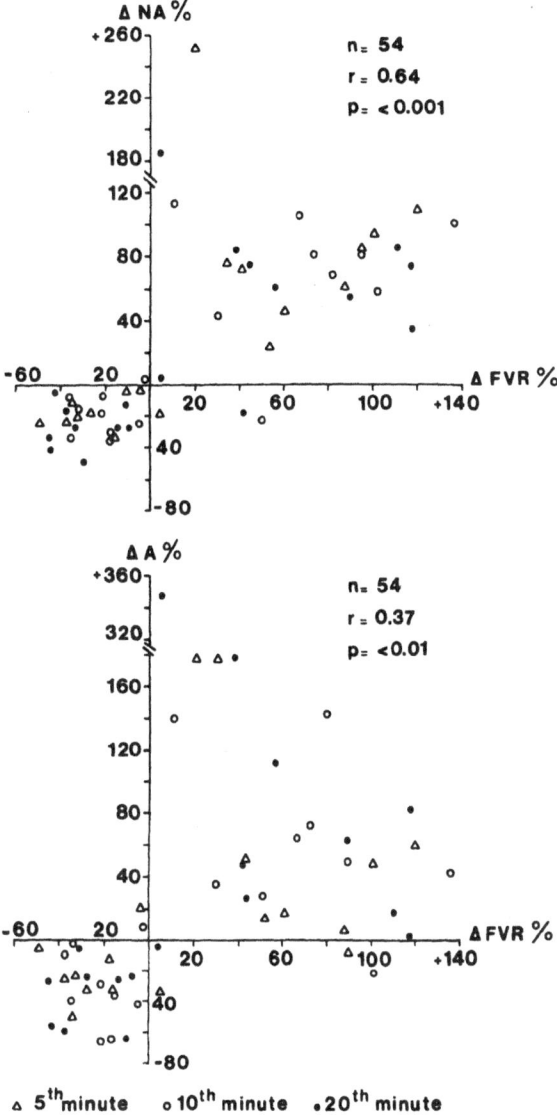

forearm induced by deactivation and stimulation, respectively, of the cardiopulmonary receptors and the simultaneously measured changes in plasma norepinephrine (Fig. 1). There is, however, no relationship between the mean levels of sympathetic nerve activity and mean levels of various blood pressure parameters in either normotensive or hypertensive subjects (Wallin and Sundlöf 1979).

It is important to recognize that the sympathetic nerve fiber discharge occurs in a highly differentiated pattern. Thus in response to reflex and central stimuli the efferent sympathetic activity varies between the different organs and tissues, and in the same organ or tissue can vary between resistance and capacitance vessels. In some instances, it may increase in some organs and decrease in others (Shepherd 1982). Hence the plasma level of norepinephrine gives no indication of the relative changes in sympathetic outflow to the different components of the cardiovascular system. To study this requires determination of norepinephrine outflow from the target organs (Folkow et al. 1983; Hjemdahl et al. 1984).

It is also important to recognize that the activity of the sympathetic nerves is not the sole determinant of the amount of norepinephrine released. This can be modified by local metabolic changes at the neuroeffector junction and by the presence of numerous receptors on the sympathetic nerve terminals. During exercise, the local metabolic changes in the active muscles, in addition to their direct relaxing action on the vascular smooth muscle of the resistance vessels, can simultaneously depress the output of norepinephrine from their nerves. Thus an increase in hydrogen or potassium ions, or in osmolality, can inhibit the exocytotic release of norepinephrine, possibly by depressing the entry of calcium ions into the nerve endings. The receptors on the sympathetic nerve endings include those that, when activated, reduce the output of norepinephrine and those that increase it. The former include alpha$_2$-adrenoceptors, muscarinic, H$_2$-histaminergic and serotonin (5-hydroxytryptamine) receptors, and P$_1$-purinergic receptors for adenosine; the latter include beta$_2$-adrenoceptors and angiotensin II receptors (Shepherd and Vanhoutte 1981; Vanhoutte et al. 1981; Verbeuren et al. 1983; Fitzgerald 1984).

In normal man, there is often a dissociation between the effects of various stimuli on the neuronal and adrenomedullary release of catecholamines. The most potent stimuli of norepinephrine release, in descencing order, were treadmill exercise, orthostasis, caffeine, the cold pressor test, sodium restriction, and handgrip exercise. By contrast, the order for plasma epinephrine was caffeine, treadmill exercise, the cold pressor test, handgrip exercise, and the Valsalva maneuver. Syncope caused a marked increase in plasma epinephrine, but only a minor increase in norepinephrine (Robertson et al. 1979).

2 Regulation of Sympathetic Outflow

The studies of Claude Bernard and Carl Ludwig between 1851 and 1866 demonstrated the importance of the sympathetic nervous system in the regulation of the circulation (Neil 1983). This, however, may only be apparent under stressful conditions. Patients with progressive autonomic failure cannot maintain their blood pressure when standing or even during mild exercise while supine (Marshall et al. 1961). If the heart is denervated but the sympathetic nerves to the systemic vessels are intact (e.g., following heart transplantation), the myocardium becomes supersensitive to circulating norepinephrine entering the bloodstream from peripheral sympathetic nerve endings. This permits the heart rate and contractility to increase appropriately to meet the demands of exercise (Donald and Shepherd 1963; Donald et al. 1968). The sympathetic outflow to the heart and the blood vessels is controlled by central nervous structures, whose activity is continuously modulated by information received from an array of peripheral sensors. The major sensory receptor groups are the arterial baroreceptors and chemoreceptors, mechano- and chemosensitive endings in the heart and lungs, and ergoreceptors in the skeletal muscles. The projections from these sensors relay at various sites, so that the pattern of the sympathetic outflow depends on complex interactions at many different levels in the central nervous system. Animal studies have shown that the baroreceptor afferents which travel in the glossopharyngeal nerves relay in the cell bodies of the petrosal and nodose ganglia respectively, and terminate in the nucleus of the solitary tract of the medulla. The nucleus of the solitary tract also receives input from the chemoreceptors and receptors in the heart and lungs subserved by vagal afferents and from receptors in the skeletal muscles. The efferent fibers of the nucleus of the solitary tract project to the cardiac vagal nuclei of the medulla (the ambiguus nucleus and the dorsal motor nucleus), to other nuclei in the reticular formation of the brain stem, including the paramedian reticular nucleus, to the sympathetic preganglionic nuclei in the interomediolateral horns of the spinal cord, and to the hypothalamus and the amygdaloid nucleus. The paramedian reticular nucleus, via connections from the fastigial nucleus of the cerebellum, has, as a consequence of input from the vestibular apparatus, an important role in the increase in sympathetic outflow that serves to maintain the arterial blood pressure when the upright position is assumed (Doba and Reis 1974). Afferents from the trigeminal nerve also pass to the brain stem and when activated in animals can cause bradycardia and hypotension.

Changes in the traffic in efferent fibers from the nucleus of the solitary tract can excite or inhibit the vagal cholinergic efferent pathway to the heart and can activate the excitatory or inhibitory bulbospinal pathways, which terminate close to the preganglionic sympathetic neurons in the intermediolateral horns of the thoracic and lumbar segments of the spinal cord. In the rat, there are neurons in the rostral portion of the ventrolateral medulla (nucleus reticularis rostroventrolateralis) which synthesize epinephrine and exert an excitatory influence on sympathetic vasomotor fibers, the adrenal medulla, and the posterior pituitary. These neurons are tonically active and thus maintain resting arterial tone; since they receive a direct projection from the nucleus of the solitary tract, they may also be an integral part of the baroreceptor reflex arc (Ross et al. 1984).

The intermediolateral horn cells are the key site of origin of the sympathetic outflow. Norepinephrine, epinephrine, 5-hydroxytryptamine, and substance P cell groups are present in the ventrolateral medulla, but the role of these different cell groups in regulating the intermediolateral horn cells is unknown. Iontophoretic application of norepinephrine and epinephrine onto preganglionic sympathetic neurons inhibits, while 5-hydroxytryptamine and substance P excite them. This suggests that 5-hydroxytryptamine and substance P are more likely to be excitatory neurotransmitters of the sympathetic preganglionic neurons than norepinephrine or epinephrine. Studies on the spontaneously hypertensive rat have suggested that substance P or a closely related peptide may be responsible for regulating the sympathetic outflow involved with tonic maintenance of vasomotor tone (Palkovits and Zaborszky 1977; Ricardo and Koh 1978; Kidd 1979; McAllen et al. 1979; Palkovits 1980, 1981; Kalia et al. 1981; Spyer 1981; Donoghue et al. 1984; Antonaccio 1984; Cechetto and Calaresu 1984; Takano et al. 1985).

The arterial baroreceptor afferents which control cardiac vagal efferents diverge in the nucleus of the solitary tract from those afferent fibers which influence sympathetic neurons. This permits selective baroreflex control of cardiac vagal and sympathetic motor neurons (Pórazász et al. 1962).

There are also projections from the higher brain centers to the nucleus of the solitary tract. This array of connections makes this nucleus a key integrating center for the reflex control of the cardiovascular system. In addition, there are the centers in the hypothalamus concerned with temperature regulation and the cardiovascular responses to emotion, and "command" centers in the cerebral cortex which modulate the sympathetic outflow during certain types of exercise. It is not surprising, in view of the complexity of the central controlling centers and their connections, that the normal response to the customary daily stresses is a differential change in sympathetic activity to the heart and the various vascular beds (Heistad et al. 1973; Abboud and Thames 1983).

The continuous registration of arterial blood pressure has demonstrated that in normal individuals this shows little fluctuation when the subject is resting quietly. However, the pressure varies widely over a 24-h period as normal activities are pursued, due to behavioral factors and to moment-to-moment changes of a respiratory and nonrespiratory nature (Miller-Craig et al. 1978; Mancia et al. 1983b); these originate in centers in the brain and cause changes in the autonomic outflow to the heart and blood vessels (Fig. 2). The continuous adjustments in autonomic activity, both

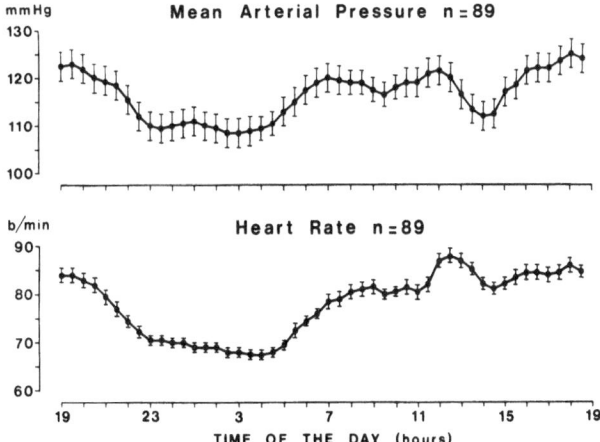

Fig. 2. The variability of arterial blood pressure and heart rate. The *points* represent the mean arterial pressure and heart rate values separately obtained for the 48 half hours of a 24-h recording period. Blood pressure was recorded by a catheter placed in a radial artery and heart rate by calculation of pulse interval valves from the blood pressure tracing. Data were analyzed beat-to-beat by a computer. Means ± SE from 89 ambulatory subjects with normal blood pressure or essential hypertension varying from mild to severe. (From Mancia et al. 1983b)

directly from brain centers and via the cardiovascular reflexes, provide the body with the perfusion pressure necessary to meet, in conjunction with the local regulation of the resistance blood vessels, the varying metabolic requirements of each of the organs and tissues of the body.

3 Methods of Studying Human Cardiovascular Reflexes

In animals it is possible to isolate specific receptor groups and to abrogate or hold constant other reflex inputs to the central nervous system. This, however, is not possible in man. Thus each technique that has been used has to be assessed critically in order to recognize the limitations of the conclusions that may be drawn from the studies (Goldstein et al. 1982; Mancia and Mark 1983; Mark and Mancia 1983). These techniques will be outlined briefly.

3.1 Arterial Baroreceptors

3.1.1 Application of Negative and Positive Pressure to the Neck

The commonly used approach to the study of the carotid baroreflex, introduced by Ernsting and Parry (1957), is to apply negative and positive pressures to the neck to cause increases and decreases respectively in carotid sinus transmural pressure, and hence resultant changes in the diameter of the carotid arteries. This alters the degree of activation of the carotid sinus baroreceptors. It may be accomplished either by enclosing the neck in a chamber or by applying a cup over the carotid sinus region which encompasses the anterior and lateral aspects of the neck (Bevegård and Shepherd 1966b; Eckberg et al. 1975; Ludbrook et al. 1977). The technique permits the establishment of the relationship between changes in carotid transmural pressure, arterial blood pressure, heart rate, and resistance in different systemic vascular beds.

All of the external pressure changes are not transmitted to the tissues adjacent to the walls of the carotid sinuses. It is, of course, the pressure inside the vessel minus the tissue pressure immediately outside it that governs the amount of stretch to which the mechanoreceptors are subjected. When positive and negative pressures are applied to the neck only 86% of the positive pressure and 64% of the negative pressure are transmitted. Fortunately, the pressure losses are similar among subjects regardless of the shape and size of the neck. Thus appropriate corrections can be applied to determine the shape of the curve relating receptor activity to hemodynamic response (Ludbrook et al. 1976, 1977). The application of positive pressure does not cause cerebral ischemia (Ludbrook et al. 1977) or trigger responses via the carotid chemoreceptors (Eckberg et al. 1975; Ludbrook et al. 1977).

The main limitation of the method is that the alterations of systemic arterial blood pressure induced by changes in the activity of the carotid sinus baroreceptors will affect in an opposite manner the aortic baroreceptors and possibly receptors in the left ventricle. Thus the blood pressure responses obtained may reflect a combination of opposing baroreceptor influences. The importance of the aortic baroreceptors in the control of arterial blood pressure in man is still unanswered. This limitation does not apply to studies of the vagally mediated response in heart rate. The much shorter latency than the sympathetically mediated responses permits the changes in heart rate to be evident prior to any significant change in arterial blood pressure; thus the aortic baroreflex is unlikely to influence the sinus node response during the first few seconds of changes in neck pressure.

The possibility that an emotional stimulus, a change in the activity of airway receptors (Trzebski et al. 1980a,b), or a reduction in venous return and hence in the activity of the low-pressure cardiopulmonary mechanoreceptors may accompany the changes in neck pressure must also be considered.

3.1.2 Occlusion of Common Carotid Arteries
and Stimulation of Carotid Sinus Nerves

Other methods that have been used infrequently include manual occlusion of the common carotid arteries (Roddie and Shepherd 1957) and electrical stimulation of the carotid sinus nerves (Carlsten et al. 1958; Neufeld et al. 1965; Epstein et al. 1969; Farrehi 1972). With the former approach, the degree of the pressure decrease in the carotid sinus is unknown and stimulus-response curves cannot be determined; with the latter, all the fibers in the carotid sinus nerve are activated simultaneously, contrary to the case in their normal recruitment, and chemo- as well as baroreceptor afferents may be stimulated.

3.1.3 Pressor and Depressor Drugs

Another approach is to use drugs to alter arterial blood pressure and to examine the resultant neurally mediated changes in heart rate. Usually, phenylephrine is used to raise the pressure and amyl nitrite, nitroglycerine, or nitroprusside to lower it. The relationship is plotted between progressive changes in systolic blood pressure and the R-R interval during the infusion. A linear relationship is found whose slope is taken as an index of the sensitivity of the arterial baroreflex control of heart rate (Smyth et al. 1969). The interpretation is complicated by the exponential relationship between R-R interval and heart rate. The alterations in R-R interval required to produce a given change in heart rate increase with increasing baseline R-R interval values. When the initial heart rate is altered, heart rate changes in response to a given stimulus to the baroreceptors may be identical. When the same experimental data are plotted using R-R intervals instead of heart rate, an inverse relationship is seen (Fig. 3). Thus two different interpretations may be made as to whether or not there is a change in sensitivity of the baroreflex heart rate response, depending on whether R-R interval or its reciprocal heart rate is used in plotting the data.

The benefit of the technique is that the same changes in pressure are sensed by the carotid and aortic baroreceptors and by other receptors in the arterial side of the circulation. Also the subjects are usually unaware of the stimulus. While the drugs used might directly alter the tension exerted by the smooth muscle cells in the wall of the carotid sinus and thus modify the stimulus to the baroreceptors (Peveler et al. 1983), these cells are scarce in man and do not appear to induce baroreceptor activation when they are exposed to topical application of epinephrine or stimulation of the cervical sympathetic trunk. The possibility does exist that the activity of other receptors may be modulated. For example, changes

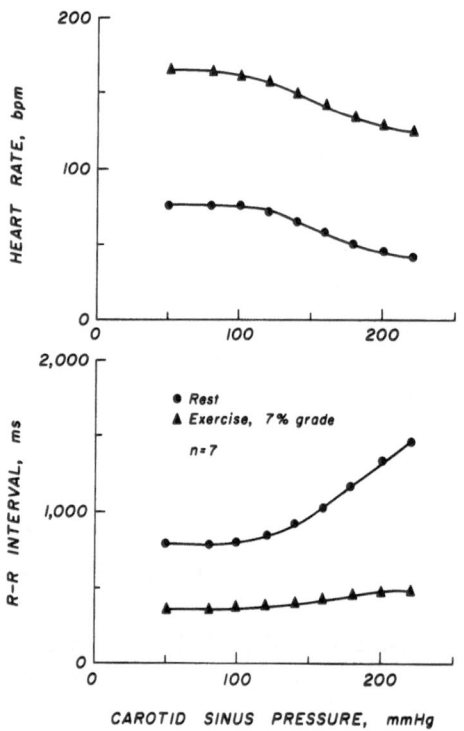

CAROTID SINUS PRESSURE, mmHg

Fig. 3. Use of heart rate versus R-R interval to determine whether or not the carotid baroreflex is reset during exercise. Dog with carotid sinuses vascularly isolated and aortic arch denervated. Heart rate and R-R interval response curves to changes in static pressure in the isolated sinuses, at rest and during treadmill exercise. The heart rate plot suggests that during exercise the setpoint of the carotid baroreflex is elevated, but the shape of the curve is unchanged. The R-R interval plot of the same data implies that the gain of the curve is much reduced. The maximal slope at rest was 6.5 ms/mmHg, compared with 1.3 ms/mmHg during exercise. Analysis of the data in this manner would have led to the conclusion that carotid baroreflex sensitivity was depressed during exercise. (Data from Melcher and Donald 1981)

in central venous pressure caused by the nitrates may affect mechanoreceptors in the atrial wall.

While such studies permit the relationship between changes in baroreceptor activity and heart rate to be examined, it is not possible to deduce from them the relationship between baroreceptor activity and arterial blood pressure (Ludbrook et al. 1980); heart rate is only one component of the many complex cardiovascular events which determine the latter.

3.2 Cardiopulmonary Mechanoreceptors

3.2.1 Negative Pressure to the Lower Body

To study reflexes from the heart and lungs the application of negative pressure to the lower body, devised by Restall and Smirk (1952) and by Brown et al. (1966), is usually the method of choice. By pulling blood from the central circulation to the periphery this reduces the cardiac filling pressure and hence the stimulus to mechanoreceptors in the low-pressure side of the central circulation. When small negative pressures are used, the central venous pressure can be reduced without change in mean arterial pressure or pulse pressure or in arterial dP/dt. Thus the activity of the arterial baroreceptors is unchanged; with higher pressure both the cardiac filling pressure and systemic arterial mean and pulse pressures are decreased, so that the activity both of the low-pressure cardiopulmonary and the arterial mechanoreceptors is reduced (Johnson et al. 1974). When 80 mmHg lower-body negative pressure was applied, the decrease in mean arterial pressure was exaggerated by beta-adrenergic blockade (Bjurstedt et al. 1977).

3.2.2 Congesting Cuffs

The application of congesting cuffs to the thighs has been used occasionally to decrease cardiac filling pressure without changing the arterial pressure, but this does not permit the same magnitude of changes in central venous pressure.

3.2.3 Tilting

Tilting the subject from the supine to the upright position also decreases central venous pressure, but this is accompanied by changes in the afferent traffic from the arterial baroreceptors and possibly also changes from the vestibular apparatus and from receptors in the postural muscles.

3.2.4 Elevation of Legs or Legs and Lower Trunk

The elevation of the legs or legs and lower trunk has been used to activate the cardiopulmonary receptors by increasing the cardiac filling pressure (Roddie and Shepherd 1956, 1958), as have head-out water immersion (Restall and Smirk 1952; Epstein 1976) and negative-pressure breathing (Roddie and Shepherd 1963).

Results obtained with these techniques and with other methods to examine the other cardiovascular reflexes in man will be summarized. For detailed analyses, the reader should consult the recent excellent reviews by Abboud and Thames (1983), Mancia and Mark (1983), Mark and Mancia (1983), Mitchell and Schmidt (1983), Lind (1983), Rowell (1983), and Ludbrook (1983).

4 Carotid Baroreflexes

The primary role of the arterial baroreceptors is the rapid adjustment of arterial blood pressure around the existing mean pressure. This is accomplished by changes in heart rate, stroke volume, cardiac contractility, total systemic vascular resistance, and venous capacitance.

4.1 Heart Rate

Increases or decreases in the activity of the carotid baroreflexes cause opposite change in heart rate. The rapid cardiac slowing with baroreceptor stimulation is mediated primarily, if not completely, by an increase in vagal activity to the sinus node (Eckberg et al. 1972; Mancia et al. 1978a). With decreased stimulation the rapid acceleration is also mediated mainly by a decrease in vagal activity; an increase in sympathetic outflow to the sinus node will contribute during a sustained deactivation of the receptors (Mancia and Mark 1983). The latency of the heart rate response to changes in baroreceptor activity is brief (Pickering and Davis 1973), averaging 260 ms with neck suction and 440 ms with neck pressure. An increase of about 10 mmHg in resting systolic pressure is sufficient to activate the reflex cardiac slowing (Eckberg 1976, 1977a,b, 1980a,b). The sinus node response in man to a single baroreceptor stimulus is biphasic, with inhibition followed by less intense but more prolonged facilitation (cardioacceleration). A single stimulus modulates sinus node function for about 7 s. These complex changes probably reflect changing temporal relationships between the arterial pulse and sinus node activity, and interactions between oscillating levels of acetylcholine and the responsiveness of the sinus node to acetylcholine (Baskerville et al. 1979). The ability of the arterial baroreceptors to change heart rate rapidly through the action of the vagus on the sinus node permits the heart rate to be regulated on a beat-to-beat basis (Eckberg 1980b). In the cat, dog, and rat chronic sinoaortic denervation is accompanied by a marked increase in arterial blood pressure variability and a marked reduction in heart rate variability (Krieger 1964; Cowley et al. 1973; Ito and Scher 1981; Norman et al. 1981; Ramirez et al. 1985).

The reflex heart rate changes in response to rising and falling arterial pressure were compared using phenylephrine to raise the pressure and amyl nitrite to lower it; the response to falling pressure, expressed as milliseconds' change in pulse interval per millimeter of mercury change in systolic pressure, was 40% of that to rising pressure. This difference was not due to different effects on pulse pressure or to different rates of change of pressure (Pickering et al. 1972a). The sinus node responses to arterial baroreceptor stimulation with phenylephrine injection or neck suction were studied in healthy men before and during changes of central venous pressure caused by lower-body negative pressure and by leg and lower trunk elevation. Variation in this pressure of between 1.1 and 9.0 mmHg did not influence arterial baroreflex-mediated bradycardia (Takeshita et al. 1979).

Fig. 4. Average tidal volume (*upper*) and maximal pulse interval (*lower*) in response to stimulation of carotid baroreceptors by neck suction in six young adults. The time of onset and the duration of each baroreflex stimulus is indicated by the *stippled bars*. Note the oscillation of baroreflex responsiveness during the respiratory cycle: responses were minimal when stimuli were applied in early and mid-inspiration, maximal when stimuli were applied in late inspiration and early expiration, and intermediate when stimuli were applied in mid- and late expiration. (From Eckberg et al. 1980)

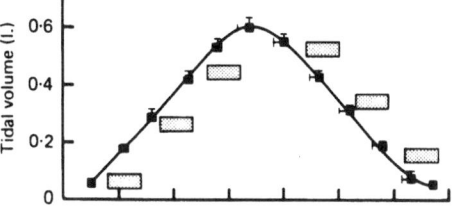

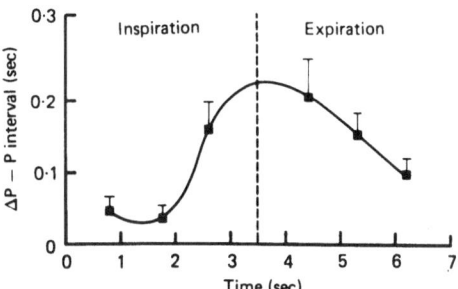

4.2 Respiratory Sinus Arrhythmia

The increase in heart rate during inspiration and the slowing during expiration, which is most marked in children and young adults, is caused by a variation in the depolarization frequency of the sinus node, caused largely by alteration in vagal activity. Four mechanisms could explain the increase in heart rate with inspiration:

1. Reflex inhibition of the cardioinhibitory center (vagal nucleus) by the slowly adapting stretch receptors in the lungs.
2. Central irradiation of inhibitory inpulses from the respiratory centers to the cardioinhibitory center.
3. Increased filling of the right atrium during inspiration, and a resultant activation of mechanoreceptors at the junction of the great veins with the right atrium; this could increase sympathetic activity to the sinus node (Linden and Kappagoda 1982).
4. A change in sensitivity of the carotid sinus baroreflex occurring in phase with respiration.

The ability of the arterial baroreceptors to control heart rate depends on the period during the respiratory cycle that the signal to the baroreceptors is changed. The sensitivity is less during early and mid-inspiration and increases in late inspiration to reach a maximum in the early expiratory phase (Fig. 4). Thus increased activation of these receptors during early and mid-inspiration and late expiration causes little sinus node inhibition; stimuli in late inspiration and early expiration cause maximal inhibition.

As breathing quickens this differential responsiveness is lost. In subjects with marked sinus arrhythmia, breath-holding does not abolish the oscillations in heart rate. The sensitivity of the carotid baroreceptor-heart rate reflex during voluntary apnea is augmented during slowing and attenuated during acceleration of the heart. The respiratory sinus arrhythmia is independent of the variations in arterial pressure with the respiratory cycle. Various explanations for this have been advanced, including central interferences from changes in input to the central nervous system from cardiopulmonary receptors, or brain centers involved in the regulation of breathing which moderate the baroreceptor—heart rate control reflex. The respiratory sinus arrhythmia may be explained best by the existence of a central respiratory oscillator, which continuously modulates the sensitivity of the arterial baroreceptor-sinus node reflex during normal quiet respiration. It seems that this central modulation is not due to a "gating" action that blocks the input from the baroreceptors, but to an additive opposition between the central neurons and the reflex input at the level of the vagal cardioinhibitory neurons (Lopes and Palmer 1976; Melcher 1976, 1980; Eckberg and Orshan 1977; Borst and Karemaker 1980; Eckberg et al. 1980).

4.3 Standing

The mechanism of the initial heart rate response to standing in normal subjects is complex, with fluctuations which last for 20—30 s. There is an immediate increase, caused by withdrawal of vagal activity to the heart which may originate directly from the higher centers in the brain and/or from ergoreceptors in the contracting postural muscles. There follows a further increase in heart rate and a rapid decrease; since these changes are accompanied by opposite changes in arterial blood pressure, it is suggested that they are mediated primarily by the arterial baroreflexes. However, the cardiopulmonary receptors may also be involved (Borst et al. 1982).

On standing following 20 min supine rest in eight normal subjects, there were also large fluctuations in blood pressure lasting for 20—30 s. There was an immediate increase in systolic blood pressure (29 ± 6 mmHg). It is suggested that this was due to compression of arteries by the contracting postural muscles. This increase was followed by fluctuation in systolic pressure to 28 ± 2 mmHg below control after 7 s and to 22 ± 2 mmHg above control after 22 s. The decrease is attributed to the passive translocation of blood from the thorax with a temporary predominance of a decrease in cardiac output prior to the reflex constriction of systemic resistance and splanchnic capacitance vessels, and to a stimulation of the arterial baroreceptors by the initial pressure increase. The recovery of blood pres-

sure is attributed to decreased stimulation both of arterial and cardiopulmonary mechanoreceptors (Borst et al. 1984). Other studies support the
view that ventricular mechanoreceptors play an important role in reflex
adjustments to orthostatic stress in humans (Ferguson et al. 1983).

4.4 Atrioventricular Conduction

The carotid baroreceptors also modulate atrioventricular conduction,
mainly through vagal cholinergic mechanisms, without affecting the ventricular portion of the atrioventricular conducting system. This modulation appears to have the role of maintaining a constant time relationship
between atrial and ventricular events during reflex changes in heart rate
(Mancia et al. 1979a). Studies using low-intensity electrical stimulation
of the carotid sinus nerves in normotensive subjects with coronary artery
disease have shown a latent period of 0.5–0.6 s for PP-interval changes
and a latency of 0.9–1.4 s for AV-interval changes. Both latencies increased
to 3 s after cholinergic blockade (Borst et al. 1974; Borst and Karemaker
1983).

4.5 Arterial Blood Pressure

In normal subjects, the myelinated fibers in the carotid sinus nerves have
been activated by electrical stimulation with frequencies of 20–200 Hz.
Judged by the absence of respiratory changes, it is unlikely that chemoreceptor afferents were activated. Changes in arterial blood pressure were
measured. The reflex effects on blood pressure increased approximately
with the number of stimulus pulses per unit time up to 80 Hz. When the
afferent impulse activity exceeded 120 Hz, frequency-dependent adaptation was observed. These studies support the view that the brain stem is
largely insensitive to the phasic, cardiac-cycle-related nature of baroreceptor afferent activity (Borst et al. 1983).

   In normal subjects the arterial blood pressure decreases when the carotid
baroreceptors are activated by application of negative pressure to the
neck and increases when their activity is lessened by positive pressure.
When a steady state is reached, and allowance is made for the loss of pressure transmission from the neck to the tissue outside the carotid sinus,
the pressor response to a given change in carotid transmural pressures is
about 50% greater than the depressor response. Thus the sensitivity of the
carotid baroreflex seems to be greater during baroreceptor deactivation
than during activation. This has been explained by the location of the set-
point (Mancia et al. 1979b). In normal subjects this is near the saturation

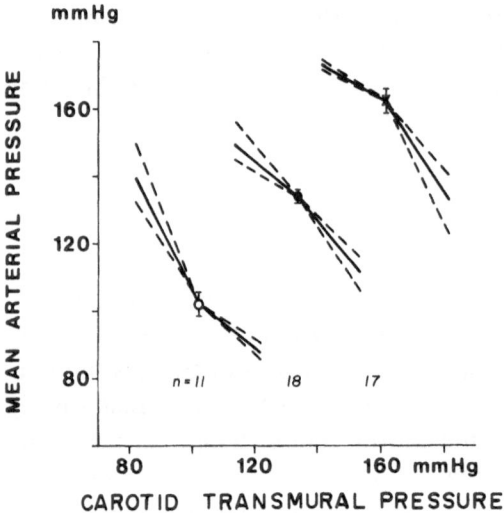

Fig. 5. Carotid baroreceptor influence on mean arterial blood pressure in 11 normo-tensive, 18 moderately hypertensive, and 17 severely hypertensive subjects. Changes in carotid transmural pressure were achieved by applying positive and negative pressures to an airtight chamber enclosing the neck. The *open circle*, the *closed circle*, and the *cross* represent the average (± standard error) mean arterial pressure during the control period in each group. The *continuous lines* represent the average of individual regression coefficients relating mean arterial pressure to increased and decreased carotid transmural pressure with respect to control values, the *dashed lines* indicating the standard errors of the regressions. The data are shown for the late or steady state effects of the neck chamber, the carotid transmural pressure being calculated as the difference between the mean arterial pressure and the tissue pressure outside the carotid sinuses. The latter was calculated from the value of the neck chamber pressure variation after application of the correction factors for loss of pressure transmission through the neck tissues. (After Mancia et al. 1978d)

limit of the curve that relates the carotid baroreflex stimulus to the changes in arterial blood pressure. This makes the reflex more effective in buffering a reduction in pressure than an increase in pressure (Fig. 5).

The decrease in arterial blood pressure with neck suction is due to various combinations of decreases in cardiac output and in total systemic vascular resistance; there is little change in stroke volume (Bevegård and Shepherd 1966b; Mancia et al. 1979b; Mancia and Mark 1983). Likewise, increases in cardiac output and/or resistance contribute to the pressor response. Since combined beta-adrenergic and parasympathetic blockade abolished the increase in cardiac output caused by a reduction of the carotid sinus transmural pressure, it seems that adjustments of cardiac output are not essential for the buffering action of the carotid sinus baroreceptors (Bjurstedt et al. 1975). Atropine does not affect the changes in blood pressure, indicating that cholinergic fibers do not participate. Also the epinephrine content in the plasma does not alter, so that the adrenal medulla is not involved. Thus alterations in total systemic vascular resistance depend on alterations in sympathetic vasoconstrictor tone (Mancia et al. 1978d). Neck suction causes a greater reduction in mean arterial pressure in young subjects because of a greater reduction in total systemic vascular resistance than is observed in older adults (Lindblad 1977).

Compared to the vagally mediated heart rate changes, the changes in arterial blood pressure mediated by the sympathetic nerves are slower. Thus, when the carotid baroreceptors are activated by negative pressure applied to the neck, a decrease in arterial pressure occurs in about 3 s, and reaches a maximum within 10–20 s; with positive pressure a steady increase in pressure is reached in 20–30 s (Eckberg et al. 1976; Mancia et al. 1979b).

In normotensive subjects with coronary artery disease, low-intensity electrical stimulation was used to activate both carotid sinus nerves. The latent period was estimated, during right atrial pacing, between the start of the stimulation and the onset of the reflex fall of diastolic arterial blood pressure. With the fixed heart rate, the arterial pressure changes started after 2–3 s. The central processing time of baroreceptor afferent activity was estimated at 0.25 s (Borst and Karemaker 1983). This long latency prevents rapid oscillations in blood pressure.

In subjects in whom arterial blood pressure was monitored continuously over 24 h, the greater or lesser tendency of blood pressure to vary within or among half-hour periods showed an inverse relationship with the sensitivity of the baroreceptor reflex as measured by the magnitude of the blood pressure response to neck suction or pressure or to the heart rate responses to phenylephrine and to trinitroglycerine injection (see Fig. 18, p. 67; Mancia et al. 1986). By contrast, heart rate variabilities within and among half hours and baroreflex sensitivities were linked by a positive relationship. These findings suggest that the arterial baroreflexes buffer spontaneous blood pressure changes in man. Baroreceptor-induced changes in heart rate, and presumably in cardiac output, may be one of the mechanisms involved. However, the spontaneous changes in blood pressure do not increase after atropinization, indicating the importance of other compensatory machanisms (Mancia et al. 1985).

4.6 Vascular Beds

4.6.1 Skeletal Muscles

The majority of the studies indicate that, contrary to the findings in animals, the carotid arterial baroreceptors may have a minor role in the control of muscle resistance vessels via the sympathetic noradrenergic

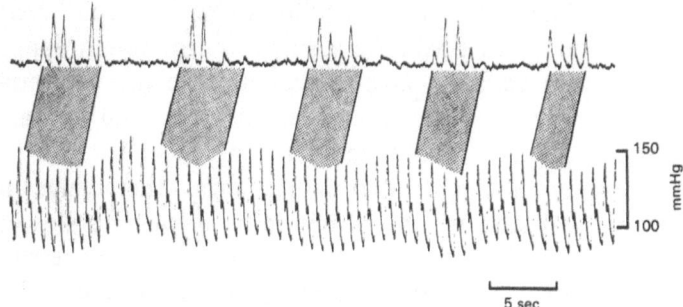

Fig. 6. Record of sympathetic efferent traffic to the leg muscle vessels of a normal human subject (peroneal nerve at the fibular head), showing more bursts occurring during a decrease than during an increase in blood pressure. *Stippled areas* indicate corresponding sequences of bursts and heart beats. Compensation made for reflex delay of 1.3 s between blood pressure and neural events. (From Sundlöf and Wallin 1978)

nerves, whereas the cardiopulmonary receptors have a major influence. This, however, may be due to the way in which the carotid baroreceptors have been activated or deactivated, i.e., by applying a constant negative pressure to the neck or by manual occlusion of both carotid arteries (Bevegård and Shepherd 1966b; Roddie and Shepherd 1957). When sine wave changes in pressure are applied to the neck, greater changes in muscle vascular resistance occur. Thus the possibility must be considered that arterial baroreceptors can exert considerable transient control over the muscle resistance vessels (Wallin et al. 1980; Lindblad et al. 1982).

Additional information has been obtained by multiunit recordings of the sympathetic impulses to the skeletal muscle vessels of the lower limbs. These are grouped in pulse-synchronous bursts, which occur during spontaneous reductions in arterial pressure and disappear during spontaneous elevations (Fig. 6). The occurrence of bursts correlates with variations of diastolic but not of systolic blood pressure; the changes in sympathetic activity per millimeter mercury blood pressure change are greater when the blood pressure is decreasing than when it is increasing. Thus acute decreases in arterial blood pressure are buffered more efficiently than acute increases, and variations of sympathetic activity to skeletal muscle are determined mainly by fluctuations in diastolic blood pressure. In normal subjects at rest there is a linear relationship between the mean level of sympathetic nerve activity to muscle and the plasma concentration of norepinephrine. With application of sinusoidal suction to the neck, there is a marked though short-lasting inhibition of sympathetic outflow in the skeletal muscle nerves. By contrast, steady suction over several minutes results in only minor changes in sympathetic activity. This suggests that the carotid baroreceptors have a greater influence in modulating the

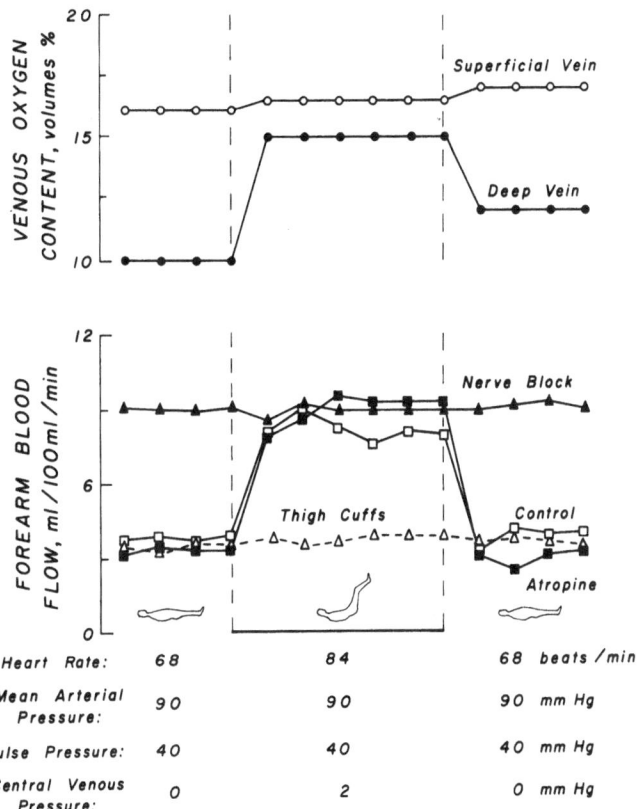

Fig. 7. Effect of passive elevation of the legs of a normal subject on the circulation through the forearm. With leg elevation the blood flow in the forearm increases. This is due to an increase in muscle blood flow, since the oxygen content of the blood draining the forearm muscles increases (*Deep Vein*) but that of the blood draining the forearm skin (*Superficial Vein*) does not. The increased flow is due to reflex dilatation, since it was prevented by blocking the sympathetic nerves to the forearm vessels with a local anesthetic. The dilatation was not affected by atropine, indicating that it was caused by decreased activity of noradrenergic fibers and not by activation of cholinergic fibers. It was caused by displacement of blood from the legs, since it was prevented by inflation of thigh cuffs to suprasystolic pressure before the leg elevation. Since arterial mean and pulse pressures were unchanged, and central venous pressure increased, the reflex dilatation is attributed to excitation of receptors somewhere in the heart and lungs. (From Donald and Shepherd 1978)

vascular resistance in the skeletal muscles during dynamic than during static changes in arterial blood pressure. The absence of a correlation between static blood pressure levels and the sympathetic activity to skeletal muscle blood vessels implies that static control of arterial blood pressure depends more on the reflex control of other parts of the vascular system (Wallin et al. 1975). This contrasts with the major influence of receptors in the low-pressure side of the cardiovascular system on the resistance vessels in human forearm skeletal muscles (Figs. 7 and 8)

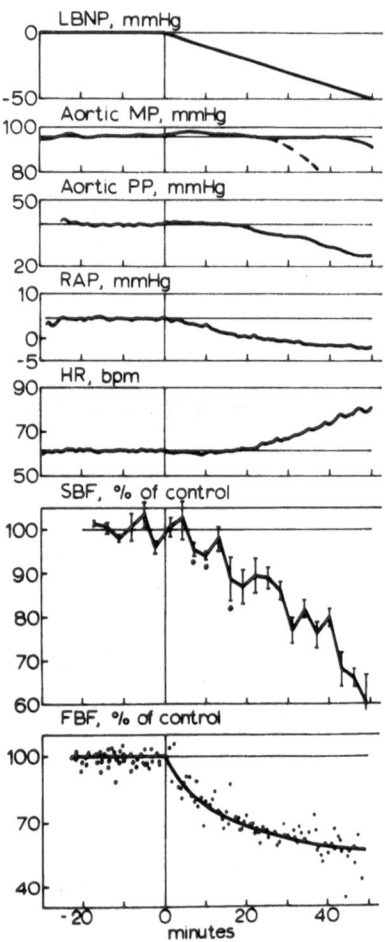

Fig. 8. Average values for responses to ramp lower-body negative pressure (*LBNP*) in nine subjects. *MP*, mean aortic pressure; *PP*, pulse pressure; *RAP*, right atrial pressure; *HR*, heart rate; *SBF*, splanchnic blood flow; and *FBF*, forearm blood flow. Ramp LBNP applied at −1 mm-Hg/min. *Bars* with entries for SBF indicate SE; values for SE for other variables were too small to be shown. *Asterisks* denote first three significant decrements in SBF below control. At 0−20 mmHg, LBNP reduces RAP without reducing aortic MP or PP and without increasing HR. This mild LBNP produced a substantial decrease in FBF and slight decreases in SBF. At 20−50 mmHg, LBNP reduced aortic PP and RAP and produced an increase in HR and further decreases in SBF and FBF. (From Johnson et al. 1974)

(Johnson et al. 1974; Wallin et al. 1975; Sundlöf and Wallin 1977; Donald and Shepherd 1978; Sundlöf and Wallin 1978b). In animal studies, by contrast, the arterial baroreflexes dominate in the control of muscle vessels, whereas the cardiopulmonary have the major role in regulating the renal vessels (Donald and Shepherd 1978).

In other studies on healthy adults, afferent carotid baroreceptor activity was modified with 30 mmHg neck suction or pressure during held inspiration, and efferent sympathetic activity was measured from peroneal nerve muscle fascicles. Although the resultant changes in muscle sympathetic responses were profound, they persisted for only 1−2 s despite continuation of neck suction or pressure (Wallin and Eckberg 1982). This is in accord with the transient vasodilatation in forearm resistance vessels during prolonged neck suction (Lindblad et al. 1982).

In contrast to these responses in the skeletal muscle sympathetic nerves, the parasympathetic outflow to the heart adapted only slightly during neck-chamber pressure changes. These results suggest that in man the input from the carotid baroreceptors is processed differently in central vagal and sympathetic networks (Wallin and Eckberg 1982).

4.6.2 Skin

Measurements of hand blood flow, which is mainly flow to the skin, indicate that these vessels are usually unresponsive to alterations in the activity of the arterial and cardiopulmonary mechanoreceptors. Thus, with a change from the recumbent to the upright position, there is no sustained constriction of human skin resistance vessels and cutaneous veins. A transient constriction may occur on standing or with a simulated postural change, but this is probably mediated by an arousal stimulus from the brain itself. This is in contrast to the sustained constriction of the muscle resistance vessels, which, together with a constriction of the splanchnic resistance and capacitance vessels, acts to maintain the arterial blood pressure on standing upright.

Recordings of sympathetic activity of skin nerves to the hand and foot show no correlation with spontaneous fluctuations of arterial blood pressure and direct stimulation of the carotid sinus nerves does not reproducibly affect their activity (Wallin et al. 1975). Thus, in contrast to the sustained activity in sympathetic fibers in muscle nerves in the upright position, there are only transient responses in the sympathetic fibers in human skin nerves. Only in circumstances where there is a marked decrease in the activity of the arterial and cardiopulmonary mechanoreceptors does a sustained constriction of skin resistance vessels contribute to the increase in systemic vascular resistance (Beiser et al. 1970; Rowell et al. 1973; Shepherd and Vanhoutte 1975; Vallbo et al. 1979; Bini et al. 1980a,b).

Following bilateral blocks with local anesthetic of the glossopharyngeal and vagus nerves in the neck of healthy subjects, there was a marked increase in muscle nerve sympathetic activity (recorded in the peroneal nerve). The normal cardiac rhythmicity was abolished and the spontaneous activity became similar to that in the sympathetic nerves to the skin but the bursts of impulses did not become synchronous (Fagius et al. 1985). At the same time there was a temporary marked increase in heart rate and arterial blood pressure consistent with the removal of the inhibitory influence of the arterial and cardiopulmonary mechanoreceptors on the vasomotor center (Guz et al. 1966; Fagius et al. 1985).

From these studies, it is suggested that baroreceptor deafferentation may explain the transient hypertension and resting tachycardia in the Guillain-Barre syndrome (Fagius and Wallin 1983; Fagius et al. 1985).

4.6.3 Splanchnic Region

Studies in animals and man have demonstrated the importance of reflexly induced changes in splanchnic blood flow for the attainment and maintenance of an appropriate arterial blood pressure (Rowell et al. 1972; Johnson et al. 1974; Tyden et al. 1979). In this the arterial baroreceptors have an important role. When lower-body suction is applied to a degree sufficient to reduce mean arterial pressure and pulse pressure, there is a marked increase in splanchnic vascular resistance (Fig. 8). The simultaneous stimulation of the carotid sinus baroreceptors during lower-body suction does little to the reflex vasoconstriction in the forearm skeletal muscles, but greatly reduces the splanchnic vasoconstriction. This illustrates again the nonuniformity in the vasomotor control by the arterial and the cardiopulmonary mechanoreceptors (Abboud et al. 1975; Abboud et al. 1979).

4.6.4 Kidney

Studies in animals have shown the importance of the arterial and cardiopulmonary mechanoreceptors in regulating the renal resistance blood vessels (Öberg and White 1970; Ott and Shepherd 1973; Mancia et al. 1975a). In man, renal blood flow declines with tilting or standing (Smith 1939). While this implies that these receptors participate in the cardiovascular reflexes, no studies are available on the alterations in the kidney circulation in response to specific changes in arterial and/or cardiopulmonary mechanoreceptor activity. Such studies are made difficult by the time required by current methods for measuring renal blood flow and by the ability of the kidney vessels to autoregulate with changes in arterial blood pressure.

4.6.5 Venous System

Studies of reflexly mediated changes in the venous system in man have been limited to the limb veins. The evidence supports the conclusion that the carotid baroreceptors do not regulate the cutaneous or other limb veins (Epstein et al. 1968b). In animals, the splanchnic capacitance vessels participate in concert with the splanchnic resistance vessels — decreasing their capacity when the arterial blood pressure decreases and vice versa. In this way they contribute importantly to the filling pressure of the heart (Shepherd and Vanhoutte 1975).

5 Aortic Baroreflexes

Of the few observations that have been made in man, one study, based on the absence of an increase in arterial blood pressure following bilateral vagal nerve block, suggested that tonic aortic baroreceptor activity is absent at pressures below the normal resting arterial blood pressure; another study suggested that these receptors were operative in the normal blood pressure range, while yet others indicated that the aortic baroreceptors are more important than the carotid in the control of heart rate (Guz et al. 1964, 1966; Mancia et al. 1977; Bath et al. 1981; Ferguson et al. 1985).

6 Sleep

Examination of the reflex bradycardia caused by injection of phenylephrine indicated that the arterial baroreflex for the control of heart rate was reset rapidly to operate around the lower blood pressure during sleep (Smyth et al. 1969; Bristow et al. 1969a). Under these conditions, the sensitivity of this reflex may be greater than when awake (Conway et al. 1983). Baroreflex sensitivity may also be related to circadian rhythms and it is suggested that there is a sleep-independent circadian rhythm of vagally mediated baroreflex sensitivity (Hossman et al. 1980). The influence of sleep on the reflex control by the arterial baroreceptors of blood pressure and vascular resistance has still to be determined (Mancia and Mark 1983).

7 Age

That baroreceptor sensitivity decreases with age is suggested by studies where the arterial blood pressure was changed by infusion of vasoactive drugs, and the heart rate changes accompanying the alterations in pressure were much less in older than in younger subjects (Gribbin et al. 1971; Korner et al. 1974; Duke et al. 1976; Randall et al. 1976, 1978). Information on baroreceptor control of systemic vascular resistance is much less conclusive. Lindblad (1977) reported that the reduction in mean arterial pressure caused by stimulation of the carotid arterial baroreceptors by neck suction was less in elderly than in younger people; since the same reduction in cardiac output occurred, this suggested that the reflex decrease in peripheral resistance was attenuated in the older subjects. However, Robinson et al. (1983) have observed no impairment of neural modulation by baroreflexes in aged subjects with autonomic dysfunction. More recently, Mancia et al. (1984a) have examined the increase in mean arterial pressure due to a reduction in carotid baroreceptor activity caused by increase in pressure in a neck chamber. The ages of the subjects ranged from less than 36 to more than 54 years. They had mild essential hypertension and the same mean arterial blood pressure. Although the pressor responses developed more slowly with aging, the overall effects of alterations in baroreceptor activity was similar in the different age groups. Thus aging may not cause an impairment of the arterial baroreceptors' control of arterial blood pressure, despite the fact that in elderly subjects the stiffer large arteries should reduce the ability of the baroreceptors to respond to changes in transmural pressure (Bader 1967; Gozna et al. 1974; Randall et al. 1976).

Plasma norepinephrine increases with aging. This may be due in part to reduced clearance of norepinephrine (Esler 1982) and to an increase in sympathetic traffic, as suggested by direct recording from muscle sympathetic nerves (Sundlöf and Wallin 1978a).

8 Exercise and Arterial Baroreflexes

During both rhythmic (dynamic) and isometric (static) exercise the carotid baroreceptors are adjusted quickly to a higher level. This permits the perfusion pressure to increase to help meet the metabolic demands of the active muscles. The shape of the curve relating the change in carotid sinus pressure to the systemic arterial pressure and to the heart rate is unchanged. Thus, as the heart rate and systemic pressure increased during supine rhythmic leg exercise, the absolute decreases in these measurements resulting from the neck suction were the same as at rest (Bevegård and Shepherd 1966b).

In normal subjects aged 31–45 years, during brief periods of exercise on a bicycle ergometer, the arterial plasma potassium increased rapidly at the start of exercise from 3.8 ± 0.3 mmol/l (mean $\pm$ SD) to plateau levels of 5.4 ± 0.1 mmol/l. It is suggested that these changes, if transmitted to the extracellular fluid, could decrease the cell transmembrane potential and as a consequence might alter the sensitivity of the arterial chemoreceptors and the arterial and cardiopulmonary mechanoreceptors (Linton et al. 1984).

Also during isometric exercise, subatmospheric neck pressure causes a similar decrease in arterial blood pressure both at rest and during the exercise (Ludbrook et al. 1978; Mancia et al. 1982b; Suarez et al. 1982). Thus exercise does not diminish the overall gain of the carotid sinus baroreflex, so that its ability to regulate blood pressure is unchanged. The reduction in arterial blood pressure with neck suction is due mainly to a decrease in cardiac output both at rest and during isometric exercise (Bevegård et al. 1977b). The age of the subject may make a difference to the response; older subjects respond mainly with a decrease in cardiac output, in younger ones there is a greater reduction in systemic vascular resistance (Lindblad 1977). The role of the carotid sinus baroreflex in the regulation of heart rate during exercise is less certain. Using intravenously injected phenylephrine to raise arterial blood pressure and studying the effect on heart interval to estimate the sensitivity of the combined carotid sinus and aortic baroreceptor reflexes, it was found that exercise caused a reduction in reflex sensitivity (Bristow et al. 1971; Pickering et al. 1971; Cunningham et al. 1972a). It was suggested that in exercise the increase in the firing rate of the cardiovascular sympathetic efferent neurons reduced the effect of the arterial baroreceptor reflex, thus permitting both heart rate and arterial blood pressure to increase. Using the variable pressure neck chamber to examine the changes in heart interval with changes in carotid sinus transmural pressure, it was found that the immediate transient increase in R-R interval that follows an acute rise in transmural pressure was diminished during isometric exercise, which agrees with the observations just cited using phenylephrine to increase the pressure. This cardiac component of the baroreflex was vagally mediated, as evidenced by its rapid onset, its abolition by atropine, and the fact that blockade of the beta-receptors by propranolol had little effect (Pickering et al. 1972b). However, the decrease in R-R interval resulting from an acute fall in carotid sinus transmural pressure was virtually unaffected (Mancia et al. 1978b; Ludbrook et al. 1978). Thus during exercise the effects on the R-R interval are evident with an increase in carotid baroreceptor stimulation, but not with a decrease. As mentioned earlier, the relation between R-R interval and heart rate is hyperbolic rather than linear. Thus if reflex changes in heart rate are the same at rest and during exercise, the corresponding changes in R-R will be smaller during exercise (Fig. 3). It seems that a reduction in arterial baroreflex sensitivity does not have an important role in the initiation of the increase in arterial blood pressure and heart rate caused by isometric exercise (Ludbrook et al. 1978). However, in contrast to blood pressure control, there may be an impairment of carotid baroreceptor control of heart rate during severe exercise. This would permit attainment of higher heart rates (Bristow et al. 1971).

Studies using dogs with chronic sinoaortic denervation have led to the conclusion that the arterial baroreceptors are not important in regulating the arterial blood pressure during running (Vanhoutte et al. 1966; Krasney et al. 1974; McRitchie et al. 1976). If, however, in dogs with chronic aortic denervation the influence of the carotid baroreceptors is acutely

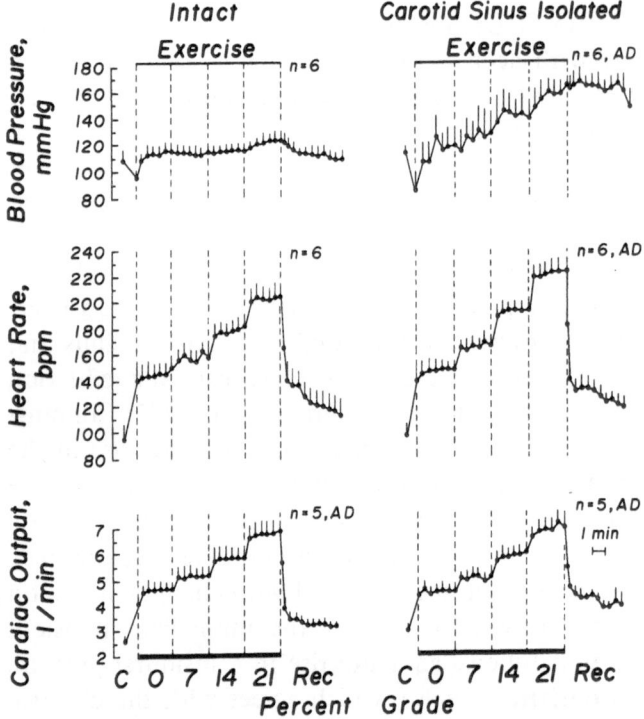

Fig. 9. Mean arterial pressure (*top panel*), heart rate (*middle panel*), and cardiac output (*bottom panel*) during continuous graded exercise in dogs with aortic arch denervation before (*left*) and after (*right*) vascular isolation and pressurization of carotid sinuses. *AD,* aortic denervation; *C,* control; *Rec,* recovery. (From Walgenbach and Donald 1984)

removed, the pattern of the arterial blood pressure during exercise is altered markedly. There is a greater decrease in blood pressure at the onset of exercise and as the severity of the exercise is increased the blood pressure rises progressively and remains elevated after the exercise ceases (Fig. 9). If the carotid sinuses are permitted to function again as soon as the exercise stops, the pressure abruptly returns to the control level. The cardiac output and heart rate changes are similar with and without the carotid sinuses operative. Thus the carotid baroreceptors regulate the overall changes in systemic vascular resistance during and after exercise, while the heart rate and cardiac output are controlled by other mechanisms (Walgenbach and Donald 1983b). The cardiopulmonary mechanoreceptors cannot compensate during exercise for the acute withdrawal of the carotid baroreflex (Walgenbach and Donald 1983a,b). However, with chronic absence of the carotid and aortic baroreflexes, the cardiopulmonary receptors do not appear to modify the blood pressure response during running; in their absence, however, the arterial blood pressure rises rapidly when the exercise ceases, at the time when the cardiac output is

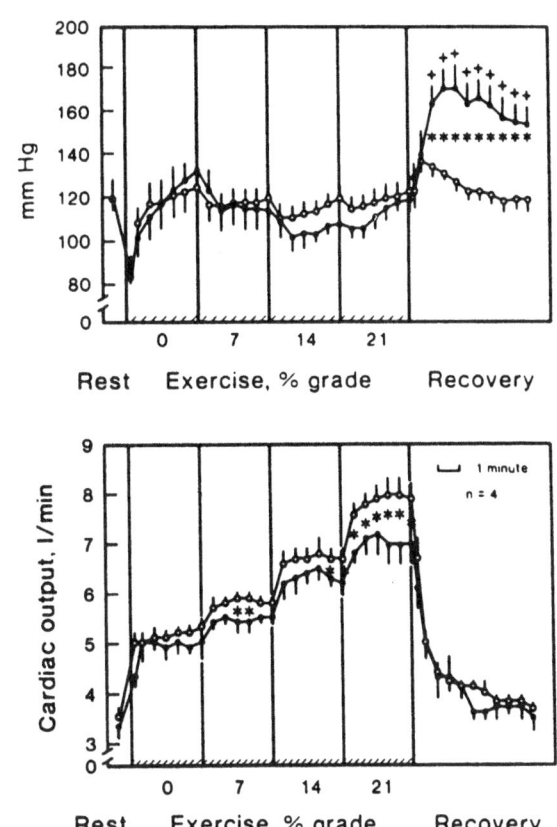

Fig. 10. Arterial blood pressure and cardiac output during 12 min graded exercise in dogs with chronic sinoaortic denervation before (O- - -O) and after (●- - -●) vagotomy. *Asterisks*, significant differences before and after vagotomy. (From Daskalopoulas et al. 1984)

decreasing rapidly to the preexercise level. Thus when the exercise ceases, mechanoreceptors in the heart and lungs are necessary, in the chronic absence of the arterial baroreflexes, to quickly reduce the total systemic vascular resistance (Daskalopoulos et al. 1984; Walgenbach and Shepherd 1984) (Fig. 10).

In young normal subjects, resting supine epinephrine was infused intravenously to produce plasma levels similar to those found in the forearm veins during bicycle exercise. The exercise increased the mean heart rate from a resting value of 64 to 177 beats/min at the end of the exercise. The infusion caused bronchodilation and elevation of plasma glucose, but no significant change in heart rate. It is suggested that during rhythmic exercise circulating epinephrine makes little contribution to the hemodynamic responses (Warren and Dalton 1983).

9 Chemoreflexes

In man little information is available on the role of the aortic and carotid chemoreceptors in controlling the circulation. This is because it is not possible to limit the effects of changes in oxygen and carbon dioxide tension and hydrogen ion concentration to these receptors. Studies in subjects who had bilateral carotid body resection, without baroreceptor denervation, for bronchial asthma suggested that the carotid bodies are essential for normal pressor responses during hypoxia, but not for the tachycardia of hypoxia or the cardiovascular responses to hypercapnia (Lugliani et al. 1973). Using the alpha-sympathomimetic drug phenylephrine to cause transient rises in blood pressure and relating changes in pulse interval to the pressure changes, it was found that hypercapnia with hyperoxia at rest was associated with a rise in systolic pressure and a fall of baroreflex sensitivity with little change in pulse interval; hypoxia with hypocapnia at rest was associated with tachycardia, but no significant change in pressure or in reflex sensitivity (Bristow et al. 1974; Cunningham et al. 1972b). In young hypertensive subjects there are increased ventilatory and pressor responses to acute hypoxia as compared to age-matched normotensive controls, suggesting that in the former the arterial chemoreceptor drive is augmented (Trzebski et al. 1982).

10 Diving Reflex

Mammals, including man, birds, and reptiles have the ability to redistribute their circulation during diving in order to utilize the nonreplenishable oxygen stores for high-priority areas. In animals this has been shown to be due to the activation of both the trigeminal and the chemoreceptor reflex. Immersion of the face in water activates the sensory endings of the trigeminal nerve, which elicits a reflex apnea, vagal induced bradycardia, and constriction of systemic vessels in the splanchnic region, skeletal muscles, and kidneys due to increased activity in the sympathetic nerves. The splanchnic and cutaneous capacitance vessels constrict. The apnea is followed quickly by a decrease in P_{O_2} and an increase in P_{CO_2} in the arterial blood; this stimulates the arterial chemoreceptors. The continued sensory input from the trigeminal nerves overrides the action of the chemoreceptors on the respiratory center. The circulatory effects of chemoreceptor stimulation augment those from the nose and face, leading to further cardiac slowing and constriction of systemic vessels. There is a reflex dilatation of the coronary arteries mediated by the vagal nerves. The arterial blood pressure is maintained, and the oxygen in the blood is made available to the most vulnerable systems, the heart and brain. Lactic acid formed and stored in muscle during a prolonged dive floods into the circulation when breathing is resumed (Angell-James and Daly 1972; Blix and Folkow 1983).

In man, a decrease in forearm and hand blood flow occurred during immersion of the face in water while breathing air or 10% oxygen. There was no change in arterial blood pressure, so that the decreased flow was due to vasoconstriction (Brick 1966; Kawakami et al. 1967). The vaso-constrictor response to facial immersion was not attenuated by the hypoxia. This peripheral vasoconstriction presumably helps in redistributing blood flow to organs more sensitive to hypoxia (Heistad and Wheeler 1970).

11 Cardiopulmonary Reflexes

Numerous studies in animals have demonstrated that mechanoreceptors in the heart and lungs, subserved both by vagal and sympathetic medul-lated and nonmedullated afferents, exert a variety of effects on the cardio-vascular system. The main effect of the vagal afferents is tonically to inhibit the vasomotor center (Mancia et al. 1976; Donald and Shepherd 1978; Malliani 1982; Bishop et al. 1983). Medullated afferents in the cervical vagus originating in large unencapsulated endings at the junctions of the venae cavae and pulmonary veins with the atria and in the atrial appendages, however, can exert a selective excitatory action on the sympa-thetic outflow to the sinus node; these receptors can also modulate renal blood flow, urinary output, and vasopressin secretion (Linden and Kappa-goda 1982). Afferent impulses from these receptors are conveyed to neurons in the medial region of the nucleus of the solitary tract (Kidd 1979).

In man only limited data are available on the cardiovascular effects exerted by different cardiopulmonary receptors. An increase or decrease in intrathoracic blood volume sufficient to alter the central venous pres-sure, but insufficient to change arterial mean or pulse pressure, results in a reflex dilatation or constriction respectively of the resistance vessels in the forearm muscles; these changes are mediated by the sympathetic nor-adrenergic nerves (Figs. 7 and 8). The vessels of the skin do not partici-pate in these reflex changes unless there is a marked decrease in the effec-tive circulatory blood volume (Samueloff et al. 1966; McNamara et al. 1969). A constriction of the splanchnic resistance vessels and tachycardia are seen only when the intrathoracic blood volume is reduced enough to cause a decrease in arterial blood pressure (Fig. 8). In the latter circum-stances the simultaneous application of negative pressure to the neck, sufficient to counteract the reduction in carotid sinus transmural pressure, prevents or attenuates the splanchnic vasoconstriction and the tachy-cardia, but does not alter the constriction of the forearm vessels. This indicates that in man low-pressure intrathoracic vascular receptors have a much greater long-term influence than the arterial baroreceptors on the forearm muscle resistance vessels (Roddie et al. 1957, 1958; Roddie and Shepherd 1958; Blair et al. 1959b; Gilbert and Stevens 1966; Zoller et al. 1972; Johnson et al. 1974; Abboud and Mark 1978; Bennett et al. 1979a; Bishop et al. 1983).

Measurement of sympathetic activity to the forearm muscle vessels showed that this was enhanced during lower body negative pressure, and that this could not be explained by changes in the activity of the arterial baroreceptors (Sundlöf and Wallin 1978b). Recent studies, however, indicate that the resistance vessels in the calf are less responsive than those in the forearm to mechanoreflexes originating in the low pressure side of the circulation (Essandoh et al. 1986). However, on changing from the lying to the standing position, there is a constriction of similar magnitude of forearm and calf resistance vessels. While the increased sympathetic outflow to the forearm muscle vessels is due predominantly to the dis-placement of blood from the thorax to the dependent parts and the con-sequent lessened stimulus to the low-pressure mechanoreceptors, that to the calf muscles is due mainly to decreased activity of the carotid sinus baroreceptors. This is brought about by the fact that in the standing posi-tion the carotid sinus is raised about 20–25 cm above heart level, so that despite the fact that there is little if any change in arterial pressure at heart level, there is a decrease in pressure at the carotid sinus barorecep-tors (Burke et al. 1977). On the reasonable assumption that this reflex constriction on standing occurs generally in the skeletal muscles which constitute about 40% of the body mass, this contributes importantly to the increase in systemic vascular resistance and hence the maintenance of the arterial blood pressure.

The effect of cardiopulmonary receptor stimulation on the renal circu-lation has not been investigated in man. When the central blood volume was reduced by application of lower-body negative pressure of 40 mmHg, sinusoidal variations in neck pressure (−10 to −40 mmHg) at a fixed frequency induced an augmented effect on the blood-pressure-regulating capacity but not on the heart rate response. Thus changes in central blood volume can modify the carotid baroreflex, presumably via low-pressure intrathoracic mechanoreceptors (Bevegård et al. 1977d).

The injection of phenyl diguanide or sodium cyanide into the main pul-monary artery of patients referred for investigation of lung function caused a stimulation of breathing, together with bradycardia and hypo-tension. Analysis of the mean injection-response time indicated that these drugs were acting on the carotid body rather than on receptors in the lungs.

In contrast, lobeline sulfate caused apnea before hyperpnea. The former was attributed to stimulation of pulmonary receptors before excitation of the carotid bodies, indicating that there are receptors in the human lung that are depolarized by chemical agents (Jain et al. 1972).

A reflex inhibition of respiration occurs in man with large lung inflations, but not at normal tidal volume. Thus the Hering-Breuer reflex appears to be less potent in humans than in animals. Electrical stimulation of vagal afferents caused a fall in arterial blood pressure. However, this was attributed to impairment of venous return and hence of cardiac output caused by the accompanying mechanical changes in the chest associated with coughing (Guz et al. 1964). In five patients studied during cardiopulmonary bypass, lung inflation had no effect on heart rate or total peripheral resistance. These negative results might be attributed to insufficient inflation of the lungs, to the halothane anesthesia, or to the low P_{CO_2} (Ott et al. 1975).

12 Axon Reflex

There is evidence both in skin and muscle that a local sympathetic axon reflex can be triggered by venous congestion. The receptor site appears to be in small veins of skin, muscle, and subcutaneous adipose tissue and the effector sites in the arterioles supplying these tissues. This axon reflex may be an important adjunct to the postural reflexes mediated through the central nervous system (Henriksen and Sejrsen 1977).

13 Renin

Animal studies have shown that release of renin is governed by intrarenal receptors that respond to alterations in renal perfusion pressure, to changes in sodium load in the macula densa, and to influences of sympathetic nerves on the juxtaglomerular cells mediated by beta-adrenergic receptors (Davis and Freeman 1976). The last mechanism is important because it can directly modify renin secretion and in addition potentiate the effects on the secretion of the local nonneural mechanisms (Di Bona 1982; Stella and Zanchetti 1984). Changes in the activity of the arterial and cardiopulmonary mechanoreceptors subserved by vagal afferents, sufficient to increase or decrease sympathetic nerve traffic to the kidney, result in increase or decrease in renin release respectively (Thames et al. 1971; Thames 1977; Mancia et al. 1975a; Thoren et al. 1976; Bevegård et al. 1977c; Thorén 1979). In the dog the control of renin release exerted by cardiopulmonary receptors is very sensitive, as shown by the fact that during hemorrhage small changes in blood volume insufficient to alter arterial pressure or heart rate cause a marked increase in renin release (Thames et al. 1978). The cardiopulmonary receptors with vagal afferents that modulate the release of renin have not been identified. Since in animals receptors in the heart and lungs with vagal myelinated or nonmyelinated afferents can reflexly alter renal sympathetic nerve activity (Mancia et al. 1975b), it is assumed that all of them can influence renin release. Much less is known of the nervous control of renin release in man. The reduction in plasma renin activity that accompanies administration of beta-adrenergic-blocking drugs suggests the existence of a tonic sympathetic stimulation of renin secretion (Conway 1983, 1984). Evidence that stimulation of renin secretion by acute renal hypotension is enhanced during the cold pressor test suggests that this reflex modulation potentiates nonneural effects (Guazzi et al. 1985). Passive leg-raising can cause a modest reduction in plasma renin activity (Grassi et al. 1985b). However, the role of cardiopulmonary receptors in control of renin release and its importance in comparison with the role of arterial baroreceptors have not been unequivocally established. Studies in subjects with normal or high renin secretion rates have shown that neck-chamber-induced changes in carotid baroreceptor activity do not alter plasma renin activity (Mancia et al. 1978c, 1981). Other studies have reported that a reduction in thoracic blood volume does not reflexly increase plasma renin activity until the reduction is sufficient to decrease arterial pressure and unload the arterial baroreceptors (Mark et al. 1978). This has been interpreted to mean that in man reflex control of renin release takes place only when cardiopulmonary and arterial baroreceptors are engaged together (Grayboys et al. 1974; Mark et al. 1978). However, evidence that a decrease in thoracic blood volume insufficient to perturb arterial baroreceptors can cause a reflex release of renin has also been reported (Kiowski and Julius 1978; Hesse et al. 1975). Using a series of maneuvers designed to examine the separate and combined effects of decreasing stretch on the high- and low-pressure receptors in normal men, Egan et al. (1983) and Julius et al. (1983) concluded that cardiopulmonary receptors interact with high-pressure baroreceptors but that in addition they have an independent role in the control of renin release. This has been supported by observations in normotensive subjects and in patients with primary hypertension; not only does a reduction in central venous pressure insufficient to alter arterial blood pressure increase renin secretion, but the increase matches the increase observed in the same subjects when central venous pressure is reduced to a similar extent by tilting (Grassi et al. 1985b) (Fig. 11). These observations suggest that cardiopulmonary receptors play in important role in the renin modulation that characterizes changes in posture (Cohen et al. 1967; Fasola and Martz 1972).

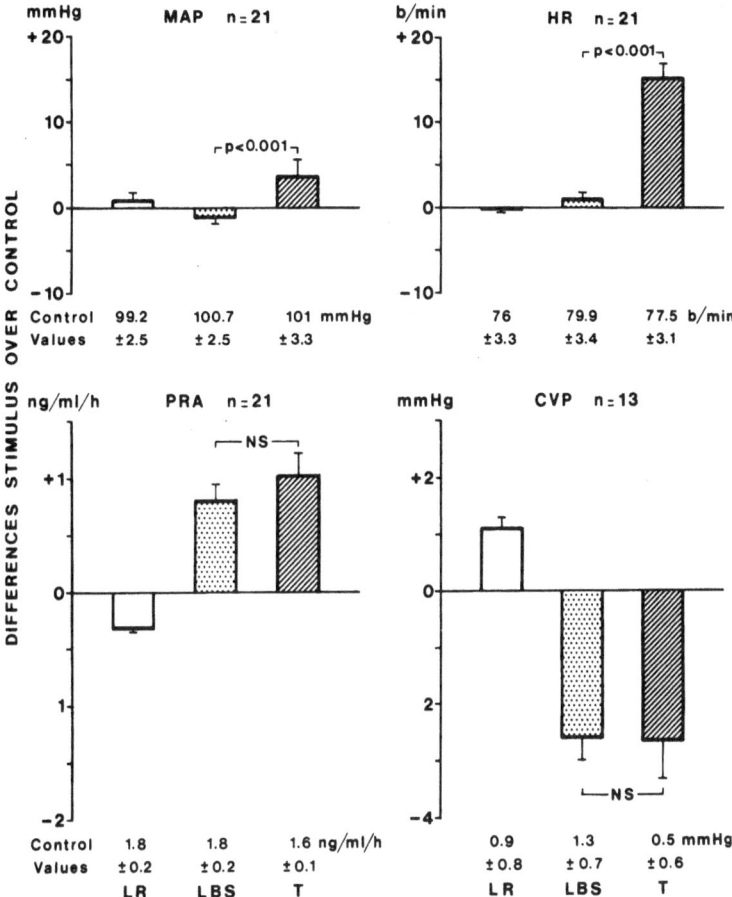

Fig. 11. Effects of a 20-min passive leg elevation (*LR*), mild lower-body suction (*LBS*), and tilting (*T*) on plasma renin activity (*PRA*), mean arterial pressure (*MAP*), heart rate (*HR*), and central venous pressure (*CVP*). Data are shown as mean (± SE) changes from control values in 21 subjects with normal blood pressure (*n* = 12) or mild essential hypertension (*n* = 9). Note that the increases in PRA induced by tilting were similar to those induced by lower-body suction, suggesting similar cardiopulmonary receptor deactivation in the two conditions. The increases in PRA were also similar, despite the fact that, during tilting, carotid baroreceptors were deactivated by the positioning of carotid sinuses above the heart. (Grassi and Mancia, unpublished data)

14 Antidiuretic Hormone, Plasma and Extracellular Fluid Volume

Under normal circumstances, the plasma and extracellular fluid volumes are maintained within narrow limits (Gauer and Henry 1976). In this the kidney has the key role by regulating the excretion of water and electrolytes, particularly sodium, which is the major determinant of the distribution of the plasma and extracellular fluid. The role of intravascular receptors in modulating water and salt excretion by the kidneys is uncertain.

It may be that extracellular fluid volume and extracellular sodium concentration are monitored by different sensors, with a central integrator to calculate total body sodium (Gilmore 1983). While the osmoreceptors signal changes in sodium concentration they cannot, with an isoosmotic volume expansion, detect an increase in total body sodium. The great distensibility of the atria and other cardiac and pulmonary structures favors the possibility that the cardiopulmonary receptors, in particular the unencapsulated endings situated in the endocardium of the atria, act as volume receptors to sense the fullness of the blood stream (Linden and Kappagoda 1982). In dogs balloon distension of the pulmonary veno-atrial junctions or the left atrial appendage causes an increase in renal blood flow and urine flow, a reduction of vasopressin secretion, and a decrease in the secretion of renin and aldosterone as a consequence of the decrease in sympathetic outflow to the kidney (Brennan et al. 1971; Mancia et al. 1976; Linden and Kappagoda 1982; Bennett et al. 1983). Increased vasopressin, renin secretion, and renal sympathetic outflow have been observed following hemorrhage or acute interruption of cardiopulmonary vagal afferents (Share 1965, 1968; Share and Levy 1962; Goetz et al. 1974; Thames and Schmid 1979). These combined humoral and nervous events will increase or reduce salt and water excretion when blood volume is increased or reduced respectively.

When normal subjects are immersed to the neck in water, about 700 ml blood is redistributed from the peripheral veins to the central circulation, and the right atrial pressure increases from −2 to +16 mmHg. The arterial blood pressure increases 10 mmHg or less. The stroke volume and cardiac output increase with no change or a decrease in heart rate. Diuresis and natriuresis take place, and there are decreases in urinary excretion of antidiuretic hormone, plasma renin activity, and plasma aldosterone level. This could be explained by stimulation of the receptors in the heart and lungs, particularly in the easily distensible atria, due to the increase in intracardiac pressures (Arborelius et al. 1972; Epstein 1976, 1978; Epstein et al. 1975, 1981a,b). However, studies in the nonhuman primate have shown that the reflex control of antidiuretic hormone is largely dependent on carotid baroreceptors and that vagal afferent pathways are not necessary for the renal responses to immersion. The challenge remains to determine whether or not large unencapsulated atrial receptors subserved by unmyelinated vagal afferents have a key role in causing the diuresis, or whether, in the primate, blood volume control is located more on the arterial side of the circulation (Gilmore 1983). The observation that in man negative pressure applied to the lower body does not increase plasma vasopressin unless arterial pressure is reduced suggests the latter possibility (Goldsmith et al. 1982). However, the mechanism(s) whereby isoosmotic volume expansion causes a decrease in plasma arginine vasopressin remain(s) incompletely defined. It seems that the decrease may occur independently of changes in the renin-angiotensin system (Epstein et al. 1981b). Studies have been conducted in cardiac transplant recipients to assess the contribution of cardiac volume receptors in the control of antidiuretic hormone release during postural changes; plasma antidiuretic hormone levels were reduced with head-down tilt, indicating that cardiac volume receptors were not involved (Convertino et al. 1984). In another study, chronic cardiac denervation inhibited the antidiuretic hormone response to acute plasma volume depletion (Grimaldi et al. 1985). When thigh cuff inflation was used to cause selective unloading of the cardiopulmonary low-pressure receptors, the plasma antidiuretic hormone levels increased (Egan et al. 1984). Thus the role of the cardiopulmonary receptors in regulating release of this hormone in man is still uncertain.

In the dog, the cardiopulmonary receptors with vagal afferents tonically inhibit the release of antidiuretic hormone (Thames et al. 1980). In humans, the reflex response to stimulation of somatic afferents (isometric exercise of one arm) is augmented when the input from the cardiopulmonary afferents is reduced by lower-body negative pressure (Walker et al. 1980; Abboud and Thames 1983). After training, the increase in plasma renin, vasopressin, and norepinephrine at the same work load are reduced; this might be explained by enhanced activity of the low-pressure intrathoracic mechanoreceptors (Convertino et al. 1983; Péronnet et al. 1981).

When the subject changes from the supine to the upright position, plasma vasopressin levels are increased, and in patients with orthostatic hypotension, tilting evoked a greater decrease in blood pressure in those subjects with a subnormal increase in plasma vasopressin. However, if the sympathetic nervous system and the renin-angiotensin system are activated normally by standing, these serve to maintain the arterial pressure in the absence of any increase in circulating vasopressin levels (Zerbe et al. 1983; Bennett and Gardiner 1985).

During spaceflight, when zero gravity is reached, the blood from the legs is displaced centrally. This is manifested by a feeling of fullness in the head and distension of the neck veins. Diuresis follows, the plasma and red cell volume decrease, and the intrathoracic blood volume is restored to normal. However, when spaceflight is prolonged, the decrease in total blood volume is such that on return to earth gravity, the central blood volume and the filling pressure of the heart are reduced as some of the blood is redistributed to the legs; as a result there is reduced tolerance for standing and fainting may occur. The zero environment thus offers unique opportunities to determine the mechanisms involved.

Recent studies have shown that the cardiocytes of mammalian atria contain specific secretory granules that synthesize and release peptides with diuretic, natriuretic and vasoactive properties. These are referred to

as atrial natriuretic factor or atriopeptins. Several of these peptides have
been purified, sequenced and synthesized (Needleman et al. 1985). Synthetic human atrial natriuretic peptide, when injected into healthy humans,
causes diuresis and natriuresis and a decrease in arterial blood pressure
(Richards et al. 1985). In the dog, the natriuresis caused by the intrarenal
infusion of synthetic atriopeptin is associated with an increase in glomerular filtration rate, a decrease in proximal tubular reabsorption and an
inhibition of renin release (Burnett et al. 1984).

It is suggested that the stimulus for release is stretch of the atrial walls
(Ledsome et al. 1985) but the physiological stimulus for atriopeptin secretion remains to be determined. The interrelationship of this system with
the blood volume control exerted by the cardiopulmonary receptors
awaits clarification. This will be facilitated by the availability of immuno-
assay techniques for measuring atriopeptins. The demonstration of atriopeptin immunoreactive neurons in regions of the brain of the rat which
have connections with the regulation of the cardiovascular system suggest
that atriopeptin might serve as a neuromodulator as well as a potential
hormone mediator of changes in body fluid composition, extracellular
volume and systemic blood pressure (Needleman et al. 1985).

15 Salt and Neural Cardiovascular Control

In animals an increase in the sodium content of the body can enhance the
vascular responses to activation of the sympathetic nerves. In the Dahl rat
prone to hypertension, chronic ingestion of an increased amount of salt
did not change the hindlimb vasoconstriction induced by infusion of norepinephrine but increased that induced by electrical stimulation of the
lumbar sympathetic chain (Takeshita and Mark 1978). In the spontaneously hypertensive rat (Okamoto 1969) a greater sodium intake produced
a more pronounced rise in blood pressure, a greater plasma concentration
of norepinephrine, and a more marked hypotensive effect of ganglio-
plegic agents (Wintermintz and Oparil 1982). Mechanoreceptors are sensitive to extracellular Na^+ and K^+ concentrations. In both normotensive and
spontaneously hypertensive rats, reduction by as little as 5% in $[Na^+]_o$
bathing baroreceptors reflexly increases arterial blood pressure and urine
output (Kunze et al. 1977; Saum et al. 1976). It is suggested, therefore,
that sodium sensitivity of baroreceptors may be an important mechanism
for prompt regulation of body fluid volume. In the deoxycorticosterone
acetate (DOCA) model of hypertension the greater sodium injection
necessary for blood pressure to rise was accompanied by evidence of a
faster turnover of norepinephrine in the sympathetic nerve terminals

(de Champlain 1977). Although less conclusive, evidence of a similar relationship between sodium and sympathetic function has been obtained in man. In healthy volunteers a short-term marked increase in the usual sodium intake (about 5 g for 4 days) increased the plasma level of norepinephrine, while epinephrine levels were not altered significantly (Nicholls et al. 1980). Feeding a healthy volunteer with 410 mEq sodium daily for 10 days increased the forearm vasoconstriction induced by cardiopulmonary receptor deactivation as compared to the response observed when only 10 mEq daily sodium were given (Mark et al. 1975). Increased pressor and tachycardic responses to mental arithmetic have been observed in a number of normotensive adolescents when fed with a diet richer in salt (Falkner et al. 1981). Also, there is a positive relationship between intralymphocytic sodium content and the pressor responses to mental arithmetic and isometric exercise (Ambrosioni et al. 1981). In young subjects with borderline hypertension moderate sodium restriction (170–100 mmol sodium/day) reduced both intralymphocytic sodium content and diastolic blood pressure at rest and the increase in pressure during handgrip. During a high-potassium diet (130 mmol potassium/day) both intralymphocytic sodium content and baseline diastolic blood pressure rose and the pressor response to handgrip was reduced by 50% (Ambrosioni et al. 1982a,b).

A reduction in sodium intake reduced the pressor response to common carotid occlusion in the unanesthetized dog (Rocchini et al. 1977). A reduction in sodium intake so small as to leave baseline blood pressure, heart rate, and plasma renin activity substantially unaltered made unanesthetized rats much less resistant to hemorrhage and development of irreversible shock (Göthberg et al. 1983). In unanesthetized rats a 5% reduction in the sodium concentration of the perfusate increased the threshold and reduced the slope of the firing rate of the arterial baroreceptors in response to a pressure stimulus. Increasing the potassium concentration of the perfusate had an opposite effect (Saum et al. 1977; Toubes and Brody 1970; Brown 1980). The effects of alteration in sodium and potassium intake on the baroreflexes have not been assessed in man. The evidence obtained in animals, however, is sufficient to suggest that some of the consequences of a reduced sodium intake may not be beneficial for cardiovascular homeostasis.

16 Static Exercise and Ergoreceptors

A strong static contraction of skeletal muscles, or rapid powerful rhythmic contractions, causes a marked increase in arterial blood pressure and

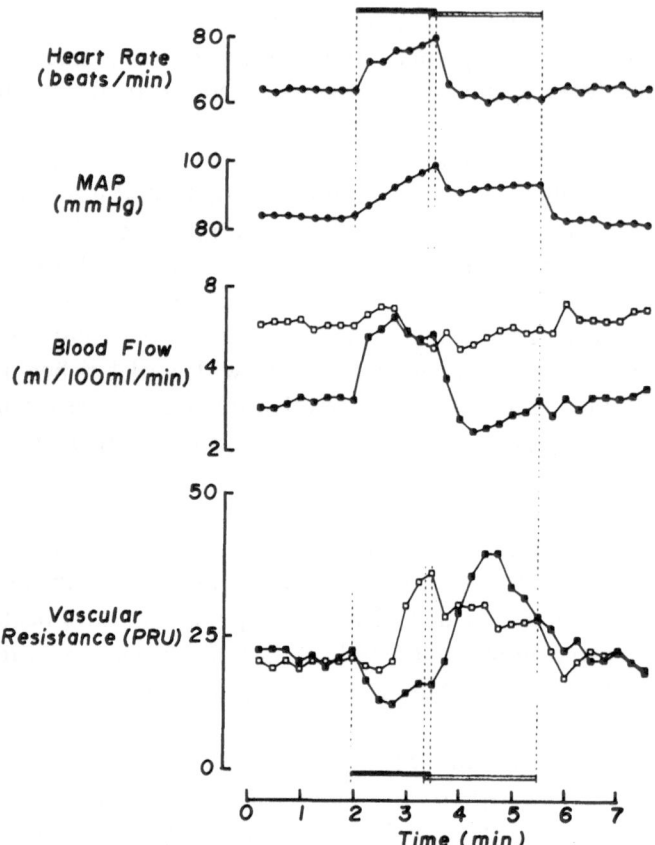

Fig. 12. Effect of isometric handgrip of the right forearm at one-third of maximal voluntary contraction, and of arterial occlusion of this forearm instituted just prior to the cessation of the contraction, on heart rate, mean arterial blood pressure (*MAP*), forearm (■—■) and calf (□—□) blood flows, and resistances. Vascular, resistance was calculated by dividing mean arterial pressure by blood flow and is expressed as peripheral resistance units (PRU). Average results for six subjects in the supine position. (From Rusch et al. (1981)

heart rate. There are resultant complex interactions with the arterial and cardiopulmonary baroreflexes (Walker et al. 1980; Abboud et al. 1981b, 1983). The increase in perfusion pressure helps to combat the mechanical compression of the vessels by the contracting muscle.

The receptors in the muscles responsible for the blood pressure rise appear to be activated by the products of muscle metabolism. A strong static contraction of the muscles of one forearm causes an increase in heart rate and blood pressure. When the circulation to this forearm is arrested by a pneumatic cuff just prior to the cessation of exercise and the arrest is continued, the blood pressure following a brief sag remains well above the control level until the circulation is restored, while the heart rate and cardiac output return to the control level. This indicates that the maintenance of the elevated pressures is due to receptors activated by metabolites trapped in the muscle by the circulatory arrest. By contrast, the rapid return of the heart rate to the control level indicates that this might be governed by muscle receptors sensitive to changes in tension (Fig. 12). The sensory fibers involved are the small myelinated (group III) and unmyelinated (group IV) afferents. These endings may serve both as metabolic receptors to initiate reflex circulatory and ventilatory changes during exercise, and as pain receptors (Mitchell and Schmidt 1983).

There does not appear to be any correlation between the cardiovascular response to static exercise and the percentage of fast-twitch muscle fibers involved (Leonard et al. 1985). Some studies have reported that static exercise of small muscle groups can evoke as large an increase in arterial blood pressure as that when large muscle masses are involved, indicating that the cardiovascular response was independent of the skeletal muscle mass active in the static contraction. Other studies, however, have shown that the cardiovascular response is greater when a larger muscle mass is involved (Lind 1983; Mitchell et al. 1980; Seals et al. 1983). The pressor response to muscle ischemia is proportional to the degree of ischemia and oxygen deficiency in dynamically exercised muscle, and the pressor responses to muscle ischemia are proportional to the mass of the ischemic muscle (Rowell et al. 1981).

In addition to the reflex from the active muscles, there is a "central command" from the cerebral cortex to the cardiovascular and respiratory centers in the brain stem (McCloskey 1981). The initial sag in the blood pressure when the exercise ceases and the circulatory arrest is maintained is indicative of the cessation of this command (Fig. 12). It seems that the appropriate cardiovascular response may be elicited via either reflex or central neural mechanisms due to their influence on the same regulatory nerve cells. It has been suggested that these two control mechanisms are redundant and that neural occlusion may be operative. In the presence of partial neuromuscular blockade, which induces a disproportion between an increase in central command and a constant or decreasing muscle tension and metabolism, the larger signal arising from central command determines the magnitude of the cardiovascular response (Mitchell and Schmidt 1983; Leonard et al. 1985). In other studies, microneurography was used to measure the sympathetic discharge in a fascicle of the peroneal nerve to muscle vessels. These suggested that central command, while increasing heart rate, inhibits increases in muscle sympathetic nerve activity. By contrast, stimulation of chemically sensitive muscle afferents increases sympathetic nerve activity but has little influence on heart rate (Mark et al. 1985b). Mark et al. (1985a) measured efferent sympathetic nerve activity to human leg muscles by inserting microelectrodes into a fascicle of the peroneal nerve. This permitted an evaluation of the role of muscle afferents and central command in the regulation of sympathetic outflow to muscle during static and rhythmic arm exercise. They found that the increase in muscle sympathetic activity in the leg during sustained handgrip was a consequence of the stimulation of chemically sensitive afferent nerves in the contracting muscles. The central command, which caused the tachycardia, did not increase but instead appeared to inhibit the sympathetic outflow to the leg muscles. These studies indicate that heart rate and muscle sympathetic nerve activity are governed by different mechanisms during static exercise. Non-fatiguing rhythmic handgrip increased heart rate and arterial blood pressure but did not increase the nerve activity, presumably because sufficient metabolites did not accumulate in the active muscles to activate the chemically-sensitive muscle afferents (Mark et al. 1986).

The increase in blood pressure with static exercise is due to an augmented sympathetic outflow, which results in cardiac acceleration, an increase in cardiac output, and only a slight increase in systemic vascular resistance. If, however, the cardiac output cannot increase normally, as occurs in healthy subjects who have been given propranolol and atropine or in patients after heart transplantation, the same increase in blood pressure is achieved by increased constriction of systemic vessels (Haskell et al. 1981; Lewis et al. 1983). Thus the prime purpose of the reflex seems to be to increase the blood flow to the contracting muscles and it calls on any combination of cardiac output and peripheral resistance to achieve this goal. At that time, the arterial and the cardiopulmonary mechanoreceptors are rendered unable to oppose the increase in blood pressure.

Dynamic as well as static exercise of small muscle groups, performed to a common end-point of local muscular fatigue, caused a marked pressor response, but only a small increase in heart rate. Cardiac output increased during the dynamic but not the static exercise, whereas the reverse is true for total peripheral resistance. The pressor response persisted after combined beta-adrenergic and parasympathetic blockade. However, the cardiac output still increased with the dynamic exercise even though the heart rate increase is absent. It is suggested that activation of the muscle pump and alpha-mediated venoconstriction enhanced the ventricular filling pressure and hence increased the stroke volume. This again indicates that the rise in systemic arterial pressure is a regulated function and is achieved by different mechanisms depending on the mode of muscular contraction (Lewis et al. 1983).

During isometric exercise of one forearm, the resistance vessels of the muscles in the other forearm and the calf respond differently. The former dilate and the latter constrict. When the circulation is arrested just before exercise ceases and the arrest is continued, the forearm vessels constrict when the exercise ceases and the calf vessels remain constricted. This implies that the muscle reflex causes constriction in muscle blood vessels but in the forearm this is opposed by the central command during exercise (Fig. 12) (Rusch et al. 1981).

17 Thermoregulation

One of the principal control mechanisms in thermoregulation is the alteration of the circulation through the skin to govern the rate of heat flow from the body core to the environment. Rapid, nervously mediated changes in skin blood flow provide the fine tuning for the control of body temperature. Along with shivering and increased sweating, this permits effective governance of the core temperature during heat stress.

The thermoregulatory effector mechanisms are controlled by receptors in the skin, sensitive to cold and to heat, and by central thermoreceptors in the hypothalamus, brain stem, spinal cord and, in sheep and rabbits, the abdominal cavity. Some of the thermosensitive nerve endings in the skin travel with the sympathetic nerves to the spinal cord. Thus the body core and the skin are inputs to the thermoregulatory control system (Hellon 1983).

The cutaneous vessels are also governed by thermoregulatory mechanisms. Local heating decreases or abolishes a reflexly induced vasoconstriction, while local cooling augments it (Zitnik et al. 1971; Shepherd et al. 1983; Cooke et al. 1984).

When resting subjects are exposed to a heat stress, there is a rapid increase in skin blood flow and in cardiac output. While the skin, like the kidney, requires only a small blood flow to meet its metabolic needs, during severe heat stress the total skin blood flow may increase from about 0.2 l/min during normothermia to about 7 l/min during severe hyperthermia, or about 50%–70% of the cardiac output. The cutaneous blood volume also increases due to the decrease in sympathetic nerve activity to the capacious subpapillary venous plexus. The increase in venous blood transit time that accompanies the increased venous volume facilitates heat exchanges with the environment. However, during the very high skin blood flow accompanying severe heat stress, the increased volume of this venous plexus may lead to a reduction of cardiac filling pressure (Roddie et al. 1956; Johnson et al. 1974; Rowell 1983).

In supine subjects at rest, the cardiac output increases by approximately 3 l/min per degree centigrade rise in the temperature of the blood in the right atrium. The increase in skin blood flow exceeds the increase in cardiac output, and this increase is compensated for by a simultaneous decrease in flow to the kidney and the splanchnic vascular beds (Rowell

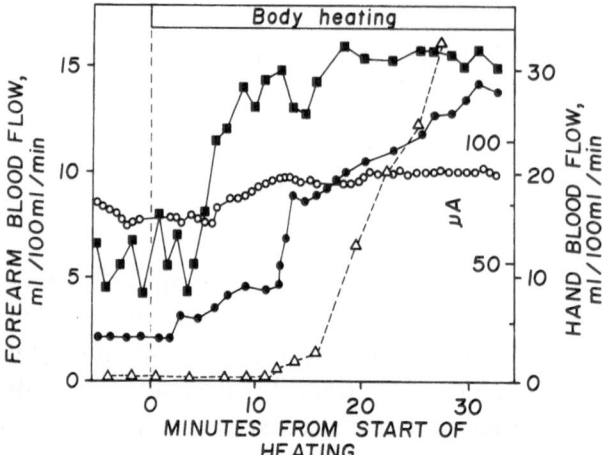

Fig. 13. Forearm and hand blood flow before and during body heating in a normal subject. ●, blood flow through the intact forearm; ○, blood flow through the nerve-blocked forearm; ■, blood flow through the hand; △, current (μα) flowing through the skin of the intact forearm: this indicates the activity of the sweat glands. Initially, the subject was lightly clad and placed in a room at $16°-18°C$ to ensure a low blood flow to the forearm. Within 5 min after the start of heating, the blood flow through the hand increased rapidly and reached 30–35' ml/100 ml forearm per minute. At the same time, there was a small increase in the blood flow to the intact forearm, which remained at this level until the onset of sweating. The flow then increased rapidly; this increase was followed by a more gradual rise until the end of the heating period, when it reached 14 ml/100 ml forearm per minute by which time the subject was sweating profusely. Since body heating dilates only the skin vessels of the forearm and the nerve block dilates both muscle and skin vessels, the greater increase in flow with body heating than with the nerve block is explained by the activity of vasodilator nerves supplying the skin of the intact forearm. Thus this experiment demonstrates that the vessels of the forearm skin are supplied with both vasoconstrictor and vaso-dilator nerves. (From Shepherd 1984)

1983). The mechanisms which lead to the constriction in these beds have not been determined. Heat stress also causes an increase in plasma renin activity (Escourrou et al. 1982). Increase in skin flow is achieved in the case of the hands and feet, where arteriovenous anastomoses are present, by a decrease in activity in the sympathetic noradrenergic nerves. In the case of the skin of the proximal part of the limbs and probably also of the skin of the face and body, the flow is much greater than that achieved by complete absence of vasoconstrictor impulses in the sympathetic nerves. When a normal subject is exposed to a generalized heat stress, the blood flow to the innervated forearm increases well above that to the opposite forearm in which the cutaneous nerves are blocked with local anesthetic. In the case of the hands no further increase occurs. This increase must therefore be due to vasodilator nerves. It is associated with the onset of sweating and can be delayed or reduced by injecting atropine into the brachial artery, indicating that it is due to a cholinergic mechanism (Fig. 13).

Microelectrodes have been used to record from cutaneous sympathetic nerves to the hands and feet. Firing in these nerves occurs in bursts. With body cooling the interval between bursts decreases and with body warming it increases (Vallbo et al. 1979; Bini et al. 1980a,b). The outflow to hands and feet is synchronized. When sympathetic nerve activity is recorded simultaneously in human nerves to muscle and skin, there is no relationship in the timing or the amplitude of the sympathetic bursts of activity. This indicates that the sympathetic ganglia contain at least two different neuronal populations with independent preganglionic inputs (Bini et al. 1980a). Recordings of sympathetic activity to different skin areas in humans have demonstrated that the reflex thermoregulatory functions in the distal glabrous skin areas are mediated mainly via changes in the activity of sympathetic vasoconstrictor fibers; sudomotor fibers only become active at relatively high temperatures. By contrast, sudomotor fibers predominate in the thermoregulatory control of the hairy skin on the dorsal side of the forearm and hand (Bini et al. 1980b). Recordings of sympathetic outflow to the skin demonstrate that during moderate warming, the activity becomes minimal, but that it increases again with the onset of sweating, presumably because of increased traffic in the sudomotor nerve fibers (Normell and Wallin 1974). It appears that with sweating an enzyme is released into the tissue spaces from the sweat glands. This enzyme acts on proteins to produce a vasodilator peptide. It has been suggested, but not proved, that a bradykinin-forming enzyme is released and that the bradykinin formed causes the vasodilatation. Thus the linkage of dilatation to the regulation of sweating, which is induced by cholinergic nerves, and the change from the dominant to the limited control by sympathetic nerves occur abruptly between the skin of the hands and feet and that of the rest of the limb (Shepherd 1963; Rowell 1981; Roddie 1983).

18 Rhythmic Exercise

With supine leg exercise the sympathetic vasoconstrictor outflow increases in proportion to the severity of the exercise (Bevegård and Shepherd 1966a, 1967). During upright exercise the arterial and cardiopulmonary mechanoreceptors are unloaded due to the effects of gravity on the cardiovascular system, leading to a shift of blood from the central circulation to the dependent parts of the body. The resultant increase in sympathetic outflow reinforces that caused by the reflex originating in the contracting muscles. Thus the blood flow to inactive or less active muscles is further reduced, as is that to the splanchnic bed and the kidney (Rowell et al. 1964). This strong constriction of the vessels outside the active muscles is reflected by the changes in cardiac output, which for identical exercise up to submaximal effort is about 2 l/min less during upright than during supine exercise (Bevegård and Shepherd 1967). There is very little rise in venous plasma epinephrine during mild or moderate treadmill exercise (Warren et al. 1984). One of the benefits of upright exercise is the action of the muscle pump, which by reducing the venous pressure in the legs leads to a marked increase in the perfusion pressure of the leg muscle vessels. It also restores the translocated blood to the thorax, which maintains the filling pressure of the heart and hence the stroke volume (Bevegård and Shepherd 1967). In patients with damage to the venous valves, this does not occur; more blood remains in the dependent parts, the cardiac filling is hampered, and the stroke volume and cardiac output are less than in normal persons (Bevegård and Lodin 1962). In the working muscles, provided the exercise is mild, a reflex excitation of sympathetic nerves can reduce their blood supply. Under these circumstances, the role of the sympathetic nerves may be to modulate the local dilator mechanisms in order to maintain the most economical ratio of blood flow to oxygen extraction. At more severe work loads the blood flow to the exercising muscles is not affected by sympathetic stimulation because the products of metabolism attenuate the action of the sympathetic nerves on the muscle resistance vessels; this permits the blood flow to reach its maximum under circumstances where maximal oxygen extraction from the blood going to the muscles is necessary to sustain their metabolism (Strandell and Shepherd 1967; Blomqvist and Saltin 1983).

In a comfortable environment, the cutaneous vessels constrict at the onset of exercise, but this is transient. As the core temperature starts to rise the skin vessels dilate. The combined stress of exercise in the upright position and of hyperthermia poses the greatest challenge to the cardiovascular system, since skin and muscles must compete for the cardiac output to satisfy simultaneously the metabolic requirements of the active muscles and the thermoregulatory requirements of the body. During heat stress the leg veins refill so rapidly during exercise, due to the great increase in skin blood flow, that the muscle pump is less efficient and more blood remains pooled in the dependent parts (Rowell et al. 1971; Rowell 1983). While there is evidence that during heat stress the removal of the inhibitory influence of the arterial and cardiopulmonary mechanoreceptor on the vasomotor center can cause a temporary constriction of the skin vessels of the forearm (Crossley et al. 1966), it is doubtful if this has any important effect on the amount of blood pooled in the dependent veins in the heated subject in an upright posture. The consequential

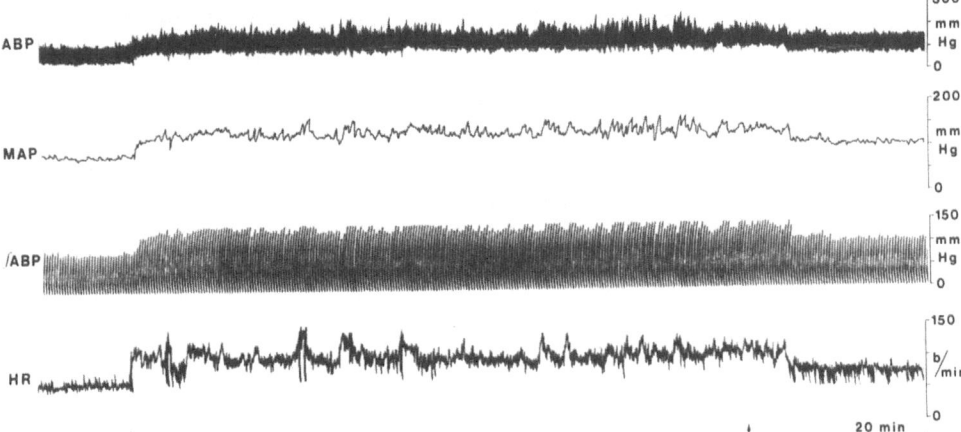

Fig. 14. Blood pressure rise during excitement. Original record taken from a 24-h blood pressure recording in a normotensive subject. The record was displayed on a polygraph after completion of the monitoring time. *ABP*, pulsatile arterial blood pressure; *MAP*, mean arterial pessure; *ʃABP*, blood pressure integration over consecutive 37.5-s intervals; *HR*, heart rate. The subject played cards (poker) during the interval *between the arrows* (about 2.5 h), during which time the blood pressure and heart rate remained elevated. (From Mancia et al. 1984)

decrease in thoracic blood volume and reduction in cardiac filling pressure limit the increase in cardiac output. Thus for the arterial blood pressure to be maintained necessitates a very marked increase in the sympathetic outflow, so that there is a maximal constriction of the splanchnic resistance and capacitance vessels and of the renal vessels. Many factors operate to cause this reflex constriction, including presumably the receptors in the active muscles, as well as the arterial baroreceptors when the arterial pressure tends to fall.

19 Emotion

Direct recordings of arterial blood pressure in subjects living in their normal environment have confirmed that a marked and sustained rise in arterial blood pressure can occur during emotional events (Fig. 14) (Mancia et al. 1984b). Studies of the cardiovascular responses to acute emotional stress elicited in the laboratory have sometimes had conflicting results, in part because the responses may be transient and in part because the same subject may at different times have different responses to the same stress (Parati et al. 1985). When normal reclining subjects were subjected to stressful mental arithmetic so that they became tense, nervous, and embarrassed, the mean arterial blood pressure and the cardiac output increased. The forearm blood flow increased, while forearm vascular resistance decreased (Blair et al. 1959a). When Stroop's color word test was used to cause mental stress in healthy volunteers, the resultant increase in arterial blood pressure (average 29/14 mmHg) was caused by an increase in cardiac output, since total systemic vascular resistance decreased. Arterial plasma norepinephrine is increased but peripheral venous levels remain unchanged (Hjemdahl et al. 1984). Changes in renal blood flow are difficult to interpret, since the duration of the stimulus is too brief to permit the collection of urine samples adequate for determination of para-aminohippurate clearance, but it is probable that the flow is decreased. Values calculated for total systemic resistance were sometimes increased and sometimes decreased, but there was thought to be no basic difference between these two types of response. The changes in total resistance depend on the balance between vasodilatation in muscle and vasoconstriction in the kidney and splanchnic circulation; when the former is dominant, the value for total systemic resistance is reduced, and vice versa (Brod et al. 1959). Cerebral vascular resistance was unchanged during the performance of mental arithmetic, but decreased during anxiety (Sokoloff et al. 1955). The cutaneous veins constrict; there are no data on other capacitance vessels.

Emotionally induced hyperemia of the forearm muscles is thought to result from the combined effects of increased levels of circulating epinephrine and activation of cholinergic vasodilator fibers. The cholinergic pathway, which has been described in the cat and the dog, originates in the motor cortex. There is a discrete descending pathway that relays in the hypothalamus and continues through the ventral mesencephalon, descending via the medullary lateral spinothalamic tracts (Üvnas 1967). The cholinergic nerves run with sympathetic noradrenergic nerves to the resistance blood vessels of skeletal muscle; their activity is reduced or abolished by atropinization. The relative importance of these two factors is controversial (Greenfield 1966). Evidence for the role of the cholinergic vasodilator fibers comes from the demonstration that the magnitude of the hyperemia is diminished by stellate ganglion block and by infusion of atropine into the artery of the limb being studied (Blair et al. 1959a; Barcroft et al. 1960).

Surprisingly, emotional stimuli may not be accompanied by any change in vascular resistance in the calf (Rusch et al. 1981). There is as yet no explanation for this. It has been suggested that the norepinephrine released from the sympathetic nerves to the forearm activates both postjunctional alpha- and beta-adrenoceptors, with the former predominating, whereas in the calf there is only stimulation of alpha-adrenoceptors (Eklund and Kaijser 1976). However, there is no evidence of neurogenically mediated beta-adrenergic receptor activation in the human forearm (Brick et al.

1968), and the vessels of the forearm and calf dilate to the same extent when exposed to isoproterenol (Whelan 1967). Another explanation is that the cholinergic vasodilator nerves are distributed to the forearm, but not to the calf vessels. That both forearm and calf vessels are capable of simultaneous reflex constriction is shown by the decrease in both forearm and calf blood flows following the Valsalva maneuver (Rusch et al. 1981). It is reported that the Valsalva maneuver and the application of ice to the forehead simultaneously stimulate adrenergic vasoconstrictor and cholinergic vasodilator pathways to forearm resistance vessels. Since the former predominates, the vasodilator effect is unmasked only after adrenergic blockade (Abboud and Eckstein 1966).

Alterations in reflex function during mental stress are difficult to determine. In cats activation of hypothalamic structures involved in the regulation of cardiovascular emotional responses impairs arterial baroreflex control of heart rate more than baroreceptor control of vascular resistance and blood pressure (Mancia and Zanchetti 1981). Inhibition of the cardiac vagal component of the arterial baroreflex during emotional stress occurs in the baboon (Stephenson et al. 1981) and monkey (Engel et al. 1982). The smaller increase in heart rate, cardiac output, and mesenteric resistance observed during naturally elicited fighting behavior in sinoaortic denervated as compared to intact cats can also be explained by central inhibition of the arterial baroreflexes (Baccelli et al. 1981).

In man phenylephrine has been used to increase arterial blood pressure and to stimulate the arterial baroreceptors during stressful mental arithmetic. During the latter the slowing of the heart caused by the drug-induced increase in arterial pressure was attenuated (Sleight et al. 1978). In another study, steady neck suction was used to decrease arterial blood pressure and heart rate; the imposition of mental stress did not alter the response to the carotid baroreceptor activation (Lindblad 1980).

The stimulus of cold results in hemodynamic changes resembling those caused by psychological factors. Keatinge et al. (1964) studied the response of normal subjects, sitting and supine, to a shower of ice-cold water ($0° - 2.5°C$) run over the chest and abdomen for 2 min at 6 1/min. The respiratory rate and the tidal volume increased, while the arterial pCO_2 decreased. The cardiac output, systemic arterial pressure, and heart rate in the seated subjects increased. The transmural atrial pressure increased. There was no significant change in plasma epinephrine levels during the showers, and the concentration of norepinephrine increased by an average of only 0.32 $\mu g/l$ plasma. The intense hyperventilation which accompanied the cold showers appeared to play no part in causing the hypertension, although it may have contributed to the tachycardia.

These changes can be explained by activation of the sympathetic adrenergic nerves to the heart and blood vessels with an increase in the force and frequency of cardiac contraction, rather than by release of catecholamines from the adrenal medulla. The reflex could be mediated through the tegmentum of the midbrain and the hypothalamus, since stimulation of these areas in cats caused similar changes (Abrahams et al. 1960). Application of cold also causes reflex venoconstriction in the limbs; if this were widespread, it could help to increase the filling pressure of the heart and thus contribute to the increase in cardiac output.

20 Fainting

Foster (1888) believed that the decrease in blood pressure during fainting resulted from vagal inhibition of the heart. Lewis (1932) found that, although atropine abolished the bradycardia, it had little effect upon the hypotension or the level of consciousness. Therefore, he regarded vasodilatation as the primary cause of the decrease of arterial pressure, and introduced the term "vasovagal syndrome" to emphasize the combined etiological role of alterations in the behavior of peripheral vessels and in vagal activity.

Fainting associated with loss of blood has been extensively studied. After removal of a large volume of blood from a normal subject by venesection, the cardiac output and right atrial pressure are decreased. The systemic arterial pressure is at first maintained through reflex constriction of resistance vessels due to the combined withdrawal of the inhibition exerted by the arterial and cardiopulmonary mechanoreceptors. With the onset of fainting, the cardiac output does not usually decrease further, but there is a marked reduction in systemic vascular resistance. The blood flow through the skeletal muscles of the forearm increases (Barcroft et al. 1944; Barcroft and Edholm 1945). Since there is simultaneously a decrease in arterial blood pressure, the increased flow is due to vasodilatation. Vasodilatation does not occur in the forearm of patients following cervicodorsal sympathectomy. Therefore, it is mediated by sympathetic nerves rather than by humoral factors such as endogenous epinephrine. The vasodilatation may be mediated by release of sympathetic vasoconstrictor tone or alternatively, as Barcroft and Edholm (1945) suggested, by increased activity of sympathetic vasodilator fibers. In one subject who fainted after standing up, there was a cessation of sympathetic bursts in peroneal muscle nerve fascicles during the initial bradycardia. This suggests that the syncope was associated with an interruption of normal baroreflex feedback between blood pressure and sympathetic outflow (Burke et al. 1977).

Bradycardia, vasodilatation in skeletal muscle, and decreased arterial blood pressure also occur during spontaneous or emotional fainting (Brigden et al. 1950; Greenfield 1951).

Some of the effects resulting from acute hemorrhage during light general anesthesia differ from those observed in conscious subject (de Wardener et al. 1953). Loss of up to about 1500 ml blood led to decreases in cardiac output and constriction of forearm resistance vessels. Hemorrhage of this order causes vagovagal fainting, with vasodilatation in skeletal muscle in the majority of conscious subjects lying supine. In the anesthetized patients, the vessels in muscle were constricted even in the rare instance where a subject had severe hypotension and bradycardia. This suggested that posthemorrhagic hypotension during anesthesia was due mainly to the decrease in cardiac output and not, as in conscious subjects, to dilatation of vessels in muscle.

In contrast to the increased blood flow to the skeletal muscle of unanesthetized subjects during fainting, flow to the splanchnic bed (Bearn et al. 1951) and kidneys (de Wardener and McSwiney 1951) is reduced. Cutaneous vasodilation may occur (Snell et al. 1955). As consciousness is lost, the electroencephalogram shows slow waves of large amplitude (Gastaut and Fischer-Williams 1957). At this stage, cerebral blood flow is reduced by about one-third (Finnerty et al. 1954). The cerebral hypoxia is mainly responsible for the hyperventilation. Öberg and Thorén (1972), from studies on cats, suggested that the increased sympathetic outflow during severe hemorrhage and lowered ventricular filling could cause excessive activation of left ventricular mechanoreceptors with vagal unmyelinated afferents. This could cause the bradycardia and peripheral vasodilatation. A similar mechanism could explain vasovagal syncope during orthostatic stress in man (Epstein et al. 1968a).

Orthostatic fainting readily occurs in the presence of conditions that interfere with the development of compensatory vasoconstriction in arteriolar beds. After spinal anesthesia, there is persistent arterial dilatation in the anesthetized portion of the body; relatively slight tipping of the body from the horizontal position may lead to an acute fall in blood pressure accompanied by vasodilatation in the forearm (Brigden et al. 1950). The exaggerated decrease in cardiac output with tipping during spinal anesthesia is contributory (Pugh and Wyndham 1950).

21 Reflex Control in Cardiovascular Disease

21.1 Autonomic Failure

Chronic autonomic failure is caused by multiple lesions of the central or peripheral autonomic pathways (Ibrahim 1975). It is characterized by an unvarying heart rate and a decrease in arterial blood pressure on standing, due to a reduced ability of the heart rate to increase and to a failure of constriction of the systemic resistance and capacitance vessels (Bannister et al. 1967; Marshall et al. 1961; White 1980). The normal heart rate response to standing has a characteristic time course, with an immediate increase due to an exercise reflex as a consequence of the muscular effort, and later changes due to alterations in the activity of the arterial and cardiopulmonary mechanoreceptors. An abrupt and large increase in heart rate on standing excludes vagal neuropathy (Wieling et al. 1983). By monitoring ambulant intraarterial blood pressure it was found that patients with autonomic failure who exhibit postural hypotension have a circadian variation in pressure. This was the inverse of the normal pattern, with the highest pressures being at night and the lowest in the morning. This is consistent with the finding of a peak incidence of the symptoms of postural dizziness in the morning (Mann et al. 1983). Recumbent hypertension may also be present. This has been attributed to loss of central control of baroreflex function, reinforced by supersensitivity of the cardiovascular system to circulating catecholamines. While most patients with orthostatic hypotension demonstrate these abnormalities in sympathetic nervous system function, different mechanisms may be involved. Those patients with histochemical and pharmacological evidence of a deficit in peripheral sympathetic nerves are classified as having idiopathic orthostatic hypotension. Those patients with a central deficit in blood pressure regulation are classified as having multiple-system atrophy (Shy-Drager syndrome). Chronic autonomic insuffiency also occurs with parkinsonism and diabetes (Shy and Drager 1960; Aminoff and Wilcox 1971; Kontos et al. 1975; Wilcox et al. 1977).

At autopsy, the most common lesion in all patients is a loss of cells in interomediolateral columns of the spinal cord. Lesions have also been reported in the locus ceruleus, the nucleus of the solitary tract, the dorsal nucleus of the vagus, and the ambiguus nucleus (Bannister 1980). In diabetic autonomic neuropathy, heart rate control may be impaired earlier than vasomotor function, which is likely to be due to the early involvement of the cardiac vagus (Bennett et al. 1979b).

Patients with autonomic dysfunction have norepinephrine depletion and a reduced plasma norepinephrine response to standing; in those with idiopathic orthostatic hypotension the plasma levels when reclining are lower than normal, whereas they are normal in those with multiple-system atrophy (Kontos et al. 1975; Cryer and Weiss 1976). The patients are not only hyperresponsive to alpha-adrenergic agonists but also have an excessive rise in arterial blood pressure in response to intravenous administration of angiotensin II (Davies et al. 1980) and vasopressin (Mohring et al. 1980). There is also increased beta-adrenoceptor responsiveness in these patients and isoproterenol can cause a marked lowering of arterial blood pressure; while there is a sixfold increase in responsiveness to its inotropic effect, there is a 17-fold increase in responsiveness to its depressor effect (Robertson et al. 1984). Studies of adrenergic receptor regulation in man have been facilitated by the identification of beta-adrenoceptors on human leukocytes and alpha-adrenoceptors on human platelets. In subjects with multiple-system atrophy with sympathetic nerve degeneration there is an increased number of beta-adrenoceptors as measured by [H^3] dihydroalprenolol ([H^3]DHA) binding to beta-receptors on lymphocytes isolated from their venous blood (Bannister et al. 1981). It is suggested that the decrease in the circulating and vascular concentrations of norepinephrine or epinephrine or both may lead to increased numbers of beta-receptors (Fraser et al. 1981a; Hui and Connolly 1981; Chobanian et al. 1982), or to their increased affinity state (Fitzgerald et al. 1981).

Another type of orthostatic hypotension, referred to as sympathotonic, does not seem to be due to abnormal sympathetic function. Patients with this disease have normal levels of norepinephrine and low mean arterial blood pressure when reclining. They also have reduced sensitivity to infused norepinephrine. On standing they have a marked tachycardia and plasma norepinephrine levels sometimes increase excessively. In these patients it seems that the effector organ response is impaired (Klein et al. 1980; Polinsky et al. 1981). Abnormalities of adrenergic physiology also occur in about 26% of patients with relatively severe diabetes. Some of these patients have hypoadrenergic postural hypotension due to diabetic adrenergic neuropathy, and hence have a lessened increase in plasma norepinephrine on standing. Other patients exhibit heightened sympathetic activity, with increased plasma norepinephrine concentrations on standing; some of them, however, have postural hypotension (hyperadrenergic postural hypotension), and it is suggested that this might be due to vascular resistance to endogenous norepinephrine (Cryer et al. 1978).

In elderly subjects with an inability to maintain their blood pressure while standing, no evidence was found of impaired baroreceptor function or of a reduction in sympathetic nervous activity. It is suggested that the orthostatic hypotension which occurs in 11%–24% of such subjects is due to mechanical factors or to failure of sympathetically mediated vasoconstriction rather than to receptor dysfunction (Robinson et al. 1983).

21.2 Cardiac Failure

Many diseases of the heart are associated with impairment of the arterial baroreflex control of heart rate. These include Chagas' disease (Amorin et al. 1968; Manco et al. 1969) and ischemic and rheumatic heart disease (Eckberg et al. 1971; Eckberg 1980b; Goldstein et al. 1975; Neto et al. 1980; Goldsmith et al. 1983a). How this occurs is a matter of speculation. In subjects with heart failure it may be due in part to reduced compliance of the aortic and carotid arterial wall as a consequence of salt and water retention.

Patients with acute heart failure due to myocardial infarction, and with radiological evicence of distension of vessels in the upper zone of the lungs, have evidence of an increase in renal function (elevation of glomerular filtration rate and an increase in urinary output). This lasts for about 2 days and could be attributed to increased cardiac filling pressure, which stimulates cardiac sensory receptors and thus inhibits the sympathetic outflow to the kidney. This helps to reduce the intravascular volume and thus the abnormally high pressure causing overdistension of the heart (Bennett et al. 1977). In congestive heart failure the usual increase in heart rate and mean arterial blood pressure during passive upright tilt are blunted, and forearm and hepatic vascular resistances do not increase (Goldsmith et al. 1983a). The circulating levels of norepinephrine, renin, angiotensin II and vasopressin are increased and there is an exaggerated sympathoadrenal response to exercise (Chidsey et al. 1962; de Champlain et al. 1963; Kramer et al. 1968; Thomas and Marks 1978; Cohn et al. 1981; Goldsmith et al. 1983b; Levine et al. 1983; Mark 1983). This can be explained by decreased tonic inhibition of the vasomotor centers by the cardiac mechanoreceptors, and can account in part for the retention of sodium and water. Impaired ability of the arterial baroreceptors to modulate the sympathetic outflow can contribute. During the relatively mild exercise of which these patients are capable, there is a greater rise in the plasma levels of norepinephrine than with similar exercise in normal subjects (Chidsey et al. 1962), and more reflex constriction of systemic vascular beds outside the active muscles. This also may be due to the decreased ability of the cardiac mechanoreceptors to buffer the excitatory reflex from the active muscles.

In patients with congestive heart failure, the forearm muscle resistance vessels do not dilate normally in response to exercise of the forearm or in response to a period of circulatory arrest (Zelis and Longhurst 1975). This is probably due to an increased stiffness of the vessels as a consequence of an increased sodium content and perhaps partly by increased tissue pressure related to interstitial edema. In patients with borderline hypertension, unlike normal subjects, excessive sodium intake causes abnormal increases in forearm vascular resistance, neurogenic vasoconstriction, and arterial blood pressure. The reasons for this are not known (Mark et al. 1975).

The normal reflex constriction of the forearm blood vessels which occurs with upright tilting of normal subjects may change to dilatation in patients in left heart failure (Brigden and Sharpey-Schafer 1950). The mechanism responsible is unknown. It is suggested that it may result from paradoxical stimulation of ventricular mechanoreceptors. Thus, during the application of lower-body negative pressure in patients with left ventricular dysfunction, the cardiac size may be reduced; this could allow for a greater degree of cardiac fiber shortening with a resultant mechanical activation of cardiac mechanoreceptors and a consequential peripheral vasodilatation (Ferguson et al. 1984). After the intravenous administration of a digitalis glycoside, the patients who showed a dilatation of the forearm resistance vessels with lower-body negative pressure prior to the drug now had a vasoconstriction. This might be explained by a direct sensitizing effect of digitalis on cardiopulmonary and/or arterial baroreceptors, or to an indirect mechanical (inotropic) effect on cardiac mechanoreceptors (Abboud et al. 1981a; Ferguson et al. 1984; Quest and Gillis 1974). Digitalis may result in a potentiation of the arterial baroreflexes; in subjects without heart failure the intravenous administration of lanatoside C enhances the bradycardic and hypotensive effects of baroreceptor stimulation by phenylephrine and neck suction respectively (Ferrari et al. 1981; 1983). The lengthening of atrioventricular conduction time induced by baroreceptor stimulation at a fixed heart rate was also potentiated markedly after administration of a single dose of digitalis (Fig. 15). Interestingly, the effect of arterial baroreceptor deactivation (i.e., tachycardia, hypertension, and shortening of atrioventricular conduction time) were not affected, a finding that may be in accordance with a sensitizing action of the drug on the receptors themselves. While sodium restriction and diuresis cause increases in renin and angiotensin levels in normal subjects, they cause decreases in patients with heart failure (Genest et al. 1968). It seems that the afferent activity from the heart, which is impaired in chronic heart failure, increases again during sodium restriction, diuresis, and upright exercise.

When patients with heart failure exercise, the reflex constriction of resistance vessels in vascular beds outside the active muscles occurs at lower work loads than in normal subjects. This is due to an earlier and more intense activation of the sympathetic outflow, which serves to redistribute the limited cardiac output to the vital organs and to maintain the arterial blood pressure (Chidsey and Harrison 1962; Francis et al. 1982). The stimulus for this and the receptors concerned have not yet been elucidated.

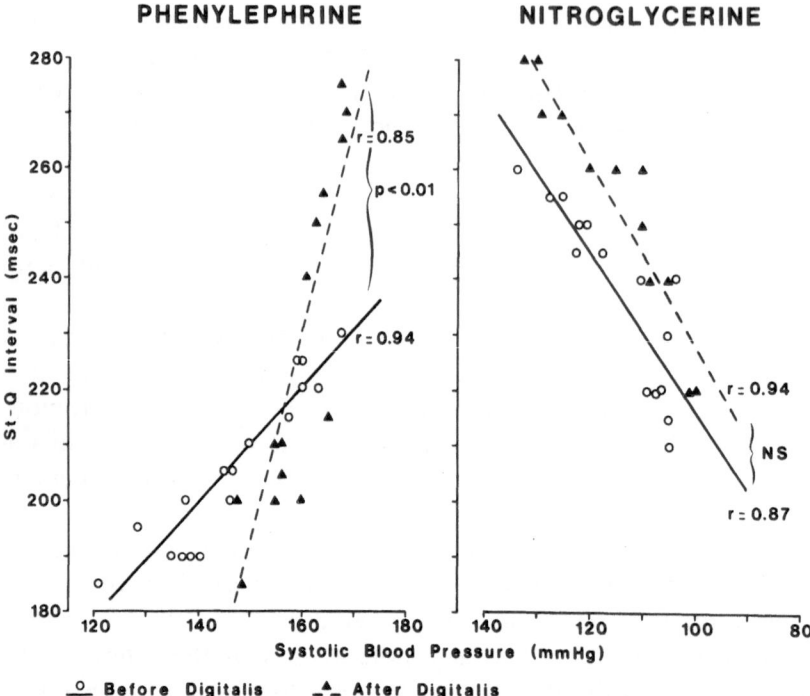

Fig. 15. Cardiac glycosides and reflex control of atrioventricular conduction. Linear regressions relating the increases and reductions in systolic blood pressure (phenyl-ephrine and nitroglycerine techniques) to lengthening and shortening respectively of St-Q interval. Data from one subject studied at a constant heart rate by atrial pac-ing, namely in the condition required to observe changes in atrioventricular conduc-tion time following arterial baroreceptor stimulation and deactivation. r values refer to correlation coefficients of the regressions, p values and NS (not significant) to statis-tical comparisons (covariant analysis) of the regressions before and after intravenous administration of 0.8 mg lanatoside C. (From Ferrari et al. 1983)

The increased sympathetic outflow does not impair the vasodilatation in the active skeletal muscles (Wilson et al. 1985).

21.3 Aortic Stenosis

Patients with aortic stenosis are prone to syncope on exertion. For a long time this has been attributed to an inability to increase the cardiac output or to a dysrhythmia. In normal subjects and patients with mitral stenosis undertaking supine leg exercise, there is a reflex constriction of the fore-arm resistance blood vessels. By contrast, this constriction is inhibited or reversed in patients with severe aortic stenosis and a history of exertional syncope. After aortic valve replacement, the normal constriction of the forearm vessels occurs. Such studies suggest that marked activation of left ventricular mechanoreceptors may be responsible for the reflex vasodilatation and as a consequence the syncope during exercise in patients with aortic stenosis (Mark et al. 1973).

21.4 Myocardial Ischemia

Experimental myocardial ischemia in dogs is accompanied by an increased number of beta-adrenoceptors in the ischemic tissue and a decrease in the norepinephrine content. This increase persists for at least 8 h. When these receptors are stimulated by isoproterenol, a biological response occurs as indicated by an increase in cyclic AMP content and phosphorylase activation (Mukherjee et al. 1982).

Coronary occlusion may simultaneously excite a wide variety of receptors that exert different reflex effects on the cardiovascular system. The early hemodynamic changes which accompany acute myocardial infarction in man may result, therefore, not only from cardiac impairment but also from changes in activity of different receptor systems. In patients seen within 30 min after the onset of infarction, only 17% have a normal heart rate and blood pressure, 48% have evidence of parasympathetic overactivity manifested by bradyarrhythmia and hypotension, and 35% show evidence of sympathetic overactivity with sinus tachycardia or elevated blood pressure or both (Pantridge et al. 1975; Pantridge 1978). It is likely that these different manifestations are the result of different degrees of activation and interaction between the vagal and sympathetic afferent input from cardiac receptors to the central nervous system and the input from the high-pressure receptors.

Experimental reduction in coronary blood flow activates cardiac receptors with vagal and sympathetic myelinated and unmyelinated fibers. Left ventricular receptors with unmyelinated vagal afferents situated in the regions supplied by the occluded coronary artery increase their activity in concert with systolic bulging of the ischemic myocardium, suggesting that excitation is mechanical rather than chemical (Thorén 1979). These inhibitory cardiac receptors have a preferential distribution to the inferior wall of the left ventricle (Thames et al. 1978; Walker et al. 1978). Activation of such receptors is accompanied by cardiac slowing and hypotension (Kezdi et al. 1974). Their sensitivity is increased by cardiac glycosides (Thames et al. 1980, 1982). These changes are usually transient. Nausea and vomiting may also occur, due to reflex gastric dilatation, and are mediated by a vagal noncholinergic pathway (Abrahamson and Thorén 1973; Johannsen et al. 1981; Sleight 1981). In patients with myocardial infarction, nausea and vomiting have been reported to occur more frequently when this involves the inferior than when it involves the anterior

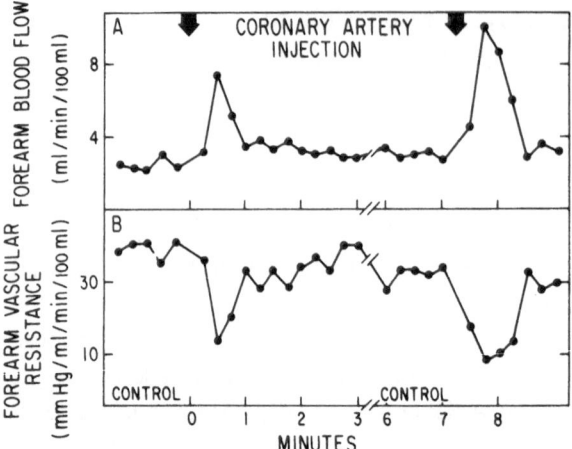

Fig. 16. Changes in **A** forearm blood flow and **B** forearm vascular resistance produced by injection of 8 ml Hypaque-M 75% into left coronary artery. *Arrows* indicate injections. Coronary injection of contrast medium caused prompt increases in forearm blood flow and decreases in forearm vascular resistance. (From Zelis et al. 1976)

wall of the left ventricle (Ahmed et al. 1978). The injection of radiographic contrast media into the coronary artery supplying the posteroinferior wall of the heart is more often accompanied by bradycardia, vasodilatation, and hypotension than when the contrast medium is injected into the artery supplying the anterior wall (Carson and Lazzara 1970; Eckberg et al. 1974; Perez-Gomez and Cardia-Aguado 1977). The bradycardia is not due to a direct depressant effect of the contrast media on the sinus node, but to activation of sensory receptors in the ventricle. The hypotension is due to the vasodilatation rather than the bradycardia, since atrial pacing to prevent the cardiac slowing only slightly reduces the decrease in blood pressure (Eckberg et al. 1974). The reflex vasodilatation in the forearm during coronary angiography is blocked by atropine, suggesting that it is due to activation of sympathetic cholinergic fibers (Fig. 16) (Zelis et al. 1976).

In dogs and cats coronary occlusion and myocardial infarction are followed by a reduction in the ability of arterial baroreceptors to modulate heart rate and sympathetic activity, possibly because of interference of cardiac receptors with the arterial baroreflex mechanisms (Toubes and Brody 1970; Felder and Thames 1979; Takeshita et al. 1980). A similar interference may take place in man; during or immediately after a myocardial infarction there is often little tachycardia when the blood pressure is reduced by vasodilator agents (Thomas et al. 1966; Smith et al. 1954). Also, little bradycardia has been reported when, 4–13 days after the myocardial necrosis, blood pressure was increased by phenylephrine, and little forearm vasoconstriction when the pressure was reduced by lower-body suction; some of the baroreflex impairment appeared to regress 1 or 2 months after the acute episode (Imaizumi et al. 1984).

With anterior myocardial infarction there may be tachycardia and elevated arterial blood ressure, and this may be attributed to increased activation of cardiac receptors with sympathetic afferents. An increase in arterial blood pressure may occur several hours after coronary bypass surgery. It is caused by a rise in total systemic vascular resistance. The heart rate shows no consistent change. The hypertension is abolished by local anesthetic of one stellate ganglion, due mainly to a reduction in the total systemic vascular resistance. It is suggested, therefore, that the hypertension is due to a reflex increase in sympathetic activity, due to stimulation of spinal (sympathetic) afferents from the heart (Tarazi et al. 1978; Estefanous and Tarazi 1980; Wallach et al. 1980; Malliani 1982).

A cardiogenic hypertensive chemoreflex might contribute to bradycardia and hypertension during myocardial ischemia. Studies in dogs have shown that chemoreceptors with a blood supply from the initial portion of the left coronary artery may be activated by release of 5-hydroxytryptamine from platelet thrombi (James et al. 1979).

The cardiopulmonary injection of prostacyclin or arachidonic acid in anesthetized or conscious dogs causes bradycardia and hypotension. This is due to the prostaglandins stimulating the C-fiber vagal afferents with chemosensitive endings located predominantly in the posterior wall of the left ventricle. The hypotension is due mainly to reflex vasodilatation in the kidneys and the skeletal muscles. Thus the possibility exists that an excessive intracardiac production of prostaglandins could activate these chemosensitive fibers (Hintze and Kaley 1984). Bradykinin, which is present in the coronary sinus blood of the ischemic heart, can excite both vagal (Kimura et al. 1973; Kaufman et al. 1980) and sympathetic (Uchida and Murao 1974; Nishi et al. 1977; Baker et al. 1980; Coleridge and Coleridge 1980· Lombardi et al. 1981) cardiac sensory endings. In conscious dogs the intracoronary injection of bradykinin initiates pressor reflexes in the presence of intact buffering mechanisms. The dogs showed no behavioral responses indicative of pain reaction (Pagani et al. 1985).

The cardiac pain associated with angina pectoris or myocardial infarction is the result of neural activity carried by sympathetic efferent fibers that enter the spinal cord via the $T_1 - T_4$ rami communicantes (White 1957). The increased activity in these fibers activates spinothalamic tract neurons, which transmit information to regions of the brain involved in the perception of pain. Stimulation of vagal afferent fibers inhibits the spinothalamic tract neuronal activity elicited by activating of cardiopulmonary sympathetic afferents by electrical stimulation or by intracardiac injection of bradykinin. This suggests a role for the vagus in the sensation of cardiac pain and may offer an explanation for the so-called silent myocardial infarction (Ammons et al. 1983).

21.5 Sudden Death from Coronary Artery Disease

The simultaneous activation of both vagal and sympathetic afferents could cause an increase in both vagal and sympathetic efferent activity to the heart, instead of the usual reciprocity between these two systems. This has been demonstrated during myocardial ischemia in animals (Gillis 1971). Activation of both noradrenergic and vagal fibers to the heart may increase the susceptibility of the patient to life-threatening arrhythmias (Bergamaschi 1978). However, in the normal anterior left ventricular myocardium of the dog, activation of muscarinic cholinergic receptors by vagal stimulation prolongs the effective refractory period by antagonism of sympathetic activity. This raises the possibility that prolongation of the effective refractory period in the ventricular myocardium may prevent or interact with arrhythmias (Martins and Zipes 1980).

Spasm of the coronary arteries is now considered one of the common causes of sudden death. The hemodynamic consequences of the spasm are due not only to diminished function of the ventricular muscle, but also to activation of sensory receptors in the heart (Perez-Gomez et al. 1979). After cardiac denervation the heart is protected from life-threatening arrhythmias following myocardial ischemia, indicating the importance of the cardiac nerves in their development (Kliks et al. 1975).

Studies in dogs have demonstrated that myocardial ischemia causes an increase in beta-adrenoceptors in the ischemic zone in association with the reduction in tissue norepinephrine (Mukherjee et al. 1979). It is possible that these changes have a role in the generation of cardiac arrhythmias during myocardial ischemia (Watanabe 1983).

Studies on patients have shown that reperfusion of acutely ischemic myocardium after intracoronary thrombolysis may be associated with transient bradycardia and hypotension, especially when the acute infarction involves the inferior wall of the left ventricle. It is likely that this cardioinhibitory and vasodepressor reflex is due to stimulation of the ventricular mechanoreceptors with vagal C-fiber afferents (Wei et al. 1983). Also, changes in $alpha_1$-adrenoceptor density might be related to the enhanced alpha-adrenoceptor responsiveness that appears to cause the lethal arrhythmias during reperfusion (Sheridan et al. 1980; Corr et al. 1981).

In dogs the depression in the baroreceptor ability to reduce heart rate caused by myocardial infarction was associated with a greater susceptibility to exercise-induced ventricular fibrillation (Billman et al. 1982). Increases in baroreflex sensitivity by training greatly improved resistance of the animals to the arrhythmogenic effect of exercise (Billman et al. 1984). Thus preservation of the baroreflex may afford protection after a coronary attack, possible via maintenace of the beneficial effect of the vagus on cardiac electrical instability. Whether this holds true in man is unknown.

21.6 Hypertension

The increase in blood pressure in patients with essential hypertension is due mainly to an increase in total systemic vascular resistance (Pickering 1936). In some patients with borderline or mild hypertension an increase in cardiac output, due to an increase in heart rate, may contribute to the rise in pressure. However, as the duration and/or the severity of hypertension increases, any initial rise in cardiac output disappears, so that in the established phase virtually all essential hypertensive subjects have a normal cardiac output, a reduction to subnormal values being frequently observed as the hypertension becomes more severe. Stroke volume and blood volume are normal throughout the course of essential hypertension, with the exception of the initial stage, in which an increase in total or central blood volume can sometimes be observed (Frohlich et al. 1970; Julius and Esler 1975; Conway 1984). In patients with secondary hypertension due to chronic renal artery stenosis, the hemodynamic pattern is similar to that in essential hypertension. In those with secondary hypertension due to primary aldosteronism and pheochromocytoma it is different in that the blood volume is frequently increased and reduced respectively (Cohn 1977; Levenson et al. 1980). Continuous recordings of intraarterial pressure have been made in patients with uncomplicated essential hypertension. The major factors influencing variability were the level of pressure and the intensity of physical activity. The variations in systolic pressure increased with progressive impairment of sinoaortic baroreflexes. Diastolic pressure increased with the level of sympathetic activity as reflected by plasma norepinephrine levels. Age of the patients was related to blood pressure variability but plasma catecholamine levels were not (Clement et al. 1978; Watson et al. 1980).

In normotensive subjects, and in those with moderate or severe hypertension, negative and positive pressures have been applied to the neck to alter the transmural pressure at the carotid sinuses (and thus respectively deactivate and stimulate the carotid baroreceptors) and the resultant changes in arterial blood pressure and heart rate have been measured. With reductions in transmural pressure the increase in arterial blood pressure is reduced in moderate and severe hypertension, whereas with increased transmural pressure there is a greater decrease in arterial pressure. These studies indicate that the usual asymmetry of the carotid baroreflex is altered in hypertension and that the set-point is displaced, as the hypertension increases in severity, from a lower to a higher point on the stim-

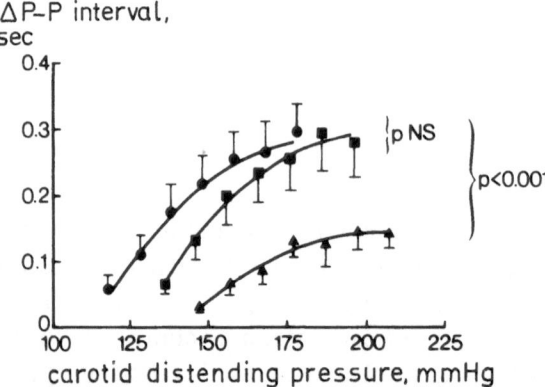

Fig. 17. Sinus node responses of subjects with normal arterial blood pressure (*circles*) and milder (*squares*) and more severe (*triangles*) borderline hypertension to brief (0.6-s) neck suction. These curves depict the combined average responses of the ten subjects in each group. Carotid distending pressure was considered to be the absolute sum of systolic arterial pressure and neck chamber pressure. *Brackets* indicate SEM. (From Eckberg 1979)

ulus-response curve (Fig. 5). The mechanism by which this occurs is unknown. It may be due to central modifications of the reflex arc, or to a progressive stiffening of the arterial walls in hypertension, so that the stimulus to the baroreceptors is lessened. However, these mechanisms are not peculiar to primary hypertension because a similar resetting occurs in renovascular hypertensive subjects (Mancia et al. 1982a). When anti-hypertensive drugs are given, the stimulus-response curves for both the arterial blood pressure and heart rate arterial baroreceptor control can be shifted to the left. This can occur within 1 h and suggests that functional rather than anatomical properties of the reflex are involved in the resetting process (Mancia et al. 1979c, 1980a,b).

The baroreceptor control of heart rate in hypertension has been studied by both the neck chamber and the vasoactive drug techniques. It has been shown that in patients with borderline hypertension the baroreceptor–heart rate responses are reset to function at higher pressure levels than normal, but the reflex sensitivity is unaltered (Randall et al. 1978). In patients with mild, moderate, and severe hypertension the reset of the baroreflex toward the elevated blood pressure values is progressively more pronounced. In addition, however, there is a progressive reduction in the baroreflex sensitivity, expressed either as changes in heart period in response to changes in blood pressure or carotid transmural pressure, or as actual changes in heart rate (Fig. 17) (Eckberg 1979; Brown 1980; Korner 1983; Bristow et al. 1969c; Gribbin et al. 1971; Mancia et al. 1978d; Takeshita et al. 1975, 1978; Korner et al. 1979). This reduction is not related to the age of the hypertensive subjects (Goldstein 1983).

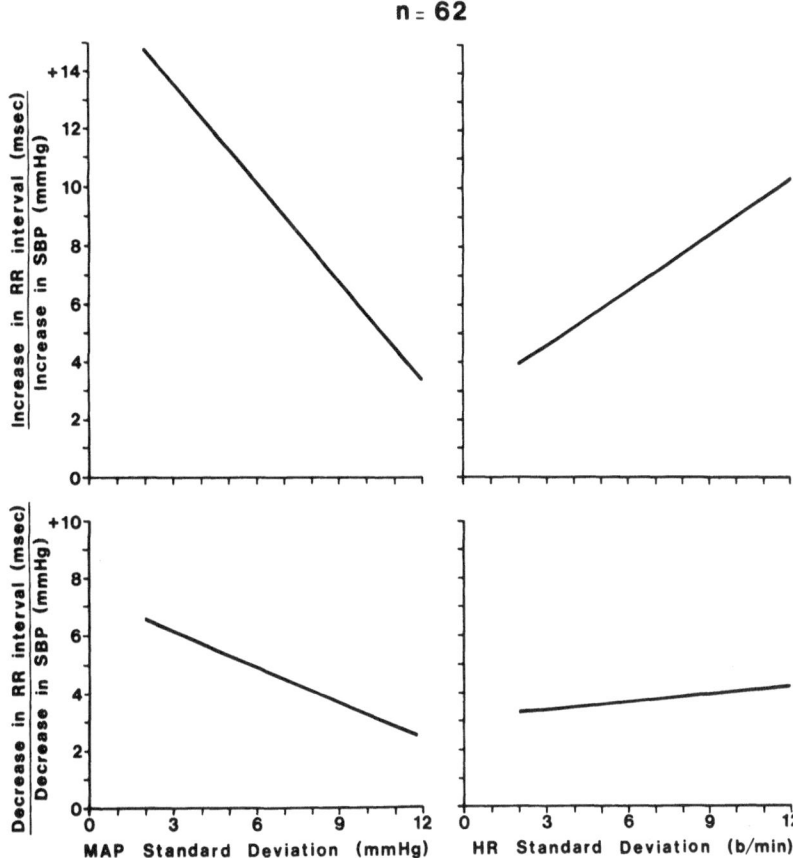

Fig. 18. Linear relationships between mean arterial pressure (*MAP*) and heart rate (*HR*) variabilities and baroreflex sensitivity in 62 ambulant subjects with normal blood pressure or essential hypertension in which a 24-h continuous blood pressure recording was performed. For each subject MAP and HR variabilities were expressed as the average of the standard deviation calculated for each of the 48 half hours of the recording. Baroreflex sensitivity was assessed by the slopes of the increase in R-R interval that accompanied a phenylephrine-induced increase in systolic blood pressure (*SBP*) (*upper panels*) and by the slopes of the reduction in R-R interval that accompanied a nitroglycerine-induced reduction in SBP (*lower panels*) respectively. The relationships were significantly negative ($p < 0.01$) for MAP variability and significantly positive ($p < 0.01$) for HR variability. (After Mancia et al. 1986)

Thus there are differences in the effects of hypertension on the arterial baroreceptor control of blood pressure and heart rate (Abboud 1982; Mancia et al. 1978d). In each case the reflex is reset; in the case of the latter it is not only reset, but its sensitivity is reduced over the total range of the stimulus-response curve (Fig. 18). Thus arterial baroreceptor control of blood pressure may be better preserved in hypertension than arterial baroreceptor control of heart rate (Bevegård et al. 1977a; Mancia et al. 1978d). These findings are in accord with those reported in experiments on animals (Angell-James and George 1978; Nosaka and Wang 1972) that have avoided some of the pitfalls inherent to the techniques for studying baroreceptor reflexes in humans.

A challenging issue that has been widely addressed is that of the mechanisms responsible for this greater loss in sensitivity of the baroreceptor control of heart rate as compared to that of blood pressure. Studies in animals have shown that a rise in blood pressure causes acute and chronic alterations in the baroreceptors and the arterial walls that increase the threshold of the baroreceptor firing and reduce its ability to respond normally to a pressure stimulus (Angell-James 1973, 1974; Angell-James and Lumley 1974; Sleight et al. 1977; Krieger et al. 1982). This, however, would impair all reflex cardiovascular influences, and therefore cannot explain the differential reduction between the cardiac and blood pressure components. This might be due to differences in reflex control of heart rate and blood pressure by the arterial baroreceptor, the former depending primarily on changes in vagal and the latter on sympathetic activity. Alternatively, the hypertension may itself lead to altered responsiveness of the sinus node and the resistance blood vessels. Cardiac hypertrophy may result in a reduced ability of the sinus node to respond to neural stimuli, whereas hypertrophy of the wall of the resistance blood vessels in hypertension will amplify the changes in resistance that occur in response to contraction of the smooth muscle (Folkow 1973, 1978, 1982; Folkow et al. 1973). In spontaneously hypertensive rats there is a reduced bradycardia but a normal suppression of sympathetic activity following arterial baroreceptor stimulation. It appears, therefore, that at least some of the differential changes in baroreceptor control of heart rate and peripheral circulation that characterize hypertension originate centrally (Ricksten and Thorén 1981). The effect of beta-adrenoceptor antagonists on sinoaortic baroreflex activity is debatable. In one study, in patients with essential hypertension the sensitivity of the baroreflex, as determined by changes in the pulse interval with injections of phenylephrine, was increased in patients under 40 years of age after chronic treatment (Watson et al. 1979).

Comparison of cardiopulmonary reflexes in normotensive and hypertensive subjects is complicated by the difficulty of properly matching the receptor stimuli in the two populations. Nevertheless, there is evidence that cardiopulmonary receptor control of renin remains effective in subjects with high blood pressure (Mancia et al. 1978c, 1981; London et al. 1983; Grassi et al. 1985b). There is also evidence that reflexes originating from heart and lung receptors sensitive to changes in blood volume may also be altered in hypertension. Thus in mild and moderately essentially hypertensive young men, the vasoconstrictor response in the forearm to deactivation of cardiopulmonary receptors (by applying lower-body negative pressure) is greater than in normal subjects (Mark and Kerber 1982)

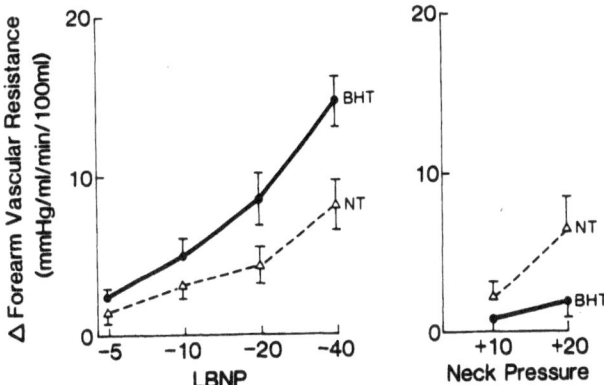

Fig. 19. Forearm vascular responses to lower-body negative pressure (*LBNP*) and neck pressure in borderline hypertensive (*BHT*) and normotensive (*NT*) subjects. Forearm vasoconstrictor responses to LBNP were augmented, whereas forearm vasoconstrictor response to neck pressure was impaired in borderline hypertensive subjects. This indicates that the cardiopulmonary baroreflexes are augmented but the carotid baroreflexes are impaired in young men with borderline hypertension. (From Mark and Mancia 1983)

(Fig. 19). This suggests that in the initial stage hypertension may be accompanied by an increase in the tonic inhibitory influence of the cardiopulmonary receptors, possibly in concomitance with the resetting, and thus the reduced tonic inhibition, of the arterial baroreflex. Later, as cardiac hypertrophy develops, it is possible that cardiopulmonary reflexes become impaired. At this stage the decreased ability of cardiopulmonary receptors to inhibit sympathetic outflow, combined with the enhanced constriction of the systemic resistance vessels due to hypertrophied vascular smooth muscle, may help in perpetuating and enhancing the high blood pressure state (Mark and Kerber 1982). By contrast, Bovy et. al. (1983) concluded that the cardiopulmonary baroreflex control of the forearm vascular resistance is normal in borderline hypertension and that the elevation of the basal forearm vascular resistance is not due to a sympathetic hyperactivity. They also conclude that the arterial baroreflex is acting normally in borderline hypertension.

In the early stages of experimental hypertension, the release of norepinephrine in the blood vessel wall is greater than normal during sympathetic nerve stimulation. This could play a role in the etiology of the disease (Collis et al. 1980; Vanhoutte et al. 1981). The greater-than-normal release could be due to an inherent defect of the adrenergic neuron, although modulating influences of circulating agents such as angiotensin II, epinephrine, or endogenous inhibition of Na^+-K^+-ATPase could contribute. Evidence has been obtained of a plasma factor affecting Na^+ cellular transport in subjects with borderline hypertension and those who are normotensive but have a family history of hypertension (Costa et al. 1982). Beta-adrenoceptors on the neuronal membrane of adrenergic nerve endings have been demonstrated in animal and human blood vessels. When activated, they augment the exocytotic release of norepinephrine. It is suggested that these receptors may have a role in certain types of hypertension when increased amounts of epinephrine reach the adrenergic neuronal membrane (Langer 1981; Majewski and Rand 1981; Vanhoutte et al. 1981). Angiotensin II also facilitates the release of the adrenergic neurotransmitter. This, combined with the activation of central angiotensin receptor sites and the facilitation of transmission at the sympathetic ganglia, can augment the amount of norepinephrine present in the vicinity of the cardiac and vascular effector cells (Zimmerman 1978, 1981).

Debate continues as to the role of the sympathetic nervous system in primary hypertension in man (Sannersted et al. 1970). This is due to the difficulties in correlating the plasma concentrations of the catecholamines with the overall sympathetic outflow and the impossibility of deducing from plasma levels difference in sympathetic outflow to the different vascular beds (Kiowski et al. 1981; Skrabal et al. 1981; Ibsen and Julius 1984). Julius et al. (1975) found that cardiac blockade with atropine and propranolol normalized the increased heart rate and cardiac output in the high-output subgroup of patients with borderline hypertension, but did not normalize the blood pressure. The increased plasma catecholamine concentrations that occur in some patients and the elevated plasma norepinephrine levels in relatively young established hypertensive patients are consistent with an abnormal and sustained increase in sympathetic outflow in this subgroup (Goldstein et al. 1983). Norepinephrine and normetanephrine concentrations were increased both in the cerebral spinal fluid and the plasma in patients with primary hypertension (aged 40 ± 2 years mean and SEM). Although in some patients this may be due to a defect in the neuronal uptake of norepinephrine (Esler et al. 1981), it is likely that both central and peripheral sympathetic activity is increased in some patients with primary hypertension, and in particular in the early stages of the disease (Dequattro et al. 1984). When the hypertension is established there is no evidence of increased sympathetic tone; the sympathetic nervous system, however, as in normal subjects, continues to play an important role in the moment-to-moment control of the arterial blood pressure (Ibsen and Julius 1984).

An increase in the sympathetic mediated response to exercise (Phillip et al. 1978; Esler and Nestle 1973; Safar et al. 1973; Cuche et al. 1974), head-up tilting, and mental stress (Falkner et al. 1979) has also been found in a fraction of the patients with primary hypertension. Thus a comparison of three groups of human adolescents, normotensive, borderline hypertensive, and hypertensive, demonstrated that those with hypertension had higher mean values for plasma norepinephrine when resting, and that those with significant hypertension had higher peak heart rates and plasma norepinephrine levels during a 90-s maximal voluntary hand-grip contraction. During dynamic exercise higher plasma norepinephrine values were observed in the group with borderline hypertension than in normotensive controls. Also, adolescents with significant hypertension had higher values for plasma norepinephrine during dynamic exercise and this had a highly significant correlation with the maximal heart rate (Klein et al. 1984). It is suggested that these higher values may be a consequence of less efficient baroreflexes in hypertensive subjects (Sleight et al. 1979).

A greater pressor response to an intravenous infusion of both norepinephrine and angiotensin II has been reported in patients with borderline hypertension resting supine than in age-matched normal subjects. The plasma levels of norepinephrine and plasma renin activity were similar in the two groups, as were exchangeable body sodium and blood volumes. The patients had a slight increase in plasma epinephrine. It is suggested that these findings indicate that borderline hypertension may be maintained by inappropriately increased cardiovascular responses to norepinephrine and angiotensin II in the presence of normal sympathetic and renin activity and body sodium volume state (Meier et al. 1981).

Since the plasma renin activity correlated with plasma norepinephrine in primary but not in renal hypertension, it is suggested that the sympathetic nervous system contributes to basal renin release only in the former but not in the latter. In renal hypertension, the basal renin release is proably maintained by intrarenal mechanisms (Skrabal et al. 1981).

Multiunit recordings of sympathetic activity in skin and muscle nerves have been made in healthy and hypertensive subjects. In both groups, the sympathetic activity recorded at rest appeared in bursts, with different temporal patterns. The bursts in the muscle nerves followed the pulse rhythm and random elevations of blood pressure above a certain level suppressed them. This inhibitory blood pressure level was higher in the hypertensive subjects, consistent with an elevated baroreceptor working range in hypertension. No other difference in muscle nerve activity between the two groups was noted. However, with the multiunit recording, a quantitative comparison could not be made of the absolute strength of the sympathetic outflow in the normotensive and hypertensive groups (Wallin et al. 1973).

In chronic hypertension there are structural changes in the blood vessels which limit the maximal degree of dilatation of which the resistance blood vessels are capable. Such limitations in the ability to achieve the same maximal dilatation of the forearm vessels as control normal subjects have been demonstrated not only in patients with intermittent or mild arterial blood pressure elevation, but also in normotensive subjects with a family history of hypertension (Takeshita and Mark 1979, 1980). The application of ice to the forehead of these latter subjects elicited a greater increase in arterial blood pressure and a greater decrease in forearm blood flow (Takeshita et al. 1982); thus other factors in addition to the arterial pressure elevation may contribute to the structural vascular changes.

21.7 Autonomic Hyperreflexia

Autonomic hyperreflexia or autonomic dysreflexia can occur in about three-quarters of patients with quadriplegia. It is manifested by generalized sympathetic hyperreactivity characterized by paroxysmal hypertension, severe headache, sweating, nausea, and bradycardia. The most common cause is distension of the urinary bladder, resulting in a spinal cord reflex (Guttman and Whitteridge 1947; Arief et al. 1962; Corbett et al. 1971a,b). The rise in arterial pressure activates the arterial baroreceptors and results in the vagally mediated cardiac slowing. However, no compensatory changes occur in the sympathetic outflow as a consequence of the spinal lesion, so the spinal reflex continues unabated and results in increased inotropism and arterial and venous constriction accompanied by a marked augmentation of plasma catecholamines (Wurster and Randall 1975; Mathias et al. 1976). Above the level of the lesion in the spinal cord, vasodilatation occurs (Pryor 1971; Mew 1981).

Central or peripheral lesions of the sympathetic nervous system and the use of sympatholytic agents cause orthostatic hypotension. In contrast, in patients with cervical spinal cord transection the arterial blood pressure does not fall during tilting (Corbett et al. 1971a,b). This has been ascribed to the widespread vasoconstriction caused by increased release of renin (Mendelsohn and Johnston 1971). However, it may also reflect the ability of an isolated but intact spinal cord to integrate homeostatic cardiovascular reflexes.

Microelectrode recordings of sympathetic activity to skin (peroneal skin nerve fascicles) have been made in patients with functionally complete spinal cord lesions mainly at cervical levels (Wallin and Stjernberg 1984). In contrast to the normal situation where the activity in skin nerves is characterized by spontaneously occurring irregular bursts that are composed of vasoconstrictor and sudomotor impulses (Hagbarth et al. 1972) this was absent in the spinal patients. This implies that spontaneous sympathetic discharges in normal man arise at supraspinal levels. There was no systematic change in cutaneous sympathetic nerve activity with changes in ambient temperature, consistent with an absence of spinal sympathetic thermoregulatory reflexes in man.

In the tetraplegic patients, deep breaths, abdominal pressure over the bladder and mechanical and electrical skin stimuli applied caudal to the spinal transection induced bursts of neural activity (Wallin and Stjernberg 1984). However, stronger cutaneous stimuli were required to evoke these responses than in normal man. Thus there is no permanent hyperactivity of spinal cutaneous sympathetic neurons in patients following spinal cord injury. Concerning the relation of these findings to autonomic hyperreflexia, the subcutaneous vasoconstriction following a single burst of sympathetic activity lasted longer in the tetraplegics than in normals. This might contribute to the episodes of high blood pressure in those with spinal cord lesions.

21.8 Carotid Sinus Syndrome

In 1933 Weiss and Baker described a syndrome characterized by the occurrence of dramatic cardiovascular changes in response to a slight compression or massage of a carotid sinus. In most subjects these changes consisted of a profound vagal bradycardia accompanied by marked hypotension, signs of impaired cerebral perfusion, and syncope. In a few subjects it consisted of a similarly marked hypotension due, however, to peripheral vasodilatation rather than to cardiac depression. The origin of the syndrome was identified in a hypersensitive carotid sinus capable of a massive baroreceptor activation in response to mild stimuli. Knowledge of the hypersensitive carotid sinus syndrome has not progressed much further since the original description. It has been found that this syndrome occurs more frequently in elderly subjects but that its overall incidence is very low (Walter et al. 1978; Morley et al. 1982; Leatham 1982; Sugrue et al. 1984). It is not yet known whether the hypersensitivity originates from the baroreceptors or from the stiffening of carotid and pericarotid tissues enhancing the baroreceptor discharge by mechanical stimuli. It has also been found that in several patients this condition reflects a marked depressor response of a sick sinus node or otherwise diseased heart rather than a true hypersensitive carotid sinus.

Carotid sinus syndrome is associated with carotid body tumors, extensive atherosclerotic lesions, and Takayasu's disease (Lown and Levine 1961; Sano and Aiba 1966; Takeshita et al. 1977; Thomas 1969; Walter et al. 1978). Recently, however, Dehn et al. (1984) and Morley et al. (1984) have reported that in subjects with carotid sinus syndrome the heart rate response to neck suction was increased nearly four times as compared to that of an age-matched control group, and that this was the case when the heart rate effects of phenylephrine injection were assessed. Thus the possibility of baroreceptors truly hypersensitive to physiological stimuli cannot be ruled out.

References

Abboud FM (1982) The sympathetic system in hypertension. State of the art review. Hypertension 4 (Suppl 2):208–225

Abboud FM, Eckstein JW (1966) Active reflex vasodilatation in man. Fed Proc 25: 1611–1617

Abboud FM, Mark AL (1978) Cardiac baroreceptors in circulatory control in humans. In: Hainsworth R, Kidd C, Linden RJ (eds) Cardiac receptors. Cambridge University Press, London, pp 437–461

Abboud FM, Thames MD (1983) Interaction of cardiovascular reflexes in circulatory control. In: Shepherd JT, Abboud FM (eds) Handbook of physiology, sect 2, vol III, part 2: The cardiovascular system. Peripheral circulation and organ blood flow. Am Physiol Soc, Bethesda, pp 675–753

Abboud FM, Mark AL, Heistad DD, Eckberg DL, Schmid PG (1975) Selectivity of autonomic control of the peripheral circulation in man. Trans Am Clin Climatol Assoc 86:184–197

Abboud FM, Eckberg DL, Johannson UJ, Mark AL (1979) Carotid and cardiopulmonary baroreceptor control of splanchnic and forearm vascular resistance during venous pooling in man. J Physiol (Lond) 286:173–184

Abboud FM, Thames MD, Mark AL (1981a) Role of cardiac afferent nerves in the regulation of the circulation during coronary occlusion and heart failure. In: Abboud FM, Fozzard HA, Gilmore JP, Reis DJ (eds) Disturbances in neurogenic control of the circulation. Am Physiol Soc, Bethesda, pp 65–86

Abboud FM, Mark AL, Thames MD (1981b) Modulation of the somatic reflex by carotid baroreceptors and by cardiopulmonary afferents in animals and humans. Circ Res 48 (Suppl 1):131–137

Abrahams VC, Hilton DM, Zbrozyna A (1960) Active muscle vasodilatation produced by stimulation of the brain stem: its significance in the defence reaction. J Physiol (Lond) 154:491–513

Abrahamson H, Thorén P (1973) Vomiting and reflex vagal relaxation of the stomach elicited from heart receptors in the cat. Acta Physiol Scand 88:433–439

Ahlquist RP (1948) A study of the adrenotropic receptors. Am J Physiol 153:586–600

Ahmed SS, Gupta RC, Brancato RR (1978) Significance of nausea and vomiting during acute myocardial infarction. Am Heart J 95:671–672

Ambrosioni E, Costa FE, Montebugnoli L, Borghi C, Magnani B (1981) Intralymphocytic sodium concentration as an index of response to stress and exercise in young subjects with borderline hypertension. Clin Sci 61:25s–27s

Ambrosioni E, Costa FV, Borghi C, Montelbugnoli L, Giordani MF, Magnani B (1982a) Effects of moderate salt restriction on intralymphocytic sodium and pressor response to stress in borderline hypertension. Hypertension 4:789–794

Ambrosioni E, Costa FV, Borghi C, Montebugnoli L, Giordvani MF, Vasconi L (1982b) Effects of moderate salt restriction and high potassium intake on intralymphocytic sodium contant and pressor response to stress in borderline hypertension. Clin Sci 63:231S–234S

Aminoff MJ, Wilcox CS (1971) Assessment of autonomic function in patients with Parkinsonian syndrome. Brit Med J 4:80–84

Ammons WS, Blair RW, Foreman RD (1983) Vagal afferent inhibition of spinothalamic cell responses to sympathetic afferents and bradykinin in the monkey. Circ Res 53:603–612

Amorin DS, Godoy RA, Manco JC, Tanaka A, Gallo J (1968) Effects of acute elevation in blood pressure and of atropine on heart rate in Chagas' disease. Circulation 38: 289–294

Andersen P, Saltin B (1985) Maximal perfusion of skeletal muscles in man. J Physiol 366:283–249

Angell-James JE (1973) Characteristics of single aortic and right subclavian barorecep-tor fiber activity in rabbits with chronic renal hypertension. Circ Res 32:149–161

Angell-James JE (1974) Arterial baroreceptor activity in rabbits with experimental atherosclerosis. Circ Res 34:27–39

Angell-James JE, De Burg Daly M (1972) Some mechanisms involved in the cardio-vascular adaptation to diving. Symp Soc Exp Biol 26:313–341

Angell-James JE, George MJ (1978) Carotid sinus baroreceptor reflex control of blood pressure and vascular resistance in experimental cardiovascular disease and hyper-tension. Ir J Med Sci 6:160

Angell-James JE, Lumley JSP (1974) The effects of carotid endarterectomy on the mechanical properties of the carotid sinus and carotid sinus nerve activity in athero-sclerotic patients. Br J Surg 61:805–810

Antonaccio MJ (1984) Central transmitters: physiology, pharmacology and effects on the circulation. In: Antonaccio M (ed) Cardiovascular pharmacology, 2nd edn. Raven, New York

Arborelius M Jr, Ballidin UI, Lilja B, Lundgren CEG (1972) Hemodynamic changes in man during immersion with the head above water. Aerosp Med 43:592–598

Arief AJ, Ripay EI, Pyrik SJ (1962) Acute hypertension induced by urinary bladder distension. Arch Neurol 2:248–256

Baccelli G, Albertini R, Del Bo A, Mancia G, Zanchetti A (1981) Role of sinoaortic reflexes in hemodynamic patterns of natural defense behaviors in the cat. Am J Physiol 240:H421–H429

Bader H (1967) Dependence of wall stress in the human thoracic aorta on age and pressure. Circ Res 20:354–361

Baker DG, Coleridge HM, Coleridge JCG, Nerdrum T (1980) Search for a cardiac nociceptor. Stimulation by bradykinin of sympathetic afferent nerve endings in the heart of the cat. J Physiol (Lond) 306:519–536

Bannister R (1980) Defective baroreflex function in autonomic failure. In: Sleight P (ed) Arterial baroreceptors and hypertension. Oxford University Press, Oxford, pp 117–122

Bannister R, Ardill L, Fentem P (1967) Defective autonomic control of blood vessels in idiopathic orthostatic hypotension. Brain 90:725–745

Bannister R, Boylston AW, Davies IB, Mathias CJ, Sever PS, Sudera D (1981) Beta-receptor numbers and thermodynamics in denervation supersensitivity. J Physiol (Lond) 319:369–377

Barcroft H, Edholm OG (1945) On the vasodilatation in human skeletal muscle during post-haemorrhage fainting. J Physiol (Lond) 104:161–175

Barcroft H, Bonnar W, Edholm OG, Effron AS (1943) On sympathetic vasoconstrictor tone in human skeletal muscle. J Physiol (Lond) 102:21–31

Barcroft H, Edholm OG, McMichael J, Sharpey-Schafer EP (1944) Posthaemorrhagic fainting. Study by cardiac output and forearm flow. Lancet 1:489–491

Barcroft H, Brod J, Hejl Z, Hirsjarvi EA, Kitchin AH (1960) The mechanism of the vasodilatation in the forearm muscle during stress (mental arithmetic). Clin Sci 19: 577–586

Baskerville AL, Eckberg DL, Thompson MA (1979) Arterial pressure and pulse interval responses to repetitive carotid baroreceptor stimuli in man. J Physiol (Lond) 297: 61–71

Bath E, Lindblad LE, Wallin GB (1981) Effects of dynamic and static neck suction on muscle nerve sympathetic activity, heart rate, and blood pressure. J Physiol (Lond) 311:551–564

Bearn AG, Billing B, Edholm OG, Sherlock S (1951) Hepatic blood flow and carbo-hydrate changes in man during fainting. J Physiol (Lond) 115:442–445

Beiser GD, Zelis R, Epstein SE, Mason DT, Braunwald E (1970) The role of skin and muscle resistance vessels in reflex mediated by the baroreceptor system. J Clin Invest 49:225–231

Bennett T, Gardiner SM (1985) Involvement of vasopressin in cardiovascular regulation. Cardiovasc Res 19:57–68

Bennett ED, Brooks NH, Keddie J, Lis Y, Wilson A (1977) Increased renal function in patients with acute left ventricular failure: a possible homeostatic mechanism. Clin Sci 52:43–50

Bennett T, Hosking DJ, Hampton JR (1979a) Cardiovascular responses to lower body negative pressure in normal subjects and in patients with diabetes mellitus. Cardiovasc Res 13:31–38

Bennett T, Hosking DJ, Hampton JR (1979b) Cardiovascular responses to graded reductions of central blood volume in normal subjects and in patients with diabetes mellitus. Clin Sci 58:193–200

Bennett KL, Linden RJ, Mary DASG (1983) The effect of stimulation of atrial receptors on the plasma concentration of vasopressin. Q J Exp Physiol 68:579–589

Bergamaschi M (1978) Role of the sympathetic and parasympathetic innervation in the genesis of ventricular arrhythmias during experimental myocardial ischemia. In: Schartz PJ, Brown AM, Malliani A, Zanchetti A (eds) Neural mechanisms in cardiac arrhythmias. Raven, New York, pp 139–154

Bertel O, Buhler FR, Kiowski W, Lutold BE (1980) Decreased beta-adrenoceptor responsiveness as related to age, blood pressure and plasma catecholamines in patients with essential hypertension. Hypertension 2:130–138

Bevegård S, Lodin A (1962) Postural circulatory changes at rest and during exercise in five patients with congenital absence of the valves in deep veins of the legs. Acta Med Scand 172:21–29

Bevegård BS, Shepherd JT (1966a) Reaction in man of resistance and capacity vessels in forearm and hand to leg exercise. J Appl Physiol 21:123–132

Bevegård BS, Shepherd JT (1966b) Circulatory effects of stimulating the carotid arterial stretch receptors in man at rest and during exercise. J Clin Invest 45:132–142

Bevegård BS, Shepherd JT (1967) Regulation of the circulation during exercise in man. Physiol Rev 47:178–213

Bevegård BS, Castenfors J, Danielson M (1977a) Carotid baroreceptor function in hypertensive patients. Scand J Lab Invest 37:495–501

Bevegård BS, Castenfors J, Lindblad LE (1977b) Effects of carotid sinus stimulation on cardiac output and peripheral vascular resistance during changes in blood volume distribution in man. Acta Physiol Scand 101:50–57

Bevegård BS, Castenfors J, Lindblad LE (1977c) Effect of changes in blood volume distribution on circulatory variables and plasma renin activity in man. Acta Physiol Scand 99:237–245

Bevegård BS, Castenfors J, Lindblad LE, Transejo J (1977d) Blood pressure and heart rate regulating capacity of the carotid sinus during changes in blood volume distribution in man. Acta Physiol Scand 99:300–312

Billman GE, Schwartz PJ, Stone HL (1982) Baroreceptor reflex control of heart rate: a predictor of sudden cardiac death. Circulation 66:874–880

Billman GE, Schwartz PJ, Stone HL (1984) The effects of daily exercise on susceptibility to sudden cardiac death. Circulation 69:1182–1189

Bing RJ, Hammond MM, Handelsman JC, Powers SR, Spencer FC, Eckenhoff JC, Goodale WT, Hafkenschiel JH, Kety SS (1949) The measurement of coronary blood flow, oxygen consumption and efficiency of the left ventricle in man. Am Heart J 38:1–24

Bini G, Hagbarth K-E, Hynninen P, Wallin GB (1980a) Thermoregulatory and rhythm-generating mechanisms governing the sudomotor and vasoconstrictor outflow in human cutaneous nerves. J Physiol (Lond) 306:537–552

Bini G, Hagbarth K-E, Hynninen P, Wallin BG (1980b) Regional similarities and differences in thermoregulatory vaso- and sudomotor tone. J Physiol (Lond) 306:553–565

Bishop VS, Malliani A, Thorén P (1983) Cardiac mechanoreceptors. In: Shepherd JT, Abboud RM (eds) Handbook of physiology. The cardiovascular system, sect 2, vol III, part 2. Peripheral circulation and organ blood flow. Am Physiol Soc, Bethesda, MD, pp 497–555

Bjurstedt H, Rosenhamer G, Tyden G (1975) Cardiovascular responses to changes in carotid sinus transmural pressure in man. Acta Physiol Scand 94:497–505

Bjurstedt H, Rosenhamer G, Tyden G (1977) Lower body negative pressure and effects of autonomic heart blockade on cardiovascular responses. Acta Physiol Scand 99:353–360

Blair DA, Glover WE, Greenfield ADM, Roddie IC (1959a) Excitation of cholinergic vasodilator nerves to human skeletal muscles during emotional stress. J Physiol (Lond) 148:633–647

Blair DA, Glover WE, Kidd BSL (1959b) The effect of continuous positive and negative pressure breathing upon the resistance and capacity blood vessels of the human forearm and hand. Clin Sci 18:9–16

Blix AS, Folkow B (1983) Cardiovascular adjustments to diving in mammals and birds. In: Shepherd JT, Abboud FM (eds) Handbood of physiology, the cardiovascular system, sect 2, vol III, part 2. American Physiological Society, Bethesda, MD, pp 917–945

Blomqvist CG, Saltin B (1983) Cardiovascular adaptations to physical training. Ann Rev Physiol 45:169–189

Borst C, Karemaker J (1980) Respiratory modulation of reflex bradycardia evoked by brief carotid sinus nerve stimulation: additive rather than gating mechanisms. In: Sleight P (ed) Arterial baroreceptors and hypertension. Oxford University Press, Oxford, pp 276–281

Borst C, Karemaker JM (1983) Time delays in the human baroreceptor reflex. J Auton Nerv Syst 9:399–409

Borst C, Karemaker JM, Bouman LM, Dunning AJ, Schopman FJG (1974) Optimal frequency of carotid sinus nerve stimulation in treatment of angina pectoris. Cardiovasc Res 8:674–680

Borst C, Wieling W, van Brederode JFM, Hond A, de Rijk LG, Dunning AJ (1982) Mechanisms of initial heart rate response to postural change. Am J Physiol 243: H676–H681

Borst C, Karemaker JM, Dunning AJ, Bouman LN, Wagner J (1983) Frequency limitation in the human baroreceptor reflex. J Auton Nerv Syst 9:381–397

Borst C, van Brederode JFM, Wieling W, van Montfrans GA, Dunning AJ (1984) Mechanisms of initial blood pressure response to postural change. Clin Sci 67:321–327

Bovy PH, Juchmes J, Fossion A (1983) The arterial baroreflex and the cardiopulmonary reflex in borderline hypertension. Eur Heart J 4 (Suppl G):41–45

Bradley SE, Inglefinger FJ, Bradley GP, Curry JJ (1945) Estimation of hepatic blood flow in man. J Clin Invest 24:890–897

Brennan LA, Malvin RL, Jochim KE, Roberts DE (1971) Influence of right and left atrial receptors on plasma concentration of ADH and renin. Am J Physiol 221: 273–278

Brick I (1966) Circulatory responses to immersing the face in water. J Appl Physiol 21:33–36

Brick I, Hutchinson KJ, Roddie IC (1968) Effect of adrenergic receptor blockade on the response of forearm blood vessels to circulating noradrenaline and vasoconstrictor nerve activity. In: Hudlicka O (ed) Circulation in skeletal muscle. Pergamon, New York, pp 25–34

Brigden W, Sharpey-Schafer EP (1950) Postural changes in peripheral blood flow in cases with left heart failure. Clin Sci 9:93–100

Brigden W, Howarth S, Sharpey-Schafer EP (1950) Postural changes in the peripheral blood flow of normal subjects with observations on vasovagal fainting reactions as a result of tilting, the lordotic posture, pregnancy and spinal anesthesia. Clin Sci 9: 79–91

Bristow JD, Honour AJ, Pickering TG, Sleight P (1969a) Cardiovascular and respiratory changes during sleep in normal and hypertensive subjects. Cardiovasc Res 3: 476–485

Bristow JD, Honour AJ, Pickering GW, Sleight P, Smyth HS (1969b) Diminished baroreflex sensitivity in high blood pressure. Circulation 39:48–54

Bristow JD, Prys-Roberts C, Fisher A, Pickering TG, Sleight P (1969c) Effects of anesthesia on baroreflex control of heart rate in man. Anesthesiology 31:422–428

Bristow JD, Brown EB, Cunningham DJC, Howson MG, Strange-Petersen E, Pickering TG, Sleight P (1971) Effect of bicycling on the baroreflex regulation of pulse interval. Circ Res 28:582–592

Bristow JD, Brown EB, Cunningham DJC, Howeson MG, Lee MJR, Pickering TG, Sleight P (1974) The effects of raising alveolar P_{CO_2} and ventilation separately and together on the sensitivity and resetting of the baroreceptor cardiodepressor reflex in man. J Physiol (Lond) 234:401–425

Brod J, Fencl V, Hejl Z, Jirka J (1959) Circulatory changes underlying blood pressure elevation during acute emotional stress (mental arithmetic) in normotensive and hypertensive subjects. Clin Sci 18:269–279

Brown AM (1980) Receptors under pressure: an update on baroreceptors. Circ Res 46:1–10

Brown E, Goei JS, Greenfield ADM, Plassaras GC (1966) Circulatory responses to simulated gravitational shifts of blood in man induced by exposure of the body below the iliac crests to sub-atmospheric pressure. J Physiol (Lond) 183:607–627

Brown MJ, Jenner DA, Allison DJ, Dollery CT (1981) Variations in individual organ release of noradrenaline measured by an improved radioenzymatic technique; limitations of peripheral venous measurements in the assessment of sympathetic nervous activity. Clin Sci 61:585–590

Burke D, Sundlöf G, Wallin BG (1977) Postural effects on muscle sympathetic activity in man. J Physiol (Lond) 272:399–414

Burnett JC Jr, Granger JP, Opgenorth TJ (1984) Effects of synthetic atrial natriuretic factor on renal function and renin release. Amer J Physiol 247:F863–F866

Carlsten A, Folkow B, Grimby G, Hamberger CA, Thulesius O (1958) Cardiovascular effects of direct stimulation of the carotid sinus nerves in man. Acta Physiol Scand 44:138–145

Carson RP, Lazzara R (1970) Hemodynamic responses initiated by coronary stretch receptors with special reference to coronary arteriography. Am J Cardiol 25:571–578

Cechetto DF, Calaresu FR (1984) Units in the amygdala responding to activation of carotid baro- and chemoreceptors. Am J Physiol 246:R832–R836

Chidsey CA, Harrison DC, Braunwald E (1962) Augmentation of the plasma norepinephrine response to exercise in patients with congestive heart failure. N Engl J Med 267:650–654

Chobanian AV, Tifft CD, Sackel H, Pitruzella A (1982) Alpha and beta adrenergic receptor activity in circulating blood cells of patients with idiopathic orthostatic hypotensin and pheochromocytoma. Clin Exp Hypertens (A) 2:793–806

Clement DL, Bogaert MG, Moerman EZ, de Schaepdryver AF (1978) Significance of elevated plasma noradrenaline in patients with essential hypertension. In: Birkenhager WH, Falke HE (eds) Circulating catecholamines and blood pressure. Bunge Scientific, Utrecht, p 17

Cohen EL, Conn JW, Rovner DR (1967) Postural augmentation of plasma renin activity and aldosterone excretion in normal people. J Clin Invest 46:418–428

Cohn JN, Levine TB, Francis GS, Goldsmith SR (1981) Neurohumoral control mechanisms in congestive heart failure. Am Heart J 102:509–514

Cohn JW (1977) Primary aldosteronism. In: Genest J, Koiw E, Kuchel O (eds) Hypertension. McGraw-Hill, New York, pp 768–780

Coleridge HM, Coleridge JCG (1980) Cardiovascular afferents involved in regulation of peripheral vessels. Annu Rev Physiol 42:413–427

Collis MG, DeMey C, Vanhoutte PM (1980) Renal vascular reactivity in the young spontaneously hypertensive rat. Hypertension 2:45−52

Convertino VA, Keil LC, Greenleaf JE (1983) Plasma volume, renin and vasopressin responses to graded exercise after training. J Appl Physiol 54;508−514

Convertino VA, Benjamin BA, Keil LC, Sandler H (1984) Role of cardiac volume receptors in the control of ADH release during acute simulated weightlessness in man. Physiologist (Suppl) 27(6):S51−52

Conway J (1983) Hypotensive mechanisms of beta-blockers. Eur Heart J 4 (Suppl D): 43−51

Conway J (1984) Hemodynamic aspects of essential hypertension in humans. Physiol Rev 64:617−660

Conway J, Boon N, Jones JV, Sleight P (1983) Involvement of the baroreceptor reflexes in the changes in blood pressure with sleep and mental arousal. Hypertension 5:746−748

Cooke JP, Shepherd JT, Vanhoutte PM (1984) The effect of warming on adrenergic neurotransmission in canine cutaneous vein. Circ Res 54:457−553

Cooper B, Handing RI, Young GH, Alexander RW (1978) Agonist regulation of the human platelet alpha-adrenergic receptors. Nature 274:703−706

Corbett JL, Frankel HL, Harris PJ (1971a) Cardiovascular changes associated with skeletal muscle spasm in tetraplegic man. J Physiol (Lond) 215:381−393

Corbett JL, Frankel HL, Harris PJ (1971b) Cardiovascular reflex responses to cutaneous and visceral stimuli in spinal man. J Physiol (Lond) 215:395−409

Corr PB, Shaynan JA, Kramer TB, Kipnis RJ (1981) Increased alpha-adrenergic receptors in ischemic cat myocardium: a potential mediator of electrophysiological derangements. J Clin Invest 67:1232−1236

Costa FV, Montebognoli L, Giordani MF, Vasconi L, Ambrosioni E (1982) Evidence for a plasma factor affecting Na^+ cellular transport in genetic normotensive subjects and in borderline hypertensive subjects. Clin Sci 63:53s−55s

Cowley AW Jr, Liard JF, Guyton AC (1973) Role of the baroreceptor reflex in daily control of arterial blood pressure and other variables in dogs. Circ Res 32:564−576

Crossley RJ, Greenfield ADM, Plassaras GC, Stephens D (1966) The interrelation of thermoregulatory and baroreceptor reflexes in the control of the blood vessels in the human forearm. J. Physiol (Lond) 183:628−636

Cryer PE, Weiss S (1976) Reduced plasma norepinephrine response to standing in autonomic dysfunction. Arch Neurol 33:275−277

Cryer PE, Silverberg AB, Santiago JV, Shah SH (1978) Plasma catecholamines in diabetes: the syndromes of hypoadrenergic and hyperadrenergic postural hypotension. Am J Med 64:407−416

Cuche J-L, Kuckel O, Barbeau A (1974) Autonomic nervous system and benign essential hypertension in man. II Circulatory and hormonal responses to upright posture. Circ Res 35:290

Cunningham DJC, Strange-Petersen E, Peto R, Pickering TG, Sleight P (1972a) Comparison of the effect of different types of exercise on the baroreflex regulation of heart rate. Acta Physiol Scand 86:444−455

Cunningham DJC, Strange-Petersen E, Pickering TG, Sleight P (1972b) The effects of hpyoxia, hypercapnia, and asphyxia on the Baroreceptor-Cardiac reflex at rest and during exercise in man. Acta Physiol Scand 86:456−465

Daskalopoulos DA, Shepherd JT, Walgenbach SC (1984) Cardiopulmonary reflexes and blood pressure in exercising sinoaortic denervated dogs. J Appl Physiol 57: 1417−1421

Davies IB, Bannister R, Hensby C, Sever PS (1980) The pressor actions of noradrenaline and angiotensin II in chronic autonomic failure treated with indomethacin. Br J Clin Pharmacol 10:223−229

Davis JO, Freeman RH (1976) Mechanisms regulating renin relase. Physiol Rev 56: 1−56

de Champlain J (1977) Experimental aspects of the relationship between the auto-
    nomic nervous system and catecholamines in hypertension. In: Genest J, Koiw E,
    Kuchel O (eds) Hypertension. McGraw-Hill, New York, pp 76–92
de Champlain J, Boucher R, Genest J (1963) Arterial angiotensin levels in oedematous
    patients. Proc Soc Exp Biol Med 113:932–937
Dehn TCB, Morley CA, Sutton R (1984) A scientific evaluation of the carotid sinus
    syndrome. Cardiovasc Res 18:746–751
Delius W, Hagbarth KE, Hongell A, Wallin BG (1972a) Manoeuvres affecting sympathe-
    tic outflow in human muscle nerves. Acta Physiol Scand 84:82–94
Delius W, Hagbarth KE, Hongell A, Wallin BG (1972b) Manoeuvres affecting sympa-
    thetic outflow in human skin nerves. Acta Physiol Scand 84:177–186
Delius W. Hagbarth KE, Hongell A, Wallin BG (1972c) General characteristics of sym-
    pathetic activity in human muscle nerves. Acta Physiol Scand 84:65–81
Dequattro V, Sullivan P, Minagawa R, Kopin I, Bornheimer J, Foti A, Barndt R (1984)
    Central and peripheral noradrenergic tone in primary hypertension. Fed Proc 43:
    47–51
de Wardener HE, McSwiney RR (1951) Renal hemodynamics in vaso-vagal fainting
    due to haemorrhage. Clin Sci 10:209–217
de Wardener HE, Miles BE, Lee C de J, Churchill-Davidson H, Wylie D, Sharpey-
    Schaefer EP (1953) Circulatory effects of haemorrhage during prolonged light
    anesthesia in man. Clin Sci 12:175–184
Di Bona GF (1982) The function of renal nerves. Rev Physiol Biochem Pharmacol 94:
    75–181
Dillon N, Chung S, Kelly J, O'Halley K (1980) Age and beta-adrenoceptor-mediated
    function. Clin Pharmacol Ther 27:769–772
Doba N, Reis DJ (1974) Role of the cerebellum and the vestibular apparatus in regula-
    tion of orthostatic reflexes in the cat. Circ Res 34:9–18
Donald DE, Shepherd JT (1963) Responses to exercise in dogs with cardiac denerva-
    tion. Am J Physiol 205:393–400
Donald DE, Shepherd JT (1978) Reflexes from the heart and lungs: physiological
    curiosities or important regulatory mechanisms. Cardiovasc Res 12:449–469
Donald DE, Ferguson DA, Milburn SE (1968) Effects of beta-adrenergic blockade on
    racing performance of greyhounds with normal and with denervated hearts. Circ
    Res 22:127–134
Donoghue S, Felder RB, Joradan D, Spyer KM (1984) The central projections of
    carotid baroreceptors and chemoreceptors in the cat: a neurophysiological study.
    J Physiol (Lond) 347:397–409
Duke PC, Wade JG, Hickey RF, Larson CP (1976) The effects of age on baroreceptor
    function in man. Can Anaesth Soc J 23:111–124
Dwyer EM Jr, Dell RB, Cannon PJ (1973) Regional myocardial blood flow in patients
    with residual anterior and inferior transmural infarction. Circulation 48:924–935
Eckberg DL (1976) Temporal response patterns of the human sinus node to brief
    carotid baroreceptor stimuli. J Physiol (Lond) 258:769–782
Eckberg DL (1977a) Baroreflex inhibition of the human sinus node: importance of
    stimulus intensity, duration and rate of pressure change. J Physiol (Lond) 269:
    561–577
Eckberg DL (1977b) Adaptation of the human carotid baroreceptor cardiac reflex.
    J Physiol (Lond) 269:579–589
Eckberg DL (1979) Carotid baroreflex function in young men with borderline blood
    pressure elevation. Circulation 59:632–636
Eckberg DL (1980a) Nonlinearities of the human carotid baroreceptor-cardiac reflex.
    Circ Res 47:208–216
Eckberg DL (1980b) Parasympathetic cardiovascular control in human disease: a criti-
    cal review of methods and results. Am J Physiol 239 (Heart Circ Physiol 8):H581–
    H593

Eckberg DL (1983) Human sinus arrhythmia as an index of vagal cardiac outflow. J Appl Physiol 54:961–966

Eckberg DL, Orshan CR (1977) Respiratory and baroreceptor reflex interactions in man. J Clin Invest 59:780–785

Eckberg DL, Drabinski M, Braunwald E (1971) Defective cardiac parasympathetic control in patients with heart disease. N Engl J Med 285:877–883

Eckberg DL, Fletcher GF, Braunwald E (1972) Mechanisms of prolongation of the R-R interval with electrical stimulation of the carotid sinus nerves in man. Circ Res 30:131–138

Eckberg DL, White CW, Kioschos JM, Abboud FM (1974) Mechanisms mediating bradycardia during coronary arteriography. J Clin Invest 54:1445–1461

Eckberg DL, Cavanaugh MS, Mark AL, Abboud FM (1975) A simplified neck suction device for activation of carotid baroreceptors. J Lab Clin Med 85:167–173

Eckberg DL, Abboud FM, Mark AL (1976) Modulation of carotid baroreflex responsiveness in man: effects of posture and propranolol. J Appl Physiol 41:383–387

Eckberg DL, Kifle YT, Roberts VL (1980) Phase relationship between normal human respiration and baroreflex responsiveness. J Physiol (Lond) 304:489–502

Egan BM, Julius S, Cotter C, Osterziel KJ, Ibsen H (1983) Role of cardiovascular receptors on the neural regulation of renin release in normal men. Hypertension 5: 779–786

Egan B, Grekin R, Ibsen H, Osterziel K, Julius S (1984) Role of cardiopulmonary mechanoreceptors in ADH release in normal humans. Hypertension 6:832–836

Eklund B, Kaijser L (1976) Effect of regional α- and β-adrenergic blockade on blood flow in the resting forearm during contralateral isometric handgrip. J Physiol (Lond) 262:39–50

Elliott HL, Sumner DS, McLean K, Rubin PC, Reid IL (1982) Effect of age on vascular alpha-adrenoreceptor responsiveness in man. Clin Sci 63:305s–308s

Engel BT, Joseph JA (1982) Attenuation of baroreflexes during operant cardiac conditioning. Psychophysiology 19:609–614

Epstein M (1976) Cardiovascular and renal effects of head-out water immersion in man: application of the model in the assessment of volume homeostasis. Circ Res 39:619–628

Epstein M (1978) Renal effects of head-out water immersion in man: implications for an understanding of volume homeostasis. Physiol Rev 58:529–581

Epstein SE, Stampfer M, Beiser GD (1968a) Role of the capacitance and resistance vessels in vasovagal syncope. Circulation 37:524–533

Epstein SE, Beiser GD, Stampfer M, Braunwald E (1968b) Role of the venous system in baroreceptor-mediated reflexes in man. J Clin Invest 47:139–152

Epstein SE, Beiser GD, Goldstein RE, Stampfer M, Wechsler AS, Glick G, Braunwald E (1969) Circulatory effects of electrical stimulation of the carotid sinus nerves in man. Circulation 40:269–276

Epstein M, Pins DS, Sancho J, Haber E (1975) Suppression of plasma renin and plasma aldosterone during water immersion in normal man. J Clin Endocrinol Metab 41: 618–625

Epstein M, DeNunzio AG, Loutzenhiser RD (1981a) Effects of vasopressin administration on the diuresis of water immersion in normal humans. J Appl Physiol 51: 1384–1387

Epstein M, Preston S, Weitzman RE (1981b) Isoosmotic central blood volume expansion suppresses plasma arginine vasopressin in normal man. J Clin Endocrinol Metab 52:256–262

Ernsting J, Parry DJ (1957) Some observations on the effects of stimulating the stretch receptors in the carotid artery of man (abstract). J Physiol (Lond) 137: 45P–46P

Escourrou P, Freund PR, Rowell LB, Johnson PG (1982) Splanchnic vasoconstriction in heat-stressed man: role of renin-angiotensin system. J Appl Physiol 52:1438–1443

Esler M (1982) Assessment of sympathetic nervous function in humans from nor-adrenaline plasma kinetics. Clin Sci 62:247−254

Esler M, Nestle P (1973) Sympathetic responsiveness to head-up tilt in essential hypertension. Clin Sci 44:213−226

Esler M, Jackman G, Bobik A, Leonard P, Kelleher D, Skews H, Jennings G, Korner P (1981) Norepinephrine kinetics in essential hypertension: defective neuronal uptake of norepinephrine in some patients. Hypertension 3:149−156

Esler M, Leonard P, O'Dea K, Jackman G, Jennings G, Korner P (1982) Biochemical quantification of sympathetic nervous activity in humans using radiotracer methodology; fallibility of plasma noradrenaline measurements. J Cardiovasc Pharmacol 4:S152−S157

Essandoh LK, Houston DS, Shepherd JT (1986) Differential effects of lower body negative pressure on forearm and calf blood vessels. J Appl Physiol 61(3): in press

Estefanous FG, Tarazi RC (1980) Systemic arterial hypertension associated with cardiac surgery. Am J Cardiol 46:685−694

Fagius J, Wallin BG (1983) Microneurographic evidence of excessive sympathetic outflow in the Guillain-Barré syndrome. Brain 106:589−600

Fagius J, Wallin BG, Sundlöf G, Nerhed C, Englesson S (1985) Sympathetic outflow in man after anaesthesia of the glossopharyngeal and vagus nerves. Brain 108:423−438

Falkner B, Onesti G, Angelakos ET, Fernandes M, Langman C (1979) Cardiovascular responses to mental stress in normal adolescents with hypertensive parents. Hypertension 1:23−30

Falkner B, Onesti G, Angelakos E (1981) Effect of salt loading on the cardiovascular response to stress in adolescents. Hypertension 3 (suppl II):195−199

Farrehi C (1972) Cardiovascular effects of carotid sinus nerve stimulation in resting man. Chest 61:121−142

Fasola AF, Martz BL (1972) Peripheral venous renin activity during 70° tilt and lower body negative pressure. Aerosp Med 43:713−715

Felder RB, Thames MD (1979) Interaction between cardiac receptors and sinoaortic baroreceptors in the control of efferent cardiac sympathetic nerve activity during myocardial ischemia in dogs. Circ Res 45:728−736

Ferguson DW, Thames MD, Mark AL (1983) Effects of propranolol on reflex vascular responses to orthostatic stress in humans: role of ventricular baroreceptors. Circulation 67(4):802−806

Ferguson DW, Abboud FW, Mark AL (1984) Selective impairment of baroreflex-mediated vasoconstrictor responses in patients with ventricular dysfunction. Circ 69:451−460

Ferguson DW, Kempf JS, Mark AL (1985) Importance of aortic baroreflexes in the heart rate response to dynamic increase in arterial pressure in normal man. J Am Coll Cardiol 5:416

Ferrari A, Gregorini L, Ferrari MC, Preti L, Mancia G (1981) Digitalis and baroreceptor reflexes in man. Circulation 63:279−285

Ferrari A, Bonazzi O, Gregorini L, Gardumi M, Perondi R, Mancia G (1983) Modification of the baroreceptor control of atrio-ventricular conduction induced by digitalis in man. Cardiovasc Res 17:633−641

Finnerty FA Jr, Witkin L, Fazekas JF (1954) Cerebral hemodynamics during cerebral ischemia induced by acute hypotension. J Clin Invest 33:1227−1232

Fitzgerald GA (1984) Peripheral presynaptic adrenoceptor regulation of norepinephrine release in humans. Fed Proc 43:1379−1381

Fitzgerald D, Doyle V, Kelly JG, O'Malley K (1984) Cardiac sensitivity to isoprenaline, lymphocyte beta-adrenoceptors and age. Clin Sci 66:697−699

Fitzgerald A, Robertson D, Wood AJJ (1981) Beta-adrenoceptor downregulation by dynamic exercise and upright posture in man. Trans Assoc Am Physicians 94:310−313

Folkow B (1973) Importance of adaptive changes in vascular design for establishment of primary hypertension: studies in man and in spontaneously hypertensive rats. Circ Res 32:I2–I15

Folkow B (1978) Cardiovascular structural adaptation: its role in the initiation and maintenance of primary hypertension. The 4th Volhard lecture. Clin Sc Mol Med 55:3–22

Folkow B (1982) Physiological aspects of primary hypertension. Physiol Rev 62: 347–504

Folkow B, DiBona GF, Hjemdahl P, Thorén P, Wallin BG (1983) Measurements of plasma norepinephrine concentrations in human primary hypertension. A word of caution on their applicability for assessing neurogenic contributions. Hypertension 5:399–403

Folkow B, Hallback M, Lundgren Y, Sivertsson R, Weiss L (1973) Importance of adaptive changes in vascular design for establishment of primary hypertension studied in man and spontaneously hypertensive rats. Circ Res 32 (suppl 1):2–16

Foster M (1888) A textbook of physiology, 5th edn. Macmillan, London

Fox IJ, Borrker LGS, Heseltine DW, Essex HE, Wood EH (1957) A tricarbocyanine dye for continuous recording of dilution curves in whole blood independent of variations in blood oxygen saturation. Proc Staff Meet Mayo Clin 32:478–484

Fox IJ, Wood EH (1960) Indicator dilution technics in study of normal and abnormal circulation. Medical Phys 3:163

Francis GS, Goldsmith SR, Ziesche SM, Cohn JN (1982) Response of plasma norepinephrine and epinephrine to dynamic exercise in patients with congestive heart failure. Amer J Cardiol 49:1152–1156

Fraser JA, Nadeau HJ, Robertson D, Wood AJJ (1981a) Down regulation of leukocyte beta-adrenoceptor density by circulating plasma levels of catecholamines in man. J Clin Invest 67:1777–1784

Fraser JA, Nadeau HJ, Robertson D, Wood AJJ (1981b) Regulation of human leucocyte beta-receptors by endogenous catecholamines. J Clin Invest 67:1777–1784

Frohlich E, Kozul V, Tarazi R, Dustan H (1970) Physiological comparison of labile and essential hypertension. Circ Res 27 (suppl 1):55–69

Gastaut H, Fischer-William M (1957) Electroencephalographic study of syncope. Lancet 2:1018

Gauer OH, Henry JP (1976) Neurohumoral control of plasma volume. Int Rev Physiol 19:145–190

Genest J, Granger P, de Champlain J, Boucher RL (1968) Endocrine factors in congestive heart failure. Am J Cardiol 22:35–42

Gilbert CA, Stevens PM (1966) Forearm vascular responses to lower body negative pressure and orthostasis. J Appl Physiol 21:1265–1272

Gillis RA (1971) Role of the nervous system in the arrhythmias produced by coronary occlusion in the cat. Am Heart J 91:677–686

Gillis CN, Greene NM, Cronau LH, Hammond GL (1972) Pulmonary extraction of 5-hydroxytryptamine and norepinephrine before and after cardiopulmonary bypass in man. Circ Res 30:666–674

Gilmore JP (1983) Neural control of extracellular volume in the human and nonhuman primate. In: Shepherd JT, Abboud FM (eds) Handbook of physiology, sect 2, vol III, part 2. Peripheral circulation and organ blood flow. Am Physiol Soc, Washington DC, pp 885–915

Goetz KL, Bond GC, Smith WE (1974) Effect of moderate hemorrhage in humans on plasma ADH and renin. Proc Soc Exp Biol Med 145:277–280

Goldsmith SR, Francis GS, Cowley AW, Cohn JN (1982) Response of vasopressin and norepinephrine to lower body negative pressure in humans. Am J Physiol 243: H970–H973

Goldsmith SR, Francis GS, Levine TB, Cohn JN (1983a) Regional blood flow responses to orthostasis in patients with congestive heart failure. J Am Coll Cardiol 1:1391–1395

Goldsmith SR, Francis GS, Cowley AW, Levine TB, Cohn JN (1983b) Increased plasma arginine vasopressin levels in patients with congestive heart failure. J Am Coll Cardiol 1:1385−1390

Goldstein DS (1983) Plasma catecholamines and essential hypertension. Hypertension 5:86−99

Goldstein RE, Beiser GD, Stampfer M, Epstein SE (1975) Impairment of the autonomically mediated heart rate control in patients with cardiac dysfunction. Circ Res 36:571−578

Goldstein DS, Lake CR, Chernow B, Ziegler MG, Coleman MD, Taylor AA, Mitchell JR, Kopin IJ, Keiser HR (1983) Age dependance of hypertensive-normotensive differences in plasma norepinephrine. Hypertension 5:100−104

Goldstein DS, Horwitz D, Keiser HR (1982) Comparison of techniques for measuring baroreceptor sensitivity in man. Circulation 66:432−439

Göthberg G, Lundin S, Aurell M, Folkow B (1983) Response to slow, graded bleeding in salt-depleted rats. Hypertension 1 (suppl 2):24−26

Gozna ER, Marble AE, Shaw A, Holland JG (1974) Age related changes in the mechanics of the aorta and pulmonary artery of man. J Appl Physiol 36:407−411

Grassi G, Gavazzi G, Cesura AM, Picotti GB, Mancia G (1985a) Changes in plasma catecholamines in response to reflex modulation of sympathetic vasoconstrictor tone by cardiopulmonary receptors. Clin Sci 68:503−510

Grassi G, Gavazzi C, Ramirez A, Sabadini E, Turolo L, Mancia G (1985b) Role of cardiopulmonary receptors in reflex control of renin release in man. Hypertension 2 (suppl 3):263−265

Grayboys TB, Lillie RD, Polansky BJ, Chobanian AV (1974) Effects of lower body negative pressure on plasma catecholamine, plasma renin activity and the vectorcardiogram. Aerosp Med 45:834−839

Greenacre JK, Connolly ME (1978) Resensibilization of the beta-adrenoceptor and lymphocytes from normal subjects and patients with phaeochromocytoma. Br J Clin Pharmacol 5:191−197

Greenfield ADM (1951) Emotional faint. Lancet 1:1302

Greenfield ADM (1966) Survey of the evidence for active neurogenic vasodilatation in man. Fed Proc 25:1607

Greenfield ADM, Whitney RJ, Mowbray JF (1963) Methods for the investigation of peripheral blood flow. Br Med Bull 19:101−109

Gribbin B, Pickering TG, Sleight P, Peto R (1971) Effect of age and high blood pressure on baroreflex sensitivity in man. Circ Res 29:424−431

Grimaldi A, Pruszczynski W, Thervet F, Ardaillou R (1985) Antidiuretic hormone response to volume depletion in diabetic patients with cardiac autonomic dysfunction. Clin Sci 68:545−552

Grossman SH, Davis D, Gunnells JC, Strand DG (1982) Plasma norepinephrine in the evaluation of baroreceptor function in humans. Hypertension 4:566−571

Guazzi M, Barbier P, Loaldi A, Montorsi P, Polese A, Tosi E, Fiorentini C (1985) Intrarenal beta-receptor and renal baroreceptor interaction in the control of the renin response to transient reduction of the renal perfusion pressure in man. Hypertension 3:39−45

Guttmann L, Whitteridge D (1947) Effects of bladder distension on autonomic mechanisms after spinal cord injuries. Brain 70:361−404

Guz A, Noble MIM, Trenchard D, Cochrane HL, Makey AR (1964) Studies on the vagus nerves in man: their role in respiratory and circulatory control. Clin Sci 27:293−304

Guz A, Noble MIM, Widdicombe JG, Trenchard D, Mushin WW, Makey AR (1966) The role of vagal and glossopharyngeal afferent nerves in respiratory sensation, control of breathing and arterial pressure regulation in conscious man. Clin Sci 30:161−170

Hagbarth K-E, Vallbo AB (1968) Pulse and respiratory groupings of sympathetic impulses in human muscle nerves. Acta Physiol Scand 74:96−108

Hagbarth K-E, Hallin RG, Hongell A, Torebjork HE, Wallin BG (1972) General characteristics of sympathetic activity in human skin nerves. Acta Physiol Scand 84: 164–176

Haskell WL, Savin WM, Schroeder JS, Alderman EA, Ingles NG Jr, Daughters GJ II, Stinson EB (1981) Cardiovascular responses to handgrip isometric exercise in patients following cardiac transplantation. Circ Res 48 pt II:156–161

Heistad DD, Wheeler RC (1970) Simulated diving during hypoxia in man. J Appl Physiol 28:652–656

Heistad DD, Abboud FM, Mark AL, Schmid PG (1973) Interaction of thermal and baroreceptor reflexes in man. J Appl Physiol 35:581–586

Hellon R (1983) Thermoreceptors. In: Shepherd JT, Abboud FM (eds) Handbook of physiology, sect 2. The cardiovascular system, vol III, part 2. Peripheral circulation and organ blood flow. Am Physiol Soc, Williams and Wilkins, Washington DC, pp 659–674

Henriksen O, Sejrsen P (1977) Local reflex in microcirculation in human skeletal muscle. Acta Physiol Scand 99:19–26

Hesse B, Nielsen I, Hansen JF (1975) The effect of reduction in blood volume on plasma renin activity in man. Clin Sci Mol Med 49:515–517

Hintze TH, Kaley G (1984) Ventricular receptors activated following myocardial prostaglandin synthesis initiate reflex hypotension, reduction in heart rate and redistribution of cardiac output in the dog. Circ Res 54:239–247

Hjemdahl P, Freyschuss U, Juhlin-Dannfelt A, Linde B (1984) Differential sympathetic activation during mental stress evoked by the Stroop test. Acta Physiol Scand (suppl 527):25–29

Hossmann V, Fitzgerald GA, Dollery CT (1980) Circadian rhythm of baroreflex reactivity and adrenergic vascular response. Cardiovasc Res 14:125–129

Hui KKP, Connolly ME (1981) Increased numbers of beta receptors in orthostatic hypotension due to autonomic dysfunction. N Engl J Med 304:1473–1476

Ibrahim MM (1975) Localization of lesion in patients with idiopathic orthostatic hypotension. Br Heart J 37:868–872

Ibsen H, Julius S (1984) Pharmacologic tools for the assessment of adrenergic nerve activity in human hypertension. Fed Proc 43:67–71

Imaizumi T, Takeshita A, Makino N, Ashihara T, Yamamoto K, Nakamura M (1984) Impaired baroreflex control of vascular resistance and heart rate in acute myocardial infarction. Br Heart J 52:418–421

Ito CS, Scher AM (1981) Hypertension following arterial baroreceptor denervation in the unanesthetized dog. Circ Res 48:576–586

Jain SK, Subramanian S, Julka DB, Guz A (1972) Search for evidence of lung chemoreceptors in man; study of respiratory and circulatory effects of phenyldiguanide and lobeline. Clin Sci 42:163–177

James TN, Hageman GR, Urthaler F (1979) Anatomic and physiologic consideration of a cardiogenic hypertensive chemoreflex. Am J Cardiol 44:852–859

Janig W, Sundlöf G, Wallin BG (1983) Discharge patterns of sympathetic neurons supplying skeletal muscle and skin in man and cat. J Auton Nerv Sys 7:239–256

Johannsen UJ, Summers R, Mark AL (1981) Gastric dilation during stimulation of cardiac sensory receptors. Circulation 63:960–964

Johnson JM, Rowell LB, Niederberger M, Eisman MM (1974) Human splanchnic and forearm vasoconstrictor responses to reductions of right atrial and aortic pressures. Circ Res 34:515–524

Julius S, Esler M (1975) Autonomic nervous cardiovascular regulation in borderline hypertension. Am J Cardiol 36:685–696

Julius S, Esler MD, Randall OS (1975) Role of the autonomic nervous system in mild human hypertension. Clin Sci Mol Med 48:2435–2525

Julius S, Cottier C, Egan B, Ibsen H, Kiowski W (1983) Cardiopulmonary mechanoreceptors and renin release in humans. Fed Proc 42:2703–2708

Kalia M, Mei SS, Kao FF (1981) Central projections from ergoreceptors (C fibers) in muscle involved in cardiopulmonary responses to static exercise. Circ 48 pt 2: I48–I62

Katona PG, Felix JIH (1975) Respiratory sinus arrhythmia: non-invasive measure of parasympathetic cardiac control. J Appl Physiol 39:801–805

Kaufman MP, Baker BG, Coleridge HM, Coleridge JCG (1980) Stimulation by bradykinin of afferent vagal C-fibers with chemosensitive endings in the heart and aorta of dogs. Circ Res 46:476–484

Kawakami Y, Natelson BH, DuBois AB (1967) Cardiovascular effects of face immersion and factors affecting diving reflex in man. J Appl Physiol 23:964–970

Keating WR, McIlroy MB, Goldfien A (1964) Cardiovascular responses to ice-cold showers. J Appl Physiol 19:1145

Keller U, Gerber PG, Buhler FR, Stauffacher W (1984) Role of the splanchnic bed in extracting circulating adrenaline and noradrenaline in normal subjects and in patients with cirrhosis of the liver. Clin Sci 67:45–49

Kety SS (1960) The cerebral circulation. In: Handbook of physiology, sect I, Neurophysiology, vol III. Am Physiol Soc, Washington DC, p 1751

Kezdi P, Kordenat RK, Misra SN (1974) Reflex inhibitory effects of vagal afferents in experimental myocardial infarction. Am J Cardiol 33:853–860

Kidd C (1979) Cardiac neurons activated by cardiac receptors. In: Hainsworth R, Kidd C, Linden RJ (eds) Cardiac receptor. Alden, Oxford, pp 377–403

Kimura E, Hashimoto K, Furukawa S, Hayakawa H (1973) Changes in bradykinin level in coronary sinus blood after experimental occlusion of a coronary artery. Am Heart J 85:635–647

Kiowski W, Julius S (1978) Renin response to stimulation of cardiopulmonary mechanoreceptors in man. J Clin Invest 62:656–663

Kiowski W, Bühler FR, Van Brummelen P, Amann FW (1981) Plasma norepinephrine concentration and α-adrenoceptor-mediated vasoconstriction in normotensive and hypertensive man. Clin Sci 60:483–489

Klein RL, Baggett J McC, Thureston-Klein A, Langford HG (1980) Idiopathic orthostatic hypotension: circulatory noradrenaline and ultrastructure of saphenous vein. J Auton Nerv Syst 2:205–222

Klein AA, McCrory WW, Engle MA, Rosenthal R, Ehlers KH (1984) Sympathetic nervous system and exercise tolerance response in normotensive and hypertensive adolescents. J Am Coll Cardiol 3:381–386

Kliks B, Burgess MJ, Abildskov JA (1975) Influence of sympathetic tone on ventricular fibrillation threshold during experimental coronary occlusion. Am J Cardiol 36: 45–49

Kontos HA, Richardson DW, Norvell JE (1975) Norepinephrine depletion in idiopathic orthostatic hypotension. Ann Intern Med 82:336–341

Korner PI (1983) The heart in hypertension. In: Robertson JIS (ed) Handbook of hypertension, vol 1. Clinical aspects of essential hypertension. Elsevier, Amsterdam

Korner PI, West MJ, Shaw J, Uther JB (1974) Steady-state properties of the baroreceptor-heart rate reflex in essential hypertension in man. Clin Exp Pharmacol Physiol 1:65–76

Korner PI, Tonkin AM, Uther JB (1979) Valsalva constrictor and heart rate reflexes in subjects with essential hypertension and with normal blood pressure. Clin Exp Pharmacol Physiol 6:97–110

Kramer RS, Mason DT, Braunwald E (1968) Augmented sympathetic neurotransmitter activity in the peripheral vascular bed of patients with congestive heart failure and cardiac norepinephrine depletion. Circulation 38:629–634

Krasney JA, Levitzky MG, Koehler RC (1974) Sinoaortic contribution to the adjustment of systemic resistance in exercising dogs. J Appl Physiol 36:679–685

Krieger EM (1964) Neurogenic hypertension in the rat. Circ Res 15:511–521

Krieger EM, Salgato HC, Michelini LC (1982) Resetting of the baroreceptors. Int Rev Physiol 26:119–146

Kunze DL, Saum WR, Brown AM (1977) Sodium sensitivity of baroreceptors mediates reflex changes of blood pressure and urine flow Nature 267:75–78

Lambert EH, Wood EH (1947) The use of a resistance wire strain gauge manometer to measure intra-arterial pressure. Proc Soc Exp Biol Med 64:186

Langer SZ (1981) Presynaptic regulation of the release of catecholamines. Pharacol Rev 32:337–362

Lassen NA, Henriksen O, Sejrsen P (1983) Indicator methods for measurement of organ and tissue blood flow. In: Shepherd JT, Abboud FM (eds) Handbook of physiology, sect 2. The cardiovascular system, vol III. Peripheral circulation and organ blood flow, part I. Am Physiol Assoc, Bethesda, pp 21–63

Leatham A (1982) Carotid sinus syncope. Br Heart J 47:409–410

Ledsome JR, Wilson N, Courneya CA, Rankin AJ (1985) Release of atrial natriuretic peptide by atrial distension. Can J Physiol Pharmacol 63:739–742

Leonard B, Mitchell JH, Mizuno M, Rube N, Saltin B, Secher NH (1985) Partial neuromuscular blockade and cardiovascular responses to static exercise in man. J Physiol (Lond) 359:365–379

Levenson JA, Safar ME, London GM, Simon ACh (1980) Haemodynamics in patients with phaeochromocytoma. Clin Sci 58:349–356

Levine TB, Francis GS, Goldsmith SR, Cohn JN (1983) The neurohumoral and hemodynamic responses to orthostatic tilt in patients with congestive heart failure. Circulation 67:1070–1075

Lewis T (1932) A lecture on vaso-vagal syncope and the carotid sinus mechanism with comments on Gower's and Nothnagel's syndrome. Br Med J 1:873–876

Lewis SF, Taylor WF, Bastian BC, Graham RM, Pettinger WA, Blomqvist CG (1983) Hemodynamic responses to static and dynamic handgrip before and after autonomic blockade. Clin Sci 64:593–599

Lind AR (1983) Cardiovascular adjustments to isometric contractions: static effort. In: Shepherd JT, Abboud FM (eds) Handbook of physiology, sect 2. The cardiovascular system, vol III, pt 2. Am Physiol Soc, Williams and Wilkins, Washington DC, pp 947–966

Lindblad LE (1977) Influence of age on sensitivity and effector mechanisms of the carotid baroreflex. Acta Physiol Scand 101:43–49

Lindblad LE (1980) Baroreceptor reflexes in man. A study of cardiovascular and neural effector mechanisms. Doctoral thesis, Stockholm, pp 1–62

Lindblad LE, Wallin BG, Bevegård S (1982) Transient vasodilatation in forearm on stimulation of carotid baroreceptors in man. J Auton Nerv System 5:373–379

Linden RJ, Kappagoda CT (1982) Atrial receptors. Monographs of the Physiological Soc no 39. Cambridge University Press, Cambridge

Linton RAF, Lim M, Wolff CB, Wilmshurst P, Band DM (1984) Arterial plasma potassium measured continuously during exercise in man. Clin Sci 67:427–431

Lombardi F, Della Bella P, Casati R, Malliani A (1981) Effects of intracoronary administration of bradykinin on the impulse activity of afferent sympathetic unmyelinated fibers with left ventricular endings in the cat. Circ Res 48:69–75

London GM, Levenson JA, Safar ME, Simon AC, Guerin AP, Payen D (1983) Hemodynamic effects of head-down tilt in normal subjects and sustained hypertensive patients. Am J Physiol 245 (Heart Circ Physiol 14):H194–H202

Lopes OV, Palmer JF (1967) Proposed respiratory "gating" mechanisms for cardiac slowing. Nature 264:454–456

Lown B, Levine SA (1961) The carotid sinus. Clinical value of its stimulation. Circulation 33:766–789

Ludbrook J (1983) Reflex control of blood pressure during exercise. Annu Rev Physiol 45:155–168

Ludbrook J, Mancia G, Ferrari A, Zanchetti A (1976) Factors influencing the carotid baroreceptor response to pressure changes in a neck chamber. Clin Sci Mol Med 51 (suppl 3):347a–349a

Ludbrook J, Mancia G, Ferrari A, Zanchetti A (1977) The variable-pressure neck-chamber method for studying the carotid baroreflex in man. Clin Sci Mol Med 53: 165—171

Ludbrook J, Faris IB, Iannos J, Jamieson GG, Russell WS (1978) Lack of effect of isometric handgrip exercise in the responses of the carotid sinus baroreceptor reflex in man. Clin Sci Mol Med 55:189—194

Ludbrook J, Mancia G, Zanchetti A (1980) Does the baroreceptor-heart rate reflex indicate the capacity of the arterial baroreceptors to control blood pressure? Clin Exp Pharmacol Physiol 7:499—503

Lugliani R, Whipp BJ, Wasserman K (1973) A role for the carotid body in cardiovascular control in man. Chest 63:744—750

Majewski H, Rand MJ (1981) Adrenaline mediated hypertension: a clue to the antihypertensive effect of beta-adrenoceptor blocking drugs. Trends Pharmac Sci 124—125

Malliani A (1982) Cardiovascular sympathetic afferent fibers. Rev Physiol Biochem Pharmacol 94:11—74

Mancia G, Mark AL (1983) Arterial baroreflexes in humans. In: Shepherd JT, Abboud FM (eds) Handbook of physiology, sect 2. The cardiovascular system, vol III, part 2. Am Physiol Soc, Williams and Wilkins, Washington DC, pp 755—793

Mancia G, Zanchetti A (1981) Hypothalamic control of autonomic functions. In: Pankseep PJ, Morgane J (eds) Handbook of hypothalamus. Dekker, New York, pp 147—202

Mancia G, Romero JC, Shepherd JT (1975a) Continuous inhibition of renin release by vagally innervated receptors in the cardiopulmonary region. Circ Res 36:529—535

Mancia G, Shepherd JT, Donald DE (1975b) Role of cardiac, pulmonary and carotid mechanoreceptors in the control of hindlimb and renal circulation. Circ Res 37: 200—208

Mancia G, Lorenz RR, Shepherd JT (1976) Reflex control of circulation by heart and lungs. Int Rev Physiol 9:111—144

Mancia G, Ferrari A, Gregorini L, Valentini R, Ludbrook J, Zanchetti A (1977) Circulatory reflexes from carotid and extracarotid baroreceptor areas in man. Circ Res 41:309—315

Mancia G, Ferrari A, Gregorini L, Ludbrook J, Zanchetti A (1978a) Baroreceptor control of heart rate in man. In: Schwartz PJ, Brown AM, Malliani A, Zanchetti A (eds) Neural mechanisms of cardiac arrhythmias. Raven, New York, pp 323—333

Mancia G, Iannos J, Jamieson GG, Lawrence HH, Sherman PR, Ludbrook J (1978b) The effect of isometric handgrip exercise on the carotid sinus baroreceptor reflex in man. Clin Sci Mol Med 54:33—37

Mancia G, Leonetti G, Terzoli L, Zanchetti A (1978c) Reflex control of renin release in essential hypertension. Clin Sci Mol Med 54:217—222

Mancia G, Ludbrook J, Ferrari A, Gregorini L, Zanchetti A (1978d) Baroreceptor reflexes in human hypertension. Circ Res 43:170—177

Mancia G, Bonazzi O, Pozzoni L, Ferrari A, Gardumi M, Gregorini L, Perondi R (1979a) Baroreceptor control of atrioventricular conduction in man. Circ Res 44: 752—758

Mancia G, Ferrari A, Gregorini L, Parati G, Pomidossi G, Zanchetti A (1979b) Control of blood pressure by carotid sinus baroreceptors in human beings. Am J Cardiol 44:895—902

Mancia G, Ferrari A, Gregorini L, Zanchetti A (1979c) Clonidine and carotid baroreflex in essential hypertension. Hypertension 1:362—370

Mancia G, Ferrari A, Gregorini L, Ferrari MC, Bianchini C, Terzoli L, Leonetti G, Zanchetti A (1980a) Effects of prazosin on autonomic control of circulation in essential hypertension. Hypertension 2:700—707

Mancia G, Ferrari A, Gregorini L, Bianchini C, Terzoli L, Leonetti G, Zanchetti A (1980b) Methyldopa and neural control of circulation in essential hypertension. Am J Cardiol 45:1237—1243

Mancia G, Ferrari A, Leonetti G, Gregorini L, Terzoli L, Parati G, Bianchini C, Zanchetti A (1981) Carotid sinus reflex control of renin release in hypertensive subjects
  with high renin secretion. Clin Sci 61:505–509
Mancia G, Ferrari A, Leonetti G, Pomidossi G, Zanchetti A (1982a) Carotid sinus
  baroreceptor control of arterial pressure in renovascular hypertensive subjects.
  Hypertension 4:47–50
Mancia G, Ferrari A, Gregorini L, Parati G, Pomidossi G (1982b) Effects of isometric
  exercise on the carotid baroreflex in hypertensive subjects. Hypertension 4:245–250
Mancia G, Ferrari L, Leonetti G, Parati G, Picotti GB, Ravazzani C, Zanchetti A
  (1983a) Plasma catecholamines do not invariably reflect sympathetically induced
  changes in blood pressure in man. Clin Sci 65:227–235
Mancia G, Ferrari A, Gregorini L, Parati G, Pomidossi G, Bertinieri G, Grassi G, di
  Rienzo M, Pedotti A, Zanchetti A (1983b) Blood pressure and heart rate variabil-
  ities in normotensive and hypertensive human beings. Circ Res 53:96–104
Mancia G, Grassi G, Bertinieri G, Ferrari A, Zanchetti A (1984a) Arterial baroreceptor
  control of blood pressure in man. J Aut Nerv Syst 11:115–124
Mancia G, Ferrari A, Gregorini L, Parati G, Cioffi P, Di Rienzo M, Pedotti A (1984b)
  Continuous intra-arterial blood pressure recording in human hypertension. In:
  Sambhi MD (ed) Fundamental fault in hypertension. Nijhoff, Boston, pp 235–250
Mancia G, Bertinieri G, Cavallazzi A, Di Rienzo M, Parati G, Pomidossi G, Ramirez
  AJ, Zanchetti A (1985) Mechanisms of blood pressure variability in man. Clin Exp
  Hypertens A7:167–178
Mancia G, Parati G, Pomidossi G, Casadei R, Di Rienzo M, Zanchetti A (1986) Arterial
  baroreflexes and blood pressure and heart rate variabilities in humans. Hypertension
  8:147–153
Manco JC, Gallo L, Godoy RA, Fernandes RG, Amorim DS (1969) Degeneration of
  cardiac nerves in Chagas' disease. Circulation 40:879–885
Mann S, Altman DG, Raftery EB, Bannister R (1983) Circadian variation of blood
  pressure in autonomic failure. Circulation 68:477–483
Mark AL (1983) The Bezold-Jarisch reflex revisited: clinical implications of inhibitory
  reflexes originating in the heart. J Am Coll Cardiol 1:90–102
Mark AL, Kerber RE (1982) Augmentation of cardiopulmonary baroreflex control of
  forearm vascular resistance in borderline hypertension. Hypertension 4:39–46
Mark AL, Mancia G (1983) Cardiopulmonary baroreflexes in humans. In: Shepherd
  JT, Abboud FM (eds) Handbook of physiology. The cardiovascular system, vol III.
  Peripheral circulation and organ blood flow, part 2. Am Physiol Soc, Washington
  DC, pp 795–813
Mark AL, Kioschos JM, Abboud FM, Heistad DD, Schmid PG (1973) Abnormal
  vascular responses to exercise in patients with aortic stenosis. J Clin Invest 52:
  1138–1146
Mark AL, Lawton WJ, Abboud FM, Fitz AE, Connor WE, Heistad DD (1975) Effects
  of high and low sodium intake an arterial pressure and forearm vascular resistance
  in borderline hypertension. Circ Res 36/37 (suppl 2):94–98
Mark AL, Abboud FM, Fitz AE (1978) Influence of low and high-pressure baroreceptors on plasma renin activity in humans. Am J Physiol 235 (Heart Circ Physiol 4):
  H29–H33
Mark AL, Victor RG, Nerhed C, Wallin BG (1985a) Microneurographic studies of the
  mechanisms of sympathetic nerve responses to static exercise in humans. Circ Res
  57:461–469
Mark AL, Victor RG, Nerhed C, Wallin BG (1985b) Microneurographic studies of the
  mechanisms of sympathetic nerve responses to static exercise in humans. Circ Res
  57:461–469
Mark AL, Victor RG, Nerhed C, Seals DR, Wallin BG (1986) Mechanisms of sympathetic nerve responses to static and rhythmic exercise: new insight from direct intra-
  neural recordings in humans. The sympatho-adrenal system. Physiology and patho-
  physiology. Alfred Benzon Symposium 23, Munkgsgaard Copenhagen, pp 221–233

Marshall RJ, Schirger A, Shepherd JT (1961) Blood pressure during supine exercise in idiopathic orthostatic hypotension. Circ 24:76–81

Martins JB, Zipes DP (1980) Effects of sympathetic and vagal nerves on recovery properties of the endocardium and epicardium of the canine left ventricle. Circ Res 46:100–110

Mathias CJ, Christensen NJ, Corbett JL, Frankel HL, Spalding JMK (1976) Plasma catecholamines during paroxysmal neurogenic hypertension in quadriplegic man. Circ Res 39:204–208

McAllen RM, Jordan D, Spyer KM (1979) The carotid baroreceptor input to the brain. In: Koepchen HP, Hilton SM, Trzebski A (eds) Central interactions between respiratory and cardiovascular control systems. Springer, Berlin Heidelberg New York, pp 87–92

McCloskey DI (1981) Centrally-generated commands and cardiovascular control in man. Clin Exp Hypertens 3:369–378

McNamara HI, Sikorski JM, Clarin H (1969) The effects of lower body negative pressure on hand blood flow. Cardiovasc Res 3:284–291

McRitchie RJ, Vatner SF, Boettcher D, Heyndrickx GR, Patrick TA, Braunwald E (1976) Role of arterial baroreceptors in mediating cardiovascular response to exercise. Am J Physiol 230:85–89

Meier A, Weidmann P, Grim M, Keush G, Gluck Z, Minder I, Ziegler WH (1981) Pressor factors and cardiovascular pressor responsiveness in borderline hypertension. Hypertension 3:367–372

Melcher A (1976) Respiratory sinus arrhythmia in man. A study in heart rate regulating mechanisms. Acta Physiol Scand (suppl) 435:1

Melcher A (1980) Carotid baroreflex heart rate control during the active and the assisted breathing cycle in man. Acta Physiol Scand 108:165–171

Melcher A, Donald DE (1981) Maintained ability of carotid baroreflex to regulate arterial pressure during exercise. Amer J Physiol 241:H383–H849

Mendelsohn FA, Johnston CI (1971) Renin release in chronic paraplegia. Aust NZ J Med 1:393–397

Mew LG (1981) Autonomic hyperreflexia. Ann Emerg Med 10:151–153

Millar-Craig MW, Bishop CN, Raftery EB (1978) Circadian variations of blood pressure. Lancet 1:795–797

Mitchell JH, Payne III FC, Saltin B, Schibye B (1980) The role of muscle mass in the cardiovascular response to static contractions. J Physiol (Lond) 309:45–54

Mitchell JH, Schmidt RF (1983) Cardiovascular reflex control by afferent fibers from skeletal muscle receptors. In: Shepherd JT, Abboud FM (eds) Handbook of physiology, sect 2. The cardiovascular system, vol III, part 2. Am Physiol Soc, Williams and Wilkins, Washington DC, pp 623–658

Mohring J, Glanzer K, Maciel JA Jr, Dusing R, Kramer HJ, Arbogast R, Kochweser J (1980) Greatly enhanced pressor response to antidiuretic hormone in patients with impaired cardiovascular reflexes due to idiopathic orthostatic hypotension. J Cardiovasc Pharmacol 2:367–376

Morley CA, Perrins EJ, Grant P, Chan SL, McBrien DJ, Sutton R (1982) Carotid sinus syncope treated by pacing. Analysis of persistent symptoms and role of atrioventricular sequential pacing. Br Heart J 47:411–418

Morley CA, Dehn TCB, Perrins EJ, Chan SL, Sutton R (1984) Baroreflex sensitivity measured by the phenylephrine pressor test in patients with carotid sinus and sick sinus syndromes. Cardiovasc Res 18:752–761

Mukherjee A, Wong TM, Buja LM, Lefkowitz RJ, Willerson JT (1979) Beta-adrenergic and muscarinic cholinergic receptors in canine myocardium. J Clin Invest 64:1423–1428

Mukherjee A, Bush LR, McCoy KE, Duke RJ, Hagler H, Buja LM, Willerson JT (1982) Relationship between beta-adrenergic receptor numbers and physiological response during experimental canine myocardial ischemia. Circ Res 50:735–741

Nachev C, Collier J, Robinson B (1971) Simplified method for measuring compliance of superficial veins. Cardiovasc Res 5:147–156

Nahirski SR, Bennett DB (1982) Biochemical assessment of adrenoceptor function and regulation: new directions and clinical relevance. Clin Sci 63:97–105

Needleman P, Adams SP, Cole BR, Currie MG, Getler DM, Michener ML, Saper CB, Schwartz D, Standaert DG (1985) Atriopeptins as cardiac hormones. Hypertension 7:469–482

Neil E (1983) Peripheral circulation: historical aspects. In: Shepherd JT, Abboud FM (eds) Handbook of physiology, sect 2. The cardiovascular system, vol III. Peripheral circulation and organ blood flow, part I. Am Physiol Soc, Bethesda, pp 1–19

Neto JAM, Maciel BC, Gallo L, Manco JC, Terra JF, Amorin DS (1980) Effect of lowering left atrial pressure on arterial baroreflex control of heart rate in patients with congestive heart failure. In: Sleight P (ed) Arterial baroreceptors and hypertension. Oxford University Press, Oxford, pp 499–509

Neufeld HN, Goor D, Nathan D, Fischler H, Yerusalmi S (1985) Stimulation of the carotid baroreceptors using a radiofrequency method. Isr J Med Sci 1:630–632

Nicholls MG, Kiowski W, Zweifler AJ, Julius S, Schork MA, Greenhouse J (1980) Plasma norepinephrine variations with dietary sodium intake. Hypertension 2:29–32

Nishi K, Sakanashi M, Takenaka F (1977) Activation of afferent cardiac sympathetic nerve fibers of the cat by pain producing substances and noxious heat. Pfluegers Arch 372:53–61

Norman RA, Coleman TG, Dent AC (1981) Continuous monitoring of arterial pressure indicates sinoaortic denervated rats are not hypertensive. Hypertension 3:119–125

Normell LA, Wallin BG (1974) Sympathetic skin nerve activity and skin temperature changes in man. Acta Physiol Scand 91:417–426

Nosaka S, Wang SC (1972) Carotid sinus baroceptor functions in the spontaneously hypertensive rat. Am J Physiol 222:1079–1084

Öberg B, Thorén P (1972) Increased activity in left ventricular receptors during hemorrhage or occlusion of caval veins in the cat. A possible cause of vasovagal reaction. Acta Physiol Scand 85:164–173

Öberg B, White S (1970) Circulatory effects of interruption and stimulation of cardiac vagal afferents. Acta Physiol Scand 80:383–394

Okamoto K (1969) Spontaneous hypertension in rats. Int Rev Exp Pathol 7:227–270

Ott NT, Shepherd JT (1973) Modifications of the aortic and vagal depressor reflexes by hypercapnia in the rabbit. Circ Res 33(2):160–165

Ott NT, Tarhan S, McGoon DC (1975) Circulatory effects of vagal inflation reflex in man. Z Kardiol 64:1066–1070

Pagani M, Pizzinelli P, Furlan R, Guzzetti S, Rimoldi O, Sandrone G, Malliani A (1985) Analysis of the pressor sympathetic reflex produced by intracoronary injections of bradykinin in conscious dogs. Circ Res 56:175–183

Palkovits M (1980) The anatomy of central cardiovascular neurons. In: Fuxe K, Goldstein M, Hokfelt B, Hokfelt T (eds) Central adrenaline neurons: basic aspects and their role in cardiovascular functions. Pergamon, Oxford, pp 3–17

Palkovits M (1981) Neuropeptides and biogenic amines and central cardiovascular control mechanisms. In: Buckle JP, Ferrario CM (eds) Central nervous system mechanisms in hypertension. Raven, New York, pp 73–97

Palkovits M, Zaborszky L (1977) Neuroanatomy of central cardiovascular control. Nucleus tractus solitarii: afferent and efferent neuronal connections in relation to the baroreceptor reflex arc. Prog Brain Res 47:9–34

Pantridge JF (1978) Autonomic disturbance at the onset of acute myocardial infarction. In: Schwartz PJ, Brown AM, Malliani A, Zanchetti A (eds) Neural mechanisms in cardiac arrhythmias. Raven, New York, pp 7–17

Pantridge JF, Adgey AAJ, Geddes JS, Webb SW (1975) The acute coronary attack. Grune and Stratton, New York

Parati G, Pomidossi G, Ramirez A, Cesana B, Mancia G (1985) Variability of the haemodynamic responses to laboratory tests employed in assessment of neural cardiovascular regulation in man. Clin Sci 69:533–540

Perez-Gomez F, Garcia-Aguado A (1977) Origin of ventricular reflexes caused by coronary arteriography. Br Heart J 39:967–973

Perez-Gomez F, De Dios RM, Rey J, Garcia-Aguado A (1979) Prinzmetal's angina: reflex cardiovascular response during episode of pain. Br Heart J 42:81–87

Péronnet F, Cléroux J, Perrault H, Cousineau D, De Champlain J, Nadear R (1981) Plasma norepinephrine response to exercise before and after training in humans. J Appl Physiol 51:812–815

Peveler RC, Bergel DH, Robinson JL, Sleight P (1983) The effect of phenylephrine upon arterial pressure, carotid sinus radius and baroreflex sensitivity in the conscious greyhound. Clin Sci 64:455–461

Phillip T, Distler A, Cordes U (1978) Sympathetic nervous system and blood-pressure control in essential hypertension. Lancet 2:959–963

Pickering GW (1936) The peripheral resistance in persistent arterial hypertension. Clin Sci 2:209–231

Pickering TG, Davis J (1973) Estimation of the conduction time of the baroreceptor-cardiac reflex in man. Cardiovasc Res 7:213–219

Pickering TG, Gribbin B, Strange-Petersen E, Cunningham DJC, Sleight P (1971) Comparison of the effects of exercise and posture on the baroreflex in man. Cardiovasc Res 5:582–586

Pickering TG, Gribbin B, Sleight P (1972a) Comparison of the reflex heart rate response to rising and falling arterial pressure in man. Cardiovasc Res 6:277–283

Pickering TG, Gribbin B, Strange-Petersen E, Cunninghan DJC, Sleight P (1972b) Effects of autonomic blockade on the baroreflex in man at rest and during exercise. Circ Res 30:177–185

Pitts RF (1968) Physiology of the kidney and body fluids. Year Book Medical, Chicago

Polinsky RJ, Kopin IJ, Ebert MH, Weise V (1981) Pharmacological distinction of different orthostatic hypotension syndromes. Neurology 31:1–7

Pórazász J, Barankay T, Szolcsanyi J, Gilbiszer-Porszasz K, Madarasz K (1962) Studies of the neural connection between the vasodilator and vasoconstrictor centres in the cat. Acta Physiol Hung 22:29–41

Pryor J (1971) Autonomic hyper-reflection. N Engl J Med 285:860

Pugh LGCE, Wyndham CL (1950) The circulatory effects of high spinal anesthesia in hypertensive and control subjects. Clin Sci 9:189–203

Quest JA, Gillis RA (1974) Effects of digitalis on carotid sinus baroreceptor activity. Circ Res 35:247–255

Ramirez AJ, Bertinieri G, Belli L, Cavallazzi A, Di Rienzo M, Pedotti A, Mancia G (1985) Reflex control of blood pressure and heart rate by arterial baroreceptors and cardiopulmonary receptors in the unanesthetized cat. J Hypert 3:327–335

Randall OA, Esler MD, Bullock GF, Maisel AS, Ellis CN, Zweifler AJ, Julius S (1976) Relationship of age and blood pressure to baroreflex sensitivity and arterial compliance in man. Clin Sci Mol Med 51:357s–360s

Randall OA, Esler M, Culp B, Julius S, Zweifler A (1978) Determinants of baroreflex sensitivity in man. J Lab Clin Invest 91:514–519

Restall PA, Smirk FH (1952) Regulation of blood pressure levels by hexamethonium bromide and mechanical devices. Brit Heart J 14:1–12

Ricardo JA, Koh ET (1978) Anatomical evidence of direct projections from the nucleus of the solitary tract to the hypothalamus, amygdala and other forebrain structures in the rat. Brain Res 153:1–26

Richards AM, Nicholls MG, Ikram H, Webster MWI, Yandle TG, Espiner EA (1985) Renal, haemodynamic, and hormonal effects of human alpha atrial natriuretic peptide in healthy volunteers. Lancet 1:545–549

Rickstein SE, Thorén P (1981) Reflex control of sympathetic nerve activity and heart rate from arterial baroreceptors in conscious spontaneously hypertensive rats. Clin Sci 61:169s–172s

Robertson D, Johnson GA, Robertson RM, Nies AS, Shand DG, Oates JA (1979) Comparative assessment of stimuli that release neuronal and adrenomedullary catecholamines in man. Circulation 59:637–642

Robertson D, Hollister AS, Carey EL, Tung Che-Se, Goldberg MR, Robertson RM (1984) Increased vascular beta$_2$-adrenoceptor responsiveness in autonomic dysfunction. J Am Coll Cardiol 3:850–856

Robinson BJ, Johnson RH, Lambie DG, Palmer KT (1983) Do elderly patients with an excessive fall in blood pressure on standing have evidence of autonomic failure? Clin Sci 64:587–591

Rocchini AP, Cant JR, Barger AC (1977) Carotid sinus reflex in dogs with low- to high-sodium intake. Am J Physiol 233 (Heart Circ Physiol2):H196–H202

Roddie IC (1983) Circulation to skin and adipose tissue. In: Shepherd JT, Abboud FM (eds) Handbook of physiology, sect 2. The cardiovascular system, vol III. Peripheral circulation and organ blood flow, pt I. Am Physiol Soc, Bethesda, pp 285–317

Roddie IC, Shepherd JT (1956) The reflex nervous control of human skeletal muscle blood vessels. Clin Sci 15:433–440

Roddie IC, Shepherd JT (1957) The effects of carotid artery compression in man with special reference to changes in vascular resistance in the limbs. J Physiol (Lond) 139:377–384

Roddie IC, Shepherd JT (1958) Receptors in the high pressure and low pressure vascular systems: their role in the reflex control of the human circulation. Lancet 1: 493–496

Roddie IC, Shepherd JT (1963) Nervous control of the circulation in skeletal muscle. Br Med Bull 19:115–119

Roddie IC, Shepherd JT, Whelan RF (1956) Evidence from venous oxygen saturation measurements that the increase in forearm blood flow during body heating is confined to the skin. J Physiol (Lond) 134:444–450

Roddie IC, Shepherd JT, Whelan RF (1957) Reflex changes in vasoconstrictor tone in human skeletal muscle in response to stimulation of receptors in a low-pressure area of the intrathoracic vascular bed. J Physiol (Lond) 139:369–376

Roddie IC, Shepherd JT, Whelan RF (1958) Reflex changes in human skeletal muscle blood flow associated with intrathoracic pressure changes. Circ Res 6:232–238

Ross CA, Ruggiero DA, Park DH, Joh TH, Sved AF, Fernandez-Pardal J, Saavedra JM, Reis DJ (1984) Tonic vasomotor control by the rostral ventrolateral medulla: effect of electrical or chemical stimulation of the area containing C$_1$ adrenaline neurons on arterial pressure, heart rate, and plasma catecholamines and vasopressin. J Neurosci 4:474–494

Rowell LB (1974) Measurement of hepatic-splanchnic blood flow in man by dye techniques. In: Bloomfield DA (ed) Dye curves. The theory and practice of indicator dilution. University Park Press, Baltimore, Chap 12, pp 209–229

Rowell LB (1981) Active neurogenic vasodilation in man. In: Vanhoutte PM, Leusen I (eds) Vasodilatation. Raven, New York, pp 1–17

Rowell LB (1983) Cardiovascular adjustments to thermal stress. In: Shepherd JT, Abboud FM (eds) Handbook of physiology, sect 1. The cardiovascular system, vol III, pt 2. Am Physiol Soc, Williams and Wilkins, Washington DC, pp 967–1024

Rowell LB, Blackmon JR, Bruce RA (1964) Indocyanine green clearance and estimated hepatic blood flow during mild to maximal exercise in upright man. J Clin Invest 43:1677–1690

Rowell LB, Brengelmann GL, Detry J-MR, Wyss C (1971) Venomotor responses to local and remote thermal stimuli to skin in exercising man. J Appl Physiol 30: 72–77

Rowell LB, Detry J-MR, Blackmon JR, Wyss C (1972) Importance of the splanchnic vascular bed in human blood pressure regulation. J Appl Physiol 32:213–220

Rowell LB, Wyss CR, Brengelmann GL (1973) Sustained human skin and muscle vasoconstriction with reduced baroreceptor activity. J Appl Physiol 34:639–643

Rowell LB, Freund PR, Hobbs SF (1981) Cardiovascular responses to muscle ischemia in humans. Circ Res 48 (suppl 1):37–47

Rusch NJ, Shepherd JT, Webb RC, Vanhoutte PM (1981) Different behavior of the resistance vessels of the human calf and forearm during contralateral isometric exercise, mental stress, and abnormal respiratory movements. Circ Res 48 (suppl 1): 118–130

Safar M, Weiss Y, Levenson J, London G, Milliez P (1973) Hemodynamic study of 85 patients with borderline hypertension. Am J Cardiol 31:315–319

Samueloff SL, Browse NL, Shepherd JT (1966) Response of capacity vessels in human limbs to head-up tilt and suction on lower body. J Appl Physiol 21:47–54

Sannerstedt RS, Julius S, Conway J (1970) Hemodynamic responses to tilt and beta-adrenergic blockade in young patients with borderline hypertension. Circulation 42:1057–1064

Sano K, Aiba T (1966) Pulseless disease: summary of our 62 cases. Jpn Circ J 30:63–67

Saum WR, Brown AM, Tuley FJ (1976) An electrogenic sodium pump and baroreceptor function in normotensive and spontaneously hypertensive rats. Circ Res 39: 497–505

Saum WR, Ayachi S, Brown AM (1977) Actions of sodium and potassium ions on baroreceptors of normotensive and spontaneously hypertensive rats. Circ Res 41: 768–774

Seals DR, Washburn RA, Hansen PG, Painter PL, Nagle FJ (1983) Increased cardiovascular response to static contraction of larger muscle groups. J Appl Physiol 54: 434–437

Share L (1965) Effects of carotid occlusion and left atrial distension on plasma vasopressin titer. Am J Physiol 208:219–223

Share L (1967) Role of peripheral receptors in the increased release of vasopressin in response to hemorrhage. Endocrinology 81:1140–1146

Share L (1968) Control of plasma ADH titer in hemorrhage: role of atrial and arterial receptors. Am J Physiol 215:1384–1389

Share L, Levy MN (1962) Cardiovascular receptors and blood titer of antidiuretic hormone. Am J Physiol 203:425–428

Shepherd JT (1963) Nervous control of the blood vessels in the skin. In: Physiology of the circulation in human limbs in healths and disease. Saunders, Philadelphia, pp 9–41

Shepherd JT (1982) Reflex control of arterial blood pressure. Cardiovasc Res 16: 357–383

Shepherd JT (1983) Circulation to skeletal muscle. In: Shepherd JT, Abboud FM (eds) Handbook of physiology, sect 2. The cardiovascular system, vol III. Peripheral circulation and organ blood flow, part I. Am Physiol Soc, Bethesda, pp 319–370

Shepherd JT (1984) Regulation of blood flow to human limbs. Int Angiology 3:31–45

Shepherd JT, Vanhoutte PM (1975) Veins and their control. Saunders, Philadelphia,

Shepherd JT, Vanhoutte PM (1981) Local modulation of adrenergic neurotransmission. Circulation 64:655–666

Shepherd JT, Rusch NJ, Vanhoutte PM (1983) Effect of cold on the blood vessel wall. Gen Pharmacol 14:61–64

Sheridan DJ, Penkoske PA, Sobel BE, Corr PB (1980) Alpha-adrenergic contributions to dysrhythmia during myocardial ischemia and reperfusion in cats. J Clin Invest 65:161–171

Shy GM, Drager GA (1960) A neurological syndrome associated with orthostatic hypotension. Arch Neurol 2:511–527

Skrabal F, Aubock J, Hortnagl H, Brucke T (1981) Plasma epinephrine and norepinephrine concentrations in primary and secondary human hypertension. Hypertension 3:373–379

Sleight P (1981) Cardiac vomiting. Br Heart J 46:5−7

Sleight P, Robinson JL, Brooks DE, Rees PM (1977) Characteristics of single carotid sinus baroreceptor fibers and whole nerve activity in the normotive and the renal hypertensive dog. Circ Res 41:750−758

Sleight P, Fox P, Lopez R, Brooks DE (1978) The effect of mental arithmetic on blood pressure variability and baroreflex sensitivity in man. Clin Sci Mol Med 55: 381s−382s

Sleight P, Floras JS, Hassan MO, Hones JV, Osikowska BA, Sever P, Turner KL (1979) Baroreflex control of blood pressure and plasma noradrenaline during exercise in essential hypertension. Clin Sci 57:169s−171s

Smith HW (1939) Studies in the physiology of the kidney. Porter lectures series IX, Univ Extension Division, University of Kansas, Lawrence

Smith WW, Wikler NS, Fox AC (1954) Hemodynamic studies of patients with myocardial infarction. Circulation 9:352−362

Smyth HS, Sleight P, Pickering GW (1969) Reflex regulation of arterial pressure during sleep in man: a quantitative method of assessing baroreflex sensitivity. Circ Res 24: 109−121

Snell ES, Cranston WI, Gerbrandy J (1955) Cutaneous vasodilatation during fainting. Lancet 1:693

Snellen HA, Dunning AJ, Arntzenius AC (1981) History and perspectives of cardiology. Catheterization, angiography, surgery and concepts of circular control. Leiden University Press, The Hague

Sokoloff L, Mangold R, Wechsler RL, Kennedy C, Kety SS (1955) The effect of mental arithmetic on cerebral circulation and metabolism. J Clin Invest 34:1101−1108

Spyer KM (1981) The neural organization and control of the baroreflex. Rev Physiol Biochem Pharmacol 88:24−124

Stephenson RB, Smith OA, Scher AM (1981) Baroreceptor regulation of heart rate in baboons during different behavioral states. Am J Physiol 241:R277−R285

Stella A, Zanchetti A (1984) Neural control of renin release. J Hypertension 2 (suppl 1): 83−87

Strandell T, Shepherd JT (1967) The effect in humans of increased sympathetic activity on the blood flow to active muscles. Acta Med Scand (suppl 472):146−167

Suarez DH, Messerli FH, Ventura HO, Aristimuno G, Dresbenski GR, Frohlich ED (1982) Baroreceptor stimulation and isometric exercise and normotensive and borderline hypertensive subjects. Clin Sci 62:307−309

Sugrue DD, Wood DL, McGoon MD (1984) Carotid sinus hypersensitivity and syncope. Mayo Clin Proc 59:637−640

Sundlöf G, Wallin BG (1977) The variability of muscle nerve sympathetic activity in resting recumbent man. J Physiol (Lond) 272:383−397

Sundlöf G, Wallin BG (1978a) Human muscle nerve sympathetic activity at rest. Relationship to blood pressure and age. J Physiol (Lond) 274:621−637

Sundlöf G, Wallin BG (1978b) Effect of lower body negative pressure on human muscle nerve sympathetic activity. J Physiol (Lond) 278:525−532

Swan HJC, Ganz W (1983) Hemodynamic measurements in clinical practice: a decade in review. J Am Coll Cardiol 1:103−113

Takano Y, Sawyer WB, Loewy AD (1985) Substance P mechanisms of the spinal cord related to vasomotor tone in the spontaneously hypertensive rat. Brain Res 334: 105−116

Takeshita A, Mark AL (1978) Neurogenic contribution to hindquarters vasoconstriction during high sodium intake in Dahl strain of genetically hypertensive rat. Circ Res 43 (suppl 1):186−191

Takeshita A, Mark AL (1979) Decreased venous distensibility in borderline hypertension. Hypertension 1:202−206

Takeshita A, Mark AL (1980) Decreased vasodilator capacity of forearm resistance vessels in borderline hypertension. Hypertension 2:610−616

Takeshita A, Tanaka S, Kuroiwa A, Nakamura M (1975) Reduced baroreceptor sensitivity in borderline hypertension. Circulation 51:738–742

Takeshita A, Tanaka S, Orita Y, Kanaide H, Nakamura M (1977) Baroreflex sensitivity in patients with Takayasu's aortitis. Circulation 55:803–806

Takeshita A, Tanaka S, Nakamura M (1978) Effects of propropanolol on baroreflex sensitivity in borderline hypertension. Cardiovasc Res 12:148–151

Takeshita A, Mark AL, Eckberg DL, Abboud FM (1979) Effect of central venous pressure on arterial baroreflex control of heart rate. Am J Physiol 236 (Heart Circ Physiol 5):H42–H47

Takeshita A, Matsuguchi H, Nakamura M (1980) Effect of coronary occlusion on arterial baroreflex control of heart rate. Cardiovasc Res 14:303–306

Takeshita A, Tsutomu I, Ashihara T, Yamamoto K, Hoka S, Nakamura M (1982) Limited maximal vasodilator capacity of forearm resistance vessels in normotensive young men with a family predisposition to hypertension. Circ Res 50:671–677

Tarazi RC, Estefanous FG, Fouad FM (1978) Unilateral stellate block in the treatment of hypertension after coronary by-pass surgery. Implications of a new therapeutic approach. Am J Cardiol 42:1013–1018

Thames MD (1977) Reflex supression of renin release by ventricular receptors with vagal afferents. Am J Physiol 233 (Heart Circ Physiol 2):H181–H184

Thames MD, Schmid PG (1979) Cardiopulmonary receptors with vagal afferents tonically inhibit ADH release in the dog. Am J Physiol 237 (Heart Circ Physiol 6): H299–H304

Thames MD, Zubair-Ul-Hassen, Brachett NC Jr, Lower RR, Kontos HA (1971) Plasma renin responses to hemorrhage after cardiac autotransplantation. Am J Physiol 221:1115–1119

Thames MD, Jarecki J, Donald DE (1978) Neural control of renin secretion in anesthetized dogs. Interaction of cardiopulmonary and carotid baroreceptors. Circ Res 42(2):237–245

Thames MD, Klopfenstein HS, Abboud FM, Mark AL, Walker JL (1978) Preferential distribution of inhibitory cardiac receptors with vagal afferents to the inferoposterior wall of the left ventricle activated during coronary occlusion in the dog. Circ Res 43:512–519

Thames MD, Peterson MG, Schmid PG (1980) Stimulation of cardiac receptors with veratrum alkaloids inhibits ADH secretion. Am J Physiol (Heart Circ Physiol 8) 239:H784–H788

Thames MD, Waickman LA, Abboud FM (1980) Sensitization of cardiac receptors (vagal afferents) by intracoronary acetylstrophanthidin. Am J Physiol 239 (Heart Circ Physiol 8):H628–H635

Thames MD, Miller BD, Abboud FM (1982) Sensitization of vagal cardiopulmonary baroreflex by chronic digoxin. Am J Physiol 243 (Heart Circ Physiol 12):H815–H818

Thomas JA, Marks BH (1978) Plasma norepinephrine in congestive heart failure. Am J Cardiol 41:233–243

Thomas JE (1969) Hyperactive carotid sinus reflex and carotid sinus syncope. Mayo Clin Proc 44:127–139

Thomas M, Malmcrona R, Shillingford J (1966) Circulatory changes associated with systemic hypotension in patients with acute myocardial infarction. Br Heart J 28: 108–117

Thorén P (1979) Role of cardiac vagal C-fibres in cardiovascular control. Rev Physiol Biochem Pharmacol 86:1–94

Thorén P, Donald DE, Shepherd JT (1976) Role of heart and lung receptors with nonmedullated vagal afferents in circulatory control. Circ Res 38 (suppl 2):2–9

Tohmen JF, Cryer PE (1979) Biphasic-adrenergic modulation of beta-adrenergic receptors in man. J Clin Invest 65:836–840

Toubes DB, Brody MJ (1970) Inhibition of reflex vasoconstriction after experimental coronary embolization in the dog. Circ Res 26:211–224

Trzebski A, Raczowska M, Kubin L (1980a) Influence of respiratory activity and hypocapnia on the carotid baroreflex in man. In: Sleight P (ed) Arterial baroreceptors and hypertension. Oxford University Press, Oxford, pp 282–290

Trzebski A, Raczkowska M, Kubin L (1980b) Carotid baroreceptor reflex in man, its modulation over the respiratory cycle. Acta Neurobiol Exp (Warcz) 40:807–820

Trzebski A, Tafil M, Zoltowski M, Przybylski T (1982) Increased sensitivity of the arterial chemoreceptor drive in young men with mild hypertension. Cardiovasc Res 163–172

Tyden G, Samnegard H, Thulin L (1979) The effects of changes in the carotid sinus baroreceptor activity on the splanchnic blood flow in anesthetized man. Acta Physiol Scand 106:187–189

Uchida Y, Murao S (1974) Bradykinin-induced excitation of afferent cardiac sympathetic fibers. Jpn Heart J 15:84–91

Üvnas B (1967) Cholinergic vasodilator innervation in skeletal muscles. Circ Res 20/21 (suppl 1):83–90

Valdes-Cruz L, Horowitz S, Nesel E, Sahn DJ, Fischer DC, Larson D (1984) A pulsed Doppler echocardiographic method for calculating pulmonary and systemic blood flow in atrial level shunts: validation studies in animals and initial human experience. Circulation 69:80–86

Vallbo AB, Hagbarth K-E, Torebjörk HE, Wallin BG (1979) Somatosensory, proprioceptive, and sympathetic activity in human peripheral nerves. Physiol Rev 59: 919–957

Vanhoutte P, Lacroix E, Leusen I (1966) The cardiovascular adaptation of the dog to muscular exercise: role of the arterial pressoreceptors. Arch Int Physiol Biochim 74:201–222

Vanhoutte PM, Verbeuren TJ, Webb RC (1981) Local modulation of adrenergic neuroeffector interaction in the blood vessel wall. Physiol Rev 61:151–247

Verbeuren TJ, Lorenz RR, Aarhus LL, Shepherd JT (1983) Prejunctional beta-adrenoceptors in human and canine saphenous veins. J Autonomic Nervous system 8: 261–271

von Euler US (1948) Identification of the sympathomimetic ergone in adrenergic nerve of cattle (sympathin N) with laevo-noradrenaline. Acta Physiol Scand 16:63–74

Walgenbach SC, Donald DE (1983a) Cardiopulmonary reflexes and arterial pressure during rest and exercise in dogs. Am J Physiol 244:H326–H369

Walgenbach SC, Donald DE (1983b) Inhibition by carotid baroreflex of exercise-induced increase in arterial pressure. Circ Res 52:253–262

Walgenbach SC, Shepherd JT (1984) Role of arterial and cardiopulmonary mechanoreceptors in the regulation of arterial pressure during rest and exercise in conscious dogs. Mayo Clinic Proc 59:467–475

Walker JL, Thames MD, Abboud FM, Mark AL, Klopfenstein HS (1978) Preferential distribution of inhibitory cardiac receptors in left ventricle of the dog. Am J Physiol 235 (Heart Circ Physiol 4):H188–H192

Walker JL, Abboud FM, Mark AL, Thames MD (1980) Interaction of cardiopulmonary receptors with somatic receptors in man. J Clin Invest 65:1491–1497

Wallach R, Karp RB, Reves JG, Oparil S, Smith LR, James TN (1980) Pathogenesis of paroxysmal hypertension developing during and after coronary bypass surgery: a study of hemodynamic and humoral factors. Amer J Cardiol 46:559–565

Wallin BG, Eckberg DL (1982) Sympathetic transients caused by abrupt alterations of carotid baroreceptor activity in humans. Am J Physiol 242 (Heart Circ Physiol II):H185–H190

Wallin BG, Stjernberg L (1984) Sympathetic activity in man after spinal cord injury. Brain 107:183–198

Wallin BG, Sundlöf G (1979) A quantitative study on muscle nerve sympathetic activity in resting normotensive and hypertensive subjects. Hypertension 1:67–77

Wallin BG, Delius W, Hagbarth KE (1973) Comparison of sympathetic nerve activity in normotensive and hypertensive subjects. Circ Res 33:9–21

Wallin BG, Sundlöf G, Delius W (1975) The effects of carotid sinus nerve stimulation on muscle and skin nerve sympathetic activity in man. Pflügers Arch 358:101–110

Wallin BG, Sundlöf G, Lindblad LE (1980) Baroreflex mechanisms controlling sympathetic outflow to the muscles in man. In: Sleight P (ed) Arterial baroreceptors and hypertension. Oxford University Press, Oxford, pp 101–107

Wallin BG, Sundlöf G, Eriksson B-M, Dominiak P, Grobecker H, Lindblad LE (1981) Plasma norepinephrine correlates to sympathetic muscle nerve activity in normotensive man. Acta Physiol Scand 111:69–73

Walter PF, Crawley IS, Dorney ER (1978) Carotid sinus hypersensitivity and syncope. Am J Cardiol 42:396–403

Warren JB, Dalton N (1983) A comparison of the bronchodilator and vasopressor effects of exercise levels of adrenaline in man. Clin Sci 64:475–479

Warren JB, Dalton N, Turner C, Clark TJH, Toseland PA (1984) Adrenaline secretion during exercise. Clin Sci 66:87–90

Watanabe AM (1983) Recent advances in knowledge about beta-adrenergic receptors. Application to clinical cardiology. J Am Coll Cardiol 1:82–89

Watson RDS, Stallard TS, Littler WA (1979) Effects of beta-adrenoceptor antagonists on sinaortic baroreflex sensitivity and blood pressure in hypertensive man. Clin Sci 57:241–247

Watson RDS, Stallard TJ, Flinn RM, Lutler WA (1980) Factors determining direct arterial pressure and its variability in hypertensive man. Hypertension 2:333–341

Wei JY, Markis JE, Malagold M, Braunwald E (1983) Cardiovascular reflexes stimulated by reperfusion of ischemic myocardium in acute myocardial infarction. Circ 67:796–801

Weiss S, Baker JB (1933) The carotid sinus reflex in health and disease: its role in the causation of fainting and convulsion. Medicine (Baltimore) 12:297–345

Whelan RF (1967) Control of the peripheral circulation in man. Thomas, Springfield,

White JC (1957) Cardiac pain. Circulation 16:644–659

White NJ (1980) Heart rate changes on standing in elderly patients with orthostatic hypotension. Clin Sci 58:411–413

Wieling W, Borst C, Van Brederode JFM, Van Dongen Torman MA, Van Montfrans GA, Dunning AJ (1983) Testing for autonomic neuropathy: heart rate changes after orthostatic maneuvers and static muscle contractions. Clin Sci 64:581–586

Wilcot CS, Aminiff MJ, Slater JDH (1977) Sodium homeostasis in patients with autonomic failure. Clin Sci Med 53:321–328

Wilson JR, Ferraro N, Wiener DH (1985) Effect of the sympathetic nervous system on limb circulation and metabolism during exercise in patients with heart failure. Circulation 72:72–81

Wintermintz SR, Oparil S (1982) Sodium-neural interactions in the development of spontaneous hypertension. Clin Exp Hypertens A4 (4 and 5):751–760

Wood EH (1978) Evolution of instrumentation and techniques for the study of cardiovascular dynamics from the thirties to 1980. Ann Biomed Eng 6:250–309

Wurster RD, Randall WC (1975) Cardiovascular responses to bladder distension in patients with spinal transection. Am J Physiol 228:1288–1292

Zelis R, Longhurst J (1975) The circulation in congestive heart failure. In: Zelis R (ed) Peripheral circulations. Grune and Stratton, Orlando, pp 283–314

Zelis R, Caudill CC, Baggette K, Mason DT (1976) Reflex vasodilation induced by coronary angiography in human subjects. Circulation 53:490–493

Zerbe RL, Henry DP, Robertson GL (1983) Vasopressin response to orthostatic hypotension. Am J Med 74:265–271

Zimmerman BG (1978) Actions of angiotensin on adrenergic nerve endings. Fed Proc 37:199–202

Zimmerman BG (1981) Adrenergic facilitation by angiotensin: does it serve a physiological function? Clin Sci 60:343–348

Zitnik RS, Ambrosioni E, Shepherd JT (1971) Effect of temperature on cutaneous venomotor reflexes in man. J Appl Physiol 31:507–512

Zoller RP, Mark AL, Abboud FM, Schmid PG, Heistad DD (1972) The role of low pressure baroreceptors in reflex vasoconstrictor responses in man. J Clin Invest 51: 2967–2972

Rev. Physiol. Biochem. Pharmacol., Vol. 105
© by Springer-Verlag 1986

Vascular Capacitance: Its Control and Importance

ROGER HAINSWORTH

Contents

1 Introduction. 102
   1.1   Capacitance Vessels 103
   1.2   Capacitance 104
   1.3   Assessment of a Change in Capacitance 105

2 Musculo-cutaneous Circulation. 107
   2.1   Studies in Cutaneous Veins 107
   2.1.1 Methods of Study 107
   2.1.2 Responses to Stimulation 109
   2.1.3 Effects of Temperature. 110
   2.1.4 Reflex Responses. 112
   2.1.5 Summary of Responses of Cutaneous Veins 115
   2.2   Regional Responses 116
   2.2.1 Methods of Study 116
   2.2.2 Responses to Direct Stimulation of Sympathetic Nerves 117
   2.2.3 Reflex Responses 120
   2.3   Capacitance Responses in Human Limbs 120
   2.3.1 Methods of Study 120
   2.3.2 Reflex Responses in Humans 122
   2.4   Importance of Capacitance in Muscle and Cutaneous Circulations. . . . 126
   2.4.1 Magnitude of Responses 126
   2.4.2 Physiological Significance of Lack of Capacitance Responses
        in Musculo-cutaneous Circulation 126

3 Abdominal Vascular Bed. 127
   3.1   Studies in Whole Abdominal Vascular Bed 128
   3.1.1 Methods of Study 128
   3.1.2 Responses to Direct Stimulation of Sympathetic Nerves 129
   3.1.3 Reflex Responses 130
   3.2   Studies in the Intestinal Vascular Bed 134
   3.2.1 Methods of Study 134
   3.2.2 Responses to Stimulation of Sympathetic Nerves. 134
   3.2.3 Reflex Responses 135
   3.3   The Liver. 135
   3.3.1 Methods of Study 135
   3.3.2 Responses to Direct Stimulation of Sympathetic Nerves 136
   3.3.3 Reflex Responses 137
   3.4   The Spleen. 137
   3.4.1 Methods of Study 137
   3.4.2 Responses to Direct Stimulation of Sympathetic Nerves 138
   3.4.3 Reflex Responses 138

Department of Cardiovascular Studies, University of Leeds, Leeds LS2 9JT,
Great Britain

3.5   Capacitance in the Human Splanchnic Circulation 139
3.5.1   Methods of Study 139
3.5.2   Responses in the Human Splanchnic Circulation 139
3.6   Importance of Capacitance in the Abdominal Circulation 140
3.6.1   Magnitude of Responses 140
3.6.2   Sensitivity of Abdominal Capacitance Vessels to Sympathetic
         Nerve Activity 142
3.6.3   Role of the Abdominal Circulation as a Blood Reservoir 142

4 The Pulmonary Circulation 144
4.1   Methods of Study 144
4.1.1   Constant Flow Perfusion: Measurement of Pressure Changes 145
4.1.2   Constant Flow Perfusion: Measurement of Volume Changes 145
4.1.3   Indicator Dilution Methods 145
4.2   Responses to Adrenergic Stimulation 146
4.3   Responses to Reflex Stimulation 146
4.4   Importance of Pulmonary Vascular Capacitance 146

5 Whole Body Capacitance 147
5.1   Methods of Study 147
5.1.1   Changes in Volume at Constant Flow and Constant Venous Pressure . . . 147
5.1.2   Mean Circulatory Filling Pressure 148
5.2   Responses to Infusions of Catecholamines 149
5.3   Reflex Responses 152
5.3.1   Baroreceptors 152
5.3.2   Responses to Other Interventions 153

6 Importance of Vascular Capacitance 154
6.1   Extent and Magnitude of Capacitance Changes 154
6.2   Active Versus Passive Volume Changes 155
6.3   Potential Importance of Capacitance Vessels 157
6.4   Vascular Capacitance in Perspective 158
6.5   Conclusion 160

References 161

1 Introduction

Interest in the physiology of veins has grown almost exponentially during most of this century. Hooker as long ago as 1918 showed that when isolated segments of vein were stimulated electrically there was an increase in the intravenous pressure, and Donegan (1921) found that when a segment of vein was perfused at a constant pressure and stimulated electrically there was a decrease in the flow rate. Folkow and Mellander reviewed the function of veins and venous tone in 1964 and concluded that veins were "far from being merely passive conduits of blood". They considered that veins were "at least as reactive and well controlled as any of the other compartments within the circulation and that their control formed an important 'missing link' in cardiovascular integration". This view seems to have been accepted for the most part uncritically ever since, although

Hainsworth and Linden (1979) suggested that some of the claims made for the importance of veins and vascular capacitance might have been overstated.

1.1 Capacitance Vessels

The classification of blood-vessels according to their physiological roles originated with Folkow (1959). The physiological classification does not coincide precisely with anatomical divisions. For example, resistance vessels are predominantly small arteries, arterioles and precapillary sphincters. However, all blood-vessels make some contribution to the overall hydraulic resistance to blood flow. Similarly, capacitance vessels are generally regarded as being synonomous with veins, but again, since the concept of capacitance means that blood is contained in the vessel and the volume contained is dependent on the distending pressure, then clearly all blood-vessels and the heart are to some extent capacitance vessels.

Rather crude estimates, based on anatomical studies, have indicated that approximately 70% of the blood volume is contained in veins (Green 1950; Green et al. 1963). Large changes in volume can occur in veins with only small changes in pressure, due partly to some veins being empty and collapsed and partly to veins being highly distensible at low volumes. Therefore, it is not unreasonable to consider that the vessels which provide the major capacitance function are indeed the veins.

Not only can the volume contained in veins be changed by a change in the distending pressure, but because most veins contain smooth muscle in their walls, the volume can also be changed actively by stimulation of that muscle. Due to the very large number of parallel channels on the venous side of the circulation (the total cross-sectional area of veins is about 30 times that of arterioles), a modest reduction in the overall size of the channels would have a relatively small effect on resistance but a relatively large effect on the volume. The exception to this would be in veins carrying a high flow of blood, for example those draining exercising muscle, where there may be a significant contribution to overall resistance and venous constriction would be likely to have the adverse effect of increasing capillary filtration (see page 127). Anatomical studies have shown the presence of smooth muscle in the wall of all veins greater than 0.6 mm in diameter (Burton 1954), which suggests a possible role for veins in the active regulation of capacitance. The extent to which active changes occur in response to physiological stimuli is the main subject of this article.

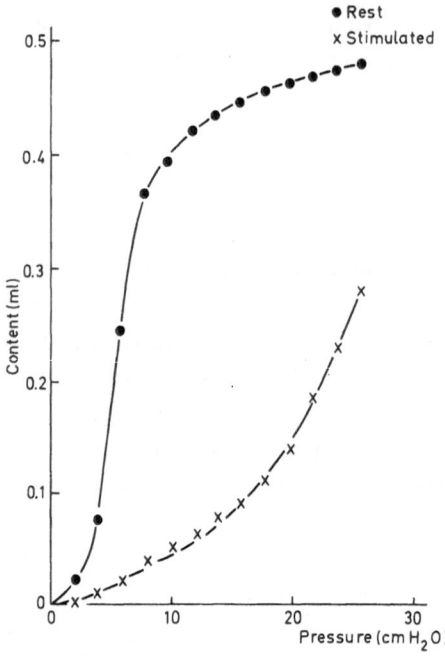

Fig. 1. Volume-pressure curves from isolated segment of saphenous vein of dog distended by injection of Krebs-Ringer solution. Curves were obtained at rest and during electrical stimulation at a frequency of 15 Hz. During stimulation the volume was reduced at all pressures and the pressure increased at any given volume. However, at pressures greater than 1 kPa (10 cmH$_2$O) the compliance of the vessel ($\Delta V/\Delta P$) was greater during stimulation. (Vanhoutte and Leusen 1969)

1.2 Capacitance

The concept of capacitance, like that of resistance, is based on an electrical analogue. Resistance is based on Ohm's law and is equal to pressure gradient divided by blood flow (electrical resistance is voltage divided by current). Electrical capacitance is the electrical charge stored divided by the potential difference across the capacitor. The relationship is linear and so the value of the capacitance of a condenser can readily be determined. In the vascular system the comparable variables are the volume of blood contained within a vessel or a region and the transmural distending pressure. However, here the analogy breaks down because the volume-pressure relationship is not linear. Therefore, it is not possible to define a single value for capacitance, but it is necessary to describe the entire volume-pressure curve (Fig. 1).

The ideal distensibility curves of veins are considered to have three components (Fig. 2). The initial portion is apparently highly distensible and represents merely the filling of the collapsed vessel so that it becomes filled and circular but its wall is not distended. The second portion corresponds to distension of the vessel wall until the limit of distension is reached due to tension developed in the collagen fibres. The volume in the vein, determined by extrapolation, at which the pressure would be zero as called the "unstressed volume". The slope of the volume-pressure curve is the "compliance" (Fig. 2). Capacitance may change either as the result of

Fig. 2. Idealized volume-pressure curve. Compliance (C) is the slope of the linear portion. Unstressed volume is determined by extrapolating the line to zero pressure and is considered to be the volume at which the vessel is filled but not distended. A change in capacitance can result in a change in compliance, unstressed volume, or both. (Rothe 1983c)

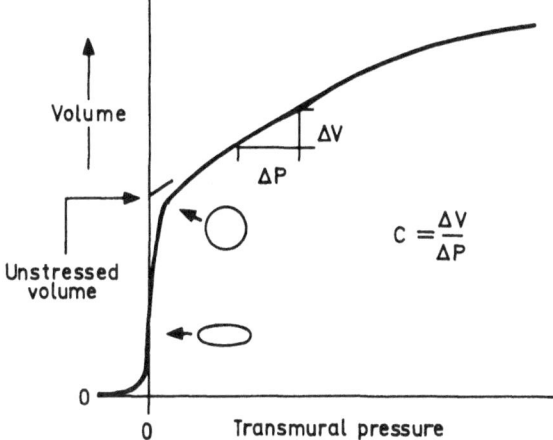

a change in unstressed volume, in which case the curve is displaced up or down, and/or as the result of a change in compliance (see Rothe 1983a,b). The idea of dividing capacitance in this way may have some conceptual value but, unless the curves are of an ideal slope, as in Fig. 2, they may be difficult to determine and impossible to interpret. For example, in Fig. 1 the unstressed volumes would be very negative in the resting state, but would increase in the stimulated state. Negative unstressed volumes do not only occur when isolated veins are considered but also can occur when one is determining the distensibility characteristics of whole organs such as the liver (e.g. Greenway et al. 1985).

The term "capacity" is sometimes used (e.g. Shepherd and Vanhoutte 1975), and although it has not been precisely defined it is usually taken to mean the same as capacitance.

1.3 Assessment of a Change in Capacitance

A change in capacitance implies that the relationship between volume and pressure has changed; there is a different curve (e.g. Fig. 1). This may be due to a change in the unstressed volume, the compliance, or both. Assessment of only a part of this curve may be misleading. Determination of changes in compliance may give erroneous results. In Fig. 1, when the compliance is assessed at distending pressures greater than 1 kPa, it will be seen that the stimulated vein is actually more compliant. This is simply because, when the vein is not stimulated, at the higher transmural pressures it is reaching the limits of distension. Similarly, as pointed out above, assessment of unstressed volume by extrapolation may be equally misleading.

To assess whether there has been an active change in vascular capacitance, due to constriction of the smooth muscle within the wall of the blood-vessel, it is necessary to demonstrate that the volume-pressure curve is different and that any change is not merely the consequence of moving along the same curve. In isolated blood-vessels, the problem is relatively simple and extensive volume-pressure curves can be drawn. In tissues or the whole body it is much more complex. The pressures distending the various blood-vessels are different and, even if precapillary vessels are ignored, it is not adequate simply to control pressure in large veins. This is because most of the blood is contained within venules and veins less than 40 μm in diameter (see Rothe 1983a), and it is very difficult to measure, let alone control, the pressure in these small vessels. If the pressure in large veins is constant and there has been a change in blood flow, there must have been a proportional change in the small venous pressure (Poiseuille's equation), and this will result in passive volume changes. The effect of this is examined in more detail later in relation to the limb and splanchnic circulations.

If it is impracticable to define complete volume-pressure curves, changes in capacitance can be inferred from a change in volume when the distending pressure is constant. This can only be achieved by vascular isolation of the region, perfusion at constant flow and draining at constant venous pressure. Furthermore, because there may be changes in capillary pressure, due to changes in pre- and postcapillary resistance (Oberg 1964), account must be taken of this and assessments should be made over relatively short periods ($<$ 2 min) to ensure that changes in volume due to changes in capillary filtration are minimized.

An alternative valid approach to the assessment of capacitance is to occlude the circulation to a region, both arterial and venous, so that volume is fixed, and to determine capacitance changes from the changes in the venous distending pressure.

In this review I will examine vascular capacitance and its control in the various regions of the body. I have adopted this approach rather than the usual one of considering the effects on capacitance of various stimuli in turn (e.g. Hainsworth and Linden 1979; Rothe 1983a,b) because, as will be shown, the responses of the various regions to all stimuli differ much more than the responses in a given organ or region differ to different stimuli. The likely importance of the various regions in their contributions to the overall capacitance response will be examined. Finally, I will attempt to put vascular capacitance into some sort of perspective and examine its likely importance for cardiovascular homeostasis.

2 Musculo-cutaneous Circulation

Because of its accessibility there have been far more studies of capacitance responses in the limb circulation than in any other region of the body. In humans, this is the only region to have been studied adequately. There are two main approaches to the study of capacitance responses in the limb circulation: responses can be studied in individual veins (usually a cutaneous vein) or in a region (either the whole limb or a part of it).

2.1 Studies in Cutaneous Veins

2.1.1 Methods of Study

Although it is not intended to give a detailed account of the methods used to study capacitance responses (for these see Hainsworth and Linden 1979 and Rothe 1983c), a short description will be given to indicate the principles of the techniques used and their limitations so that the quality of the experimental findings can be assessed. The large number of experimental approaches used testifies to the difficulties; no method is ideal. Some methods are, at best, only qualitative, whereas others are able to grade responses and to allow comparison of the results of one intervention with another.

   In order to assess a capacitance response it is necessary to know not only the volume contained within a vessel, but also the pressure distending it. One variable should be held constant while changes in the other are determined.

2.1.1.1 Direct Observation of Cutaneous Veins
In translucent tissue it is possible to visualize a vein and to estimate its diameter by use of a microscope. This technique has been used to study venous responses in a bat's wing (Wiedeman 1959). An alternative approach which can be applied to non-translucent tissue and to deeper vessels is to inject a radio-opaque dye into a vessel and to estimate its diameter radiographically (Morris et al. 1974). Although the radiographic method is claimed to be able to discriminate changes in diameter as little as 0.1 mm, it does have the disadvantage that expensive equipment is required. Furthermore, estimations of vessel diameter are only of value provided that the pressure distending the vein is known.

2.1.1.2 Measurement of Volume Changes in Isolated Venous Segments
If a segment of vein is vascularly isolated and maintained at a constant distending pressure, a change in diameter or volume can provide a measure of the change in the capacitance (Gero and Gerova 1968).

2.1.1.3 Measurement of Pressure Changes in Isolated Venous Segments

In this method the blood contained within a vein is held constant, so that a change in the distending pressure indicates a change in capacitance. By infusion and withdrawal of blood it is possible to define pressure-volume relationships.

Salzman (1957) recorded the pressure in a latex balloon inserted into a saphenous vein in the anaesthetized dog and considered that changes in pressure in the balloon signified a change in the degree of constriction of the vein. However, the method has not been validated and it is possible that changes in the flow of blood into the vein might influence its pressure. Indeed, the paper contains an illustration showing a large response to a stimulus to which later and validated methods show little or no response.

2.1.1.4 Study of Pressure Gradients Along Veins

Constriction of a blood-vessel not only decreases the volume of blood contained within it but also increases the resistance to the flow of blood along the vessel. Because the resistance to flow is a function of the fourth power of the radius and the volume is related only to its square, methods based on resistance to flow would be expected to be more sensitive. It can be calculated that if vessel circumference (and therefore vessel radius) decreases by 10%, there will be a decrease in volume of 19% but an increase in resistance of over 50%. There have been a number of studies which have attempted to assess venomotor responses by determining the pressure gradients along naturally perfused veins. However, since in such studies flow along the veins is not known, these methods are useless for assessing even volume changes, let alone active capacitance responses. In some studies the total flow to a region has been measured (e.g. Zoster and Tom 1967), but this does not necessarily imply a constant flow in the segment of the vein studied. Even when the flow along a segment of vein is known (e.g. Hall et al. 1976), there are problems in calculating venous responses from the ratio of pressure to flow when both variables are allowed to change, due to non-linearity of the pressure-flow relationship.

Abboud and Eckstein (1966) vascularly isolated a limb which was perfused through the artery at constant flow. They assessed venomotor responses from the changes in pressure gradients along segments of veins. However, during electrical stimulation of the efferent sympathetic nerves to a limb, blood becomes redistributed from cutaneous veins to muscle veins (Diana et al. 1968), so changes in pressure gradients may be due solely to changes in flow.

The errors due to changes in flow along a segment of vein can be avoided by artificially perfusing it. Donegan, as long ago as 1921, studied venomotor responses by perfusing a segment of vein at a constant pressure and recording changes in flow. More recently, Rice et al. (1966) perfused a cutaneous vein at a constant flow and assessed venomotor responses from the gradient between the perfusion pressure and a catheter with its tip a few centimetres downstream. Webb-Peploe and Shepherd (1968a) used a similar technique but measured the pressure gradient between the perfused vein and the common iliac vein. Because a change in the flow of blood into the perfused vein from collateral tributaries would be likely to influence the venous pressure gradient, they occluded the common iliac artery for the period over which observations were made. This method is a reliable and sensitive way of determining the responses of individual venous segments. However, it does have the disadvantage that arterial resistance responses cannot be determined simultaneously. Also, due to the arrested circulation, there would be likely to be a change in the external environment of the vein which would eventually influence the responses. These problems were addressed by Hainsworth et al. (1975), who vascularly isolated a dog's hind limb by the use of strong nylon ligatures, which were placed round the main muscle groups but were positioned to avoid crushing the femoral and sciatic nerves. The femoral artery and a superficial metatarsal vein were perfused at constant flows and the blood was drained from the limb into an open reservoir. With this preparation responses of total vascular resistance were assessed from changes in arterial perfusion pressure and venous responses were determined from changes in the pressure gradient between the perfused vein and the femoral vein. One possible criticism of this method is that, although total blood flow to the limb is constant, a change in the distribution of flow between the muscle and cutaneous circulations might change the flow in collateral venous tributaries and thereby change the venous perfusion pressure. This is unlikely to be of importance due to the high perfusion pressure used (usually about 4 kPa) and, furthermore, large changes in arterial flow had little or no effect on the venous pressure (R. Hainsworth and F. Karim, unpublished observations). A further criticism could be that the vein might behave abnormally due to its being perfused with arterial blood. However, there was no change in perfusion pressure if venous blood was used (R. Hainsworth and F. Karim, unpublished observations). In a further refinement to this method (Karim et al. 1980a), the circulation to the limb was separated from that to the rest of the body by the use of an artificial gas exchanger. This allowed the study of responses in the limb in the absence of changes in the concentrations of blood-borne agents from the rest of the circulation.

2.1.2 Responses to Stimulation

The sympathetic innervation to the veins in limbs of dogs, cats and rabbits was investigated by Donegan (1921). He noted that stimulation of the spinal roots between T6 and T8 resulted in constriction of forelimb veins. Hind limb veins constricted in response to stimulation of the lumbar roots between L4 and L7 or S1. Constriction of veins in response to stimulation of sympathetic nerves is not uniform. The largest responses are obtained in the more peripheral vessels and little if any change occurs in the larger, more central veins. Webb-Peploe and Shepherd (1968a) gradually withdrew a catheter so that its tip moved from a femoral vein down the saphenous vein and noted an abrupt change in pressure near the popliteal fossa. The conclusion was that only the distal veins responded to sympathetic stimulation. However, there is some evidence that the femoral vein can constrict to some extent (Gero and Gerova 1968; Morris et al. 1974), and it is possible that the abrupt change in venous pressure near the knee noted by Webb-Peploe and Shepherd (1968a) was partly due to the larger diameter of the vein and the existence of collateral veins near the knee. The rate of constriction of veins depends to some extent on the stimulus frequency and is more rapid at higher frequencies. It may require up to 4 min to reach a maximum value, but this is then sustained for at least 15 min (Gerova and Gero 1975).

There are conflicting reports as to the shape of the stimulus—response curve of cutaneous veins to sympathetic stimulation. Gero and Gerova (1968) obtained a hyperbolic relationship such that over 50% of the maximal response was obtained at a stimulus frequency of only 2 Hz. Others (Rice et al. 1966; Webb-Peploe and Shepherd 1968a) found that the venous perfusion pressure increased in a near linear way with increasing stimulus frequencies. The difference in the responses is probably at least partly related to the method of determining them. Because of the Poiseuille relationship, pressure gradient is inversely proportional to the fourth power of the radius of the vessel, whereas a change in volume is related to its square. As diameter decreases towards zero, the change in volume approaches 100% whereas the change in resistance approaches infinity.

Alpha-adrenergic drugs constrict cutaneous veins (e.g. Rice et al. 1966). However, the effects of beta-agonists are not as obvious. In resting conditions, where there is little or no vasoconstrictor tone to the vein, isoprenaline has little or no effect (Webb-Peploe and Shepherd 1969). However, if the vein is initially constricted by sympathetic stimulation or by infusion of noradrenaline, then isoprenaline does relax it.

2.1.3 Effects of Temperature

Cutaneous veins constrict in response to cold and dilate in response to warming. Shepherd and co-workers published several papers which described the response of a cutaneous vein to local and central temperature changes. Veins in both the forelimb and the hind limb in the dog have been shown to constrict progressively as the temperature of the perfusate is decreased from 47°C to 17°C (Webb-Peploe and Shepherd 1968b). The effect was dependent on sympathetic nervous activity, as it was almost abolished after alpha-adrenergic or ganglion blockade. Sympathectomy also almost abolished the response (Webb-Peploe and Shepherd 1968b).

The effects of local temperature changes are particularly marked when the sympathetic nerves are stimulated at low frequencies. Karim et al. (1980a) showed that when a vein was perfused with blood at 31°C, stimulation at frequencies as low as 0.1–0.2 Hz caused venous constriction, whereas when the blood was at 38°C it was necessary to stimulate at 0.5 Hz or above. Temperatures of 31°C are not unusual in cutaneous veins; Karim et al. (1980) recorded temperatures of 31°C–34°C in the saphenous veins of dogs immediately after exposure to ambient temperatures of about 10°C. Webb-Peploe and Shepherd (1968b) noted that the venoconstriction in response to cooling was not related to cooling of the overlying skin. However, the local effect of temperature was augmented by a central action. When the rest of the dog was cooled by passing blood through a heat exchanger, the cutaneous vein constricted and responded more powerfully to local cooling (Webb-Peploe and Shepherd 1968c). This effect was observed in cutaneous veins but not in the femoral vein or in the spleen (Webb-Peploe 1969). The responses to central cooling were attributed to hypothalamic-temperature-sensitive neurones (Hardy et al. 1964). However, no attempt was made to localize the central cooling and Hainsworth and Karim (unpublished observations) did not observe any constriction of a dog's cutaneous vein when the blood perfusing the cephalic region alone was cooled.

The effect of a local temperature change on venous smooth muscle is not uniform (Webb-Peploe and Shepherd 1968d). Vanhoutte and Lorenz (1970) found that, whilst the constrictions of helical strips of saphenous and mesenteric veins in response to electrical stimulation were enhanced by cooling, that of a strip of femoral vein was inhibited. Cooling also enhances the relaxation by the beta-agonist isoprenaline of cutaneous veins that have been stimulated to contract.

The mechanism whereby local cooling enhances the constriction of cutaneous veins has been the subject of several investigations. The response is not peculiar to sympathetic or adrenergic stimuli, but occurs also when 5-hydroxytryptamine, acetylcholine or ATP are added. However, potassium chloride or barium chloride causes a venous contraction which is reduced with cooling (Vanhoutte and Shepherd 1970). This led to the conclusion that the effect of cooling was on the muscle excitation process rather than the contraction mechanism. It was not due to depression of neuronal uptake mechanisms because it was unaffected by chronic sympathectomy or by administration of cocaine, which blocks the neuronal uptake of noradrenaline (Webb-Peploe 1969). Janssens and Vanhoutte (1978) determined the affinities of saphenous and femoral veins for noradrenaline at different temperatures and found that in the saphenous vein, in which cooling potentiated the effect of noradrenaline, the affinity of the receptor for noradrenaline was increased by cooling. In the femoral vein, however, in which no such potentiation occurs, there was no effect on the affinity of the receptor.

2.1.4 Reflex Responses

2.1.4.1 Baroreceptor Reflexes

Most studies of baroreceptor reflexes have been confined to the carotid sinuses. The relatively few studies of aortic baroreceptors have shown that, although the cardiovascular reflex responses are qualitatively similar to those from the carotid baroreceptors, they require higher blood pressures to elicit them and the reponses tend to be smaller (Hainsworth et al. 1970; Donald and Edis 1971). This is supported by electrophysiological studies which have shown the discharge from aortic baroreceptors to occur at higher pressures (Pelletier et al. 1972).

There have been several studies of the responses of cutaneous veins to the baroreceptor reflex. The earlier studies provided conflicting results (Salzman 1957; Zingher and Grodins 1964; DiSalvo et al. 1971). However, in none of these studies were possible effects from responses of precapillary vessels excluded. Brender and Webb-Peploe (1969) studied the responses of a saphenous vein which was perfused at constant flow and found no significant response at all to changes in carotid sinus pressure. Hainsworth et al. (1975) determined responses of a perfused metatarsal vein to changes in both aortic and carotid sinus pressures. Large changes in pressure in either region resulted in a small dilatation of the cutaneous vein. This response was similar to that obtained by cooling the lumbar sympathetic trunks to 5°C to block efferent sympathetic discharge. However, it was noted in these studies that, although the venous responses were significant, electrical stimulation of the sympathetic nerves at only 1 Hz resulted in responses four times as large. To induce arterial responses similar to those obtained reflexly it was necessary to stimulate at 2–5 Hz.

Those experiments provided evidence for an almost complete dissociation of sympathetic activity to resistance vessels and cutaneous veins in the limb. Because the small response might have been due to changes in circulating levels of catecholamines, Karim et al. (1980a) completely vascularly isolated a limb and perfused it in a closed circuit with blood which passed through a gas exchange unit. Using this preparation, the venous response, although still significant, was even smaller. It was also found that the reflex response, like the response to direct electrical stimulation of the sympathetic nerves, was significantly increased by cooling the perfusate.

2.1.4.2 Chemoreceptor Reflexes

The resting discharge rate from both aortic and carotid chemoreceptors at normal levels of blood gas tensions and normal blood pressures is low (Biscoe et al. 1970; Sampson and Hainsworth 1972). The discharge frequency rapidly increases with hypoxia, particularly when accompanied by hypercapnia. The effect of acidaemia on aortic chemoreceptors is less consistent (Sampson and Hainsworth 1972) and, unless the oxygen tension is low, chemoreceptors are relatively insensitive to moderate degrees of hypotension (Lee et al. 1964). Although both groups of chemoreceptors stimulate respiration, their effects on the heart are different. In the absence of a change in respiration, stimulation of carotid chemoreceptors results in negative chronotropic and inotropic responses (Daly and Scott 1958, 1962; Hainsworth et al. 1979), whereas stimulation of aortic chemoreceptors results in positive chronotropic and inotropic responses (Karim et al. 1980b).

The effect of stimulation of carotid chemoreceptors on a perfused cutaneous vein was studied by Pelletier and Shepherd (1972). They reported that, although the resistance vessels constricted, there was a dilatation of the saphenous vein. The magnitude of this response depended on the pre-existing sympathetic efferent nervous activity to the vein, and it was inhibited by giving further doses of the anaesthetic chloralose. The degree of dilatation of the vein could be graded by grading the intensity of the stimulus to the chemoreceptors (Pelletier 1972). The response was also dependent on the carotid sinus pressure (Mancia 1975). This was despite the failure of stimulation of the carotid baroreceptors to cause any direct effects on the perfused vein. Venous responses, like those of arterial resistance, were maximal at intermediate levels of carotid pressure (about 18 kPa). At lower carotid pressures, the venous responses to stimulation of chemoreceptors were attenuated and at higher pressures they were totally abolished.

There have been few satisfactory studies of responses of a cutaneous vein to stimulation of aortic chemoreceptors. In the studies of Calvelo et al. (1970) and Iizuka et al. (1970), nicotine and sodium cyanide were injected into the aortic arch to stimulate the chemoreceptors. These authors attempted to delay the effect on the carotid chemoreceptors by use of delay coils originally described by Comroe and Mortimer (1964). However, this technique is fraught with problems: firstly, due to a pressure gradient along the delay coils it is likely that some of the drug would reach the carotid or other sensory areas by a collateral route and earlier than expected; secondly, it is likely that the drug would have reached other receptors such as those perfused by the coronary circulation. More recently, Britton and Donald (1982) injected sodium cyanide into the aortic root and obtained a dilatation of a metatarsal vein when the carotid sinus regions were denervated. The responses were shown to be abolished after vagotomy or aortic nerve section. Similar responses were also obtained when the vascularly isolated aortic arch was perfused with hypoxic blood. Thus it appears that, although the effects of aortic and carotid chemo-receptors on the heart are different, stimulation of both groups of chemo-receptors dilates cutaneous veins.

2.1.4.3 Cardiac and Pulmonary Receptors

2.1.4.3.1 Atrial Receptors. Atrial receptors can be stimulated by the use of small ballons or pouches (Ledsome and Linden 1964, 1967). The only reflex cardiovascular response to this discrete stimulation is an in-crease in heart rate mediated solely by a bundle of efferent sympathetic nerves which is not affected by stimulation of baroreceptors, chemo-receptors or cutaneous or visceral nerves (Linden et al. 1982; Hassan et al. 1985). There have been no studies of responses in cutaneous veins, although Carswell et al. (1970) obtained no change in hind limb vascular resistance and Karim et al. (1972) reported that there was no change in the efferent nervous activity in the lumbar sympathetic nerves.

2.1.4.3.2 Ventricular Receptors. The ventricles are innervated mainly by unmyelinated vagal afferents (Baker et al. 1979; Thorén 1979), although there are some myelinated fibres (Paintal 1973) and afferent nerves running in the sympathetic nerves have also been described (Malliani 1979). Ven-tricular receptors can be stimulated mechanically, for example by ven-tricular distension or by injection into the coronary circulation of stimu-lant drugs, such as the Veratrum alkaloids. The resulting response is the Bezold-Jarisch reflex, which is bradycardia, hypotension and vasodilata-tion in several vascular beds.

Ford et al. (1984) studied the responses to injection of veratridine into the coronary circulation and found that, in addition to bradycardia and arterial vasodilatation, there was also a small dilatation of a perfused cutaneous vein. More recently, S. Challenger, R. Hainsworth and K.H. McGregor (to be published) increased ventricular systolic pressure and obtained a small reflex dilatation of resistance vessels in a perfused limb but no significant response of a cutaneous vein.

2.1.4.3.3 Pulmonary Receptors. The lung is very richly innervated with a variety of myelinated and non-myelinated nerves (see Widdicombe 1973; Coleridge and Coleridge 1984). The various receptors can be stimulated in several ways, such as by inhalation of irritant gases, lung inflation, pulmonary congestion and oedema. Each type of receptor can be excited by several stimuli, which makes it difficult to determine which type of receptor is responsible for a particular reflex. Lung inflation can cause tachycardia or bradycardia and vasoconstriction or vasodilatation (Anrep et al. 1936; Daly and Robinson 1968; Glick et al. 1969; Salisbury et al. 1959; Hainsworth 1974; Greenwood et al. 1980), although at moderate inflation pressures there is usually tachycardia and vasoconstriction and at higher pressures, bradycardia and vasodilatation (Hainsworth 1974; Wood et al. 1985). In the only study of responses in a cutaneous vein (Greenwood et al. 1977) there was no response to lung inflation.

2.1.4.4 Central Nervous System Effects

An increase in sympathetic discharge and a pronounced rise in arterial blood pressure occur in response to an increase in intracerebral pressure (Brown 1956), cerebral arterial hypotension and hypoxia (Sagwa et al. 1961; Downing et al. 1963; De Geest et al. 1965; Hainsworth and Karim 1973) and hypercapnia (Hainsworth et al. 1984, 1985; Ford et al. 1985b; Soladoye et al. 1985). There have been no reports of the effects of these interventions on cutaneous veins.

2.1.5 Summary of Responses of Cutaneous Veins

Cutaneous veins constrict powerfully in response to sympathetic and adrenergic stimulation and to a number of vasoconstrictor agents. Beta-adrenoreceptor stimulation causes venodilatation but only on a background of stimulation. The venoconstriction is greatly potentiated by cooling the venous perfusate by a mechanism which appears to increase the affinity of the receptor for the agonist. Central cooling has been reported to cause venoconstriction and to increase the constriction in response to local cooling.

Cutaneous veins are weakly engaged in the baroreceptor reflex, if at all. However, if they are initially constricted they do dilate slightly in response to stimulation of carotid chemoreceptors and to injections of veratridine into the coronary circulation. However, even when they do occur, responses of cutaneous veins to reflex stimuli are small and there is usually evidence of marked dissociation in the magnitude and sometimes the direction of the changes in the sympathetic activity to cutaneous veins and resistance vessels.

2.2 Regional Responses

2.2.1 Methods of Study

Changes in volume in a region can be inferred from the changes in weight or determined directly by plethysmography. Other techniques involve the use of radioactive tracers and the integration of the difference between inflow and outflow of blood to the region. An alternative approach is to occlude the vessels to the region and to determine changes in intravascular pressures. With all these methods it is important to distinguish active changes in intravascular volume from passive changes due to changes in distending pressure, or changes in extravascular tissue volume.

2.2.1.1 Occluded Limb Techniques

When an occluded limb technique is used, the region investigated, usually hindquarters, whole limb or part of a limb, is vascularly isolated from the rest of the circulation, and once the pressures have been stabilized, a change in venous pressure indicates a change in capacitance (Browse et al. 1966a—c; Verrier et al. 1969). If this method is used it is essential to be certain that there could be no unoccluded collateral vessels, as this would invalidate the results. Disadvantages of this method are that the pressures tend to be unstable and any responses decline after a few minutes of occlusion. Also, it is difficult to be certain that all vessels are in communication with each other and responses from reactive areas, for example cutaneous veins, may disproportionately influence measurements. The responses to a constrictor stimulus are dependent on the initial venous pressure, being reduced at abnormally high pressures (Samueloff and Chinyanga 1968).

2.2.1.2 Plethysmography

Mellander (1960) described a technique for determining changes in volume in the hindquarters of the cat. The region was perfused at constant pressure and changes in volume determined by use of a plethysmograph. An increase in activity in sympathetic nerves to the region results in a decrease in blood flow, indicating constriction of resistance vessels, and an initial rapid decrease in volume followed by a slower prolonged decrease in volume. The initial rapid phase was attributed to constriction of capacitance vessels and the slower phase to changes in extravascular fluid volume. Although this method has been very widely used, it cannot be regarded as being adequate for determining active responses of capacitance vessels. The two main problems are that it cannot readily distinguish slow venous responses from changes in extravascular volume and that, due to changes in flow, the pressure distending the small veins is likely to change and to result in passive changes in volume.

Plethysmography has been used with constant flow perfusion (Baum and Hosko 1965; Baker 1966). This can give reliable estimates of capacitance responses provided that vascular isolation is complete, venous pressures are constant and the effects of changes in extravascular volume are taken into account either by restricting the time for estimation to about 2 min or by estimating changes in capillary filtration by use of radioisotopes.

2.2.1.3 Weight Changes

If an organ is continuously weighed a change in weight denotes a change in its volume (Shadle et al. 1958; Haddy et al. 1961). The same constraints and limitations apply to this method as to plethysmography. Lesh and Rothe (1969) determined changes in weight in an isolated perfused muscle and assessed the intravascular and extravascular components by use of ^{51}Cr-labelled erythrocytes and by determining mean transit time (see also Rothe 1983c).

2.2.1.4 Inflow-Outflow Difference

Changes in volume in a limb can be determined by integration of the difference between the inflow and the outflow of blood to the region (Hainsworth et al. 1983c). The limitations of this method are similar to those of plethysmography but, in addition, a very high degree of vascular isolation is necessary, otherwise changes in collateral blood flow result in changes in outflow, which are indistinguishable from changes in volume.

2.2.2 Responses to Direct Stimulation of Sympathetic Nerves

The reported changes in limb blood volume differ greatly according to the experimental approach used. Browse et al. (1966a) studied responses in the occluded hind limb and reported quite large increases in venous pressure during sympathetic stimulation. These experiments provided evidence for constriction of vessels in the limb but they may have been influenced disproportionately by responses of superficial veins and the effect on venous pressure due to arterial constriction cannot readily be quantitated.

Mellander (1960) reported consistent although small changes in volume of the cat's hindquarters in response to sympathetic stimulation. In this paper he compared the relative responses of resistance and capacitance at each stimulus frequency and concluded that, at low frequencies, capacitance responses were relatively more complete. A criticism of this work, apart from the small size of the responses, is that it does not distinguish changes in volume due to active constriction of capacitance vessels from volume changes due to changes in venous distending pressures when the

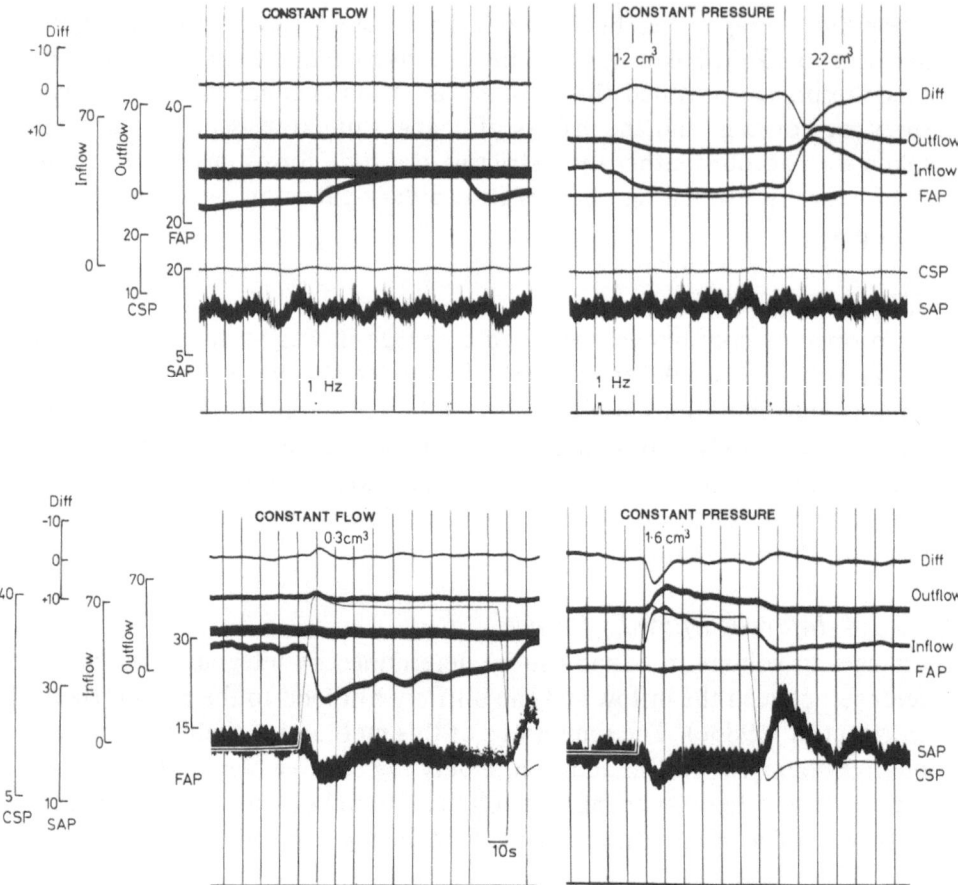

Fig. 3. Changes in volume in vascularly isolated, innervated dog's hind limb in response to electrical stimulation of efferent sympathetic nerves at 1 Hz and to large step changes in carotid sinus pressure. *Left traces* were obtained during perfusion and the limb at constant flow and *right traces* under constant pressure. Traces are of difference between inflow to the limb and outflow from limb (*Diff*), outflow, inflow, femoral arterial perfusion pressure (*FAP*), carotid sinus pressure (*CSP*), systemic arterial pressure (*SAP*). All pressures are in kPa and flows in ml·min^{-1}. Traces show that during constant flow perfusion both stimulation of the sympathetic nerves and changes in carotid sinus pressure resulted in changes in perfusion pressure between inflow and outflow). However, both interventions resulted in changes in both flow and volume during constant pressure perfusion. (Modified from Hainsworth et al. 1983c)

flow changes. Davis (1963) also reported changes in paw volume in the dog due to sympathetic stimulation but, again, the secondary effects of changes in flow were not excluded. Nervous stimulation caused a decrease in weight of the isolated perfused dog's limb (Shadle et al. 1958), although a possible effect from muscular contraction cannot entirely be ruled out.

Lesh and Rothe (1969) evaluated responses in perfused dog's gracilis muscle. They reported that there was no appreciable change in weight in

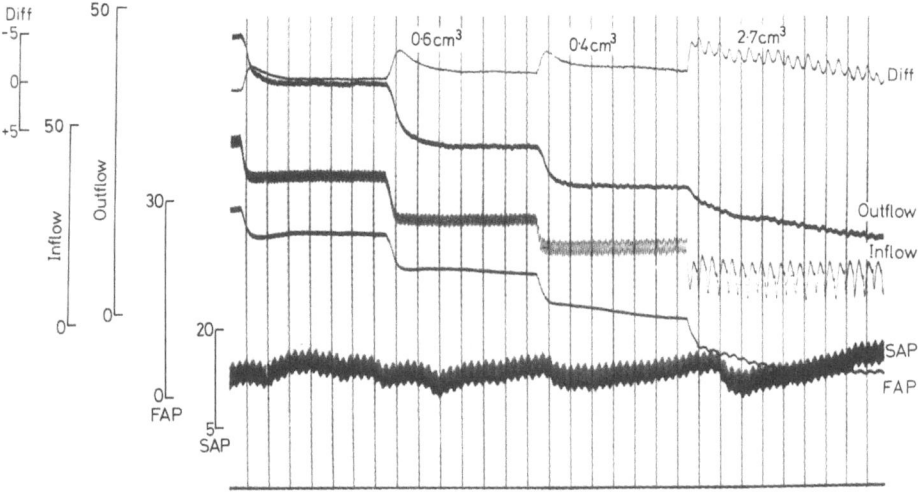

Fig. 4. Changes in volume in vascularly isolated perfused hind limb of dog in response to changes in rate of perfusion. Preparation and traces as in Fig. 3. Each step change in outflow was followed by a similar change in outflow. However, there was always a lag of the response of outflow which resulted in a change in volume similar to that which was obtained when the activity in the sympathetic nerves was changed by either direct electrical or reflex stimulation. (Hainsworth et al. 1983c)

response to sympathetic nervous stimulation, despite a large increase in total vascular resistance. There were, however, changes in weight when the muscle was perfused at constant pressure.

Recently, Hainsworth et al. (1983c) determined capacitance responses in the vascularly isolated dog's hind limb in response to sympathetic nerve stimulation during both constant pressure and constant flow perfusion. Stimulation of the sympathetic nerves supplying the limb, during constant flow perfusion, resulted in no change or a very small decrease in the limb volume (Fig. 3). During constant pressure perfusion responses, although still not large, were ten times greater. The difference in the responses during the two types of perfusion is explained by the mechanical effects of the changes in flow (Fig. 4), which passively change the distension of the venules and small veins. This conclusion is supported to some extent by the earlier findings of Oberg (1967), who attempted to control flow to a cat's hindquarter during sympathetic stimulation by releasing the partial occlusion of the artery. Although that work can be criticized due to a very low perfusion pressure, changes in volume did occur which were attributed partly to changes in flow.

2.2.3 *Reflex Responses*

Since carefully controlled studies of active capacitance responses in muscle tissue (Lesh and Rothe 1969) or the whole limb (Hainsworth et al. 1983c) have shown that responses to direct electrical stimulation of sympathetic efferent nerves were very small, it seems unlikely that large responses could occur to reflex stimuli.

The only reports of relatively large or consistent responses of limb blood volume to changes in carotid sinus pressure have come from studies in which flow was allowed to change (Oberg 1964; Hadjiminas and Oberg 1968; Hainsworth et al. 1983c). In a study in which responses during constant flow and constant pressure perfusion were compared (Hainsworth et al. 1983c), an increase in carotid pressure resulted in either no change or a small decrease in limb volume during constant flow perfusion but consistent increases in volume during constant pressure perfusion (Fig. 4). The absence of reflex venous responses in muscle was also noted by Hebert and Marshall (1985), who observed no changes in diameter, by in vivo microscopy of muscle blood-vessels of the anaesthetized rat, in response to changes in carotid sinus pressure.

Studies of responses in the occluded limb (Browse et al. 1966a,b) have also shown little or no change in venous pressure in response to changes in carotid sinus pressure, although there was a decrease in pressure in response to stimulation of carotid chemoreceptors (Browse et al. 1966b), or to aortic injections of cyanide which, amongst other things, stimulates aortic chemoreceptors (Browse and Shepherd 1966). This suggests that this method may reflect preferentially the responses of cutaneous vessels, since only these have been shown to respond to neural stimulation.

Stimulation of various areas in the brain, including the medulla, hypothalamus and midbrain, resulted in small decreases in limb volume similar to those occurring during direct stimulation of the sympathetic outflow (Baum and Hosko 1965; Manchanda and Bhattarai 1974; Manchanda et al. 1975; Meninger and Baker 1976).

2.3 Capacitance Responses in Human Limbs

2.3.1 *Methods of Study*

Because the volume of blood contained within a vein or a region is greatly influenced by factors other than the state of active constriction of the vessel, it has proved extremely difficult to develop reliable techniques for assessing capacitance responses in humans. The particular problem which has been most difficult to overcome is to distinguish active capacitance responses from effects due to changes in arterial blood pressure or blood flow.

2.3.1.1 Study of Segments of Veins

It is possible to isolate a segment of vein which does not contain tributaries by applying an external occlusion. Then, by injection and withdrawal of volumes of blood, volume-pressure curves can be defined (De Pasquale and Burch 1963). It is important to ensure that no small tributaries are present. Also, if the effect of an intervention is to be studied the same portions of the volume-pressure curves should be examined. If a segment of vein is isolated and distended with blood, changes in the distending pressure can be used to provide an index of changes in the degree of constriction of the vessel wall (Burch and Murthada 1956; Duggan et al. 1953; Page et al. 1955). An alternative approach is to assess the pressure-diameter relationship by use of a microscope focussed on the skin overlying a superficial vein (Nachev et al. 1971).

2.3.1.2 Measurement of Volume Changes in a Limb

There have been several methods used to determine the relationship between distending pressure and volume in limbs, and these techniques have been most widely used in human research. Changes in volume have been determined by air- or water-filled plethysmographs or mercury-in-rubber strain gauges (Whitney 1953). Changes in distending pressure are imposed by inflation of a pneumatic cuff to an appropriate pressure. Responses are assessed from the slope of the plot of change in volume to change in pressure (Wood and Eckstein 1958; Sharpey-Schafer 1961; Bevegård and Shephard 1965a,b; Eckstein et al. 1965; Mason and Braunwald 1964; Burki and Guz 1970). An alternative method of changing venous transmural pressure is to apply positive and negative pressures to the limb by means of a plethysmograph (Greenfield and Patterson 1956; Coles and Patterson 1957; Coles et al. 1957).

Continuous estimates of volume changes have been determined during inflation of a proximal limb cuff to a constant pressure (e.g. 4 kPa) (Ardill et al. 1968; Abboud et al. 1968).

None of these methods can be regarded as adequate. Firstly, the pressure distending the capacitance vessels is not known. It cannot be assumed to be the same as the cuff pressure or even to be a constant value above it when blood is flowing. Even when venous pressure is recorded directly there can be a wide difference in the values recorded in different veins (Brown et al. 1966; Hollenberg and Boreus 1972).

2.3.1.3 Measurement of Pressure Changes in Vascularly Isolated Limb

The measurement of pressure changes in vascularly isolated limb has been
described in connection with animal experiments. A proximal cuff
occludes both arteries and veins to a limb. Initial partial occlusion would
allow the vessels to become distended before totally occluding them.
After about 2 min of occlusion, the pressure in the vein becomes relatively
stable and responses can be determined for periods of up to 12 min
(Samueloff et al. 1966; Zitnik and Lorenz 1969). The occluded limb
technique is probably the only valid technique for assessing responses in
humans. However, it is difficult to make a quantitative assessment of the
exact significance of a change in venous pressure.

2.3.2 Reflex Responses in Humans

It is clearly impossible to obtain the same degree of control of the relevant
variables in experiments on human subjects as in those involving anaesthetized experimental animals. However, there have been a number of studies
which at least point to the probability that responses in humans do not
differ from those in animals.

2.3.2.1 Baroreceptor Stimulation

Several techniques have been used to change the stimulus to arterial baroreceptors in man. However, all methods have major disadvantages and in
none of them is it possible to localize a stimulus to the carotid baroreceptors and to prevent modification of the responses from changes in the
stimuli to other reflexogenic areas. Perhaps the most direct method is to
stimulate the sinus nerve directly. This was employed in the treatment of
anginal patients (e.g. Carlsten et al. 1958; Epstein et al. 1968). Although
this is an effective method of exciting baroreceptor afferent fibres, it does
not discriminate between baroreceptor and chemoreceptor afferents.
Also, it is not possible to quantify the stimulus. Nevertheless, it is possible
to show decreases in blood pressure, cardiac output and vascular resistance
and the absence of any consistent venous response (Epstein et al. 1968).

An alternative method of changing the stimulus to baroreceptors, which
has the advantage of affecting all arterial baroreceptors, is to inject intravenously vasoactive drugs, such as sodium nitroprusside and phenylephrine (Smyth et al. 1969). This technique is mainly useful for studying
the control of heart rate, since the drugs act directly on blood-vessels and
this precludes the study of vascular responses. However, if the region
studied is vascularly isolated by occluding the limb before the drug is
injected, reflex venous responses can be studied from the changes in
venous pressure. Using this approach Epstein et al. (1968) showed that
there was no venous response to changes in arterial pressure even though

changes did occur in response to mental arithmetic or to placing ice on the abdomen!

Carotid baroreceptors can be stimulated by enclosing the neck in a rigid airtight chamber and applying a subatmospheric pressure (Ernsting and Parry 1957; Eckberg et al. 1975). This increases the transmural pressure in the carotid sinuses and thus stimulates the baroreceptors. The advantage of this is that the stimulus is relatively well localized to the barore-ceptors. The disadvantages are that the stimulus declines as blood pressure changes; that the stimulus to other baroreceptor areas, in particular aortic receptors, is changed; and that the sudden application of suction can elicit emotional venous responses in uninitiated subjects. Distension of carotid baroreceptors in this way, both at rest and during exercise, failed to cause any venous responses in an occluded limb (Bevegård and Shepherd 1966). Other techniques which change the stimulus to baroreceptors include the application of a subatmospheric pressure to the lower part of the body and upright tilting. However, these would be likely to change the stimuli to many other reflexogenic areas and will be considered separately.

2.3.2.2 Chemoreceptor Stimulation

It is not possible in humans to apply a discrete stimulus to chemorecep-tors. Animal studies have shown that some of the cardiovascular responses from aortic and carotid chemoreceptors are different and that the primary cardiovascular responses are complicated by the effects on respiration. It is against this background that responses to hypoxia and hypercapnia must be considered.

Moderately severe hypoxia (breathing $< 11\%$ oxygen) induces forearm venous constriction (Eckstein and Horsley 1960). However, it is not clear whether this is related to hypoxia, the concomitant increase in ventilation or hypocapnia. Hintze and Thron (1961) and Weil et al. (1971) reported little or no venoconstriction in response to hypoxia when P_{ACO_2} was held constant. Thus the most potent stimulus for venoconstriction seems to be the combination of hypoxia, hypocapnia and hyperventilation (Weil et al. 1971). However, hypercapnia also causes venoconstriction (Duggan et al. 1953), and since moderate hypoxia alone has little effect on the limb vein, it seems unlikely that any venous responses can be attributed solely to the direct results of stimulation of chemoreceptors.

There have been several studies of the changes in venous distensibility in response to ascent to high altitude (Weil et al. 1969, 1971; Wood and Roy 1970; Cruz et al. 1976). Venous constriction was reported to occur in all the studies. However, these studies reported venous compliance which may be misleading due to non-linear volume-pressure curves (page 104). Also, responses could be influenced by changes in flow. They certainly do little to elucidate the role of chemoreceptors.

2.3.2.3 Deep Breaths

Taking a voluntary deep inspiration is one of the few procedures which consistently results in constriction of veins in limbs (Quiroz et al. 1960; Samueloff et al. 1966; Epstein et al. 1968; Zelis and Mason 1969; Browse and Hardwick 1969; Delius and Kellerova 1971). The mechanism responsible for inducing this response is uncertain. It may partly be related to the cortical centres responsible for taking the breath or to a change in P_{CO_2}. Browse and Hardwick (1969) found that the response did not occur to positive pressure inflation in anaesthetized and paralyzed subjects, which suggests that distension of pulmonary receptors is unlikely to be the cause of the response. Also, Burch (1975) found that there was habituation, so that the responses to repeated deep inspirations became much smaller.

2.3.2.4 Psychogenic Stimuli

Mental activity seems to have an important effect on cutaneous veins. Their distensibility increases during sleep (Watson 1962). Veins are reported to constrict during various mental or emotional stimuli, including mental arithmetic, fear and anticipation (Martin et al. 1959; Delius and Kelerova 1971; Burch 1975; Brod et al. 1976). As for the responses to deep breaths, Burch (1975) found that repeated applications of alerting stimuli resulted in habituation of the responses.

2.3.2.5 Responses to Shifts in Blood Volume

The main techniques used have been the application of a subatmospheric pressure to the body below the iliac crests and upright tilting on a tilt table. The effect of both of these procedures is to cause distension of veins in the lower abdomen and limbs and effectively to reduce the circulating blood volume. A similar redistribution of blood volume may also occur in response to the Valsalva manoeuvre (Sharpey-Schafer 1961, 1963), positive pressure breathing (Mowassaghi et al. 1969) or haemorrhage (Robinson et al. 1966).

Both upright tilting and application of lower body negative pressure result in decreases in venous return and cardiac output and increases in vascular resistance and heart rate. The responses to upright tilting by $60°-70°$ are very similar to those resulting from a negative pressure of 4 kPa applied below the iliac crests (Al-Shamma and Hainsworth 1985). However, during tilting the hydrostatic pressure distending the blood-vessels is graded along the length of the body, whereas the suction applies the same pressure to all vessels in the region. Also, upright tilting would decrease the pressure to the upper part of the body, including the carotid baroreceptors. The maximum volume shift in response to suction with 9 kPa is about 10 ml/kg body weight (Brown et al. 1966). Most authors have reported an increase in limb venous tone in response both to leg suction (Gilbert and Stevens 1966; Boreus and Hollenberg 1972; Paessler et al. 1968; Page et al. 1955) and lower body negative pressure (Gilbert and Stevens 1966; Ardill et al. 1968; Tripathi et al. 1984). However, responses were frequently small and transient (Samueloff et al. 1966; Paessler et al. 1968), and Epstein et al. (1968) noted that the response frequently preceded the stimulus, suggesting the existence of a psychogenic component. Tripathi et al. (1984) reported much larger changes in venous volume in response to changes in ambient temperature, although with this, as with most other studies, it is not possible to know the extent to which the responses were due to active constriction of capacitance vessels.

The receptive areas responsible for any responses which occur during shifts of circulating blood volume are not clear. Many authors have assumed that the changes are due to "cardiopulmonary receptors" because changes in central venous pressure were observed, but there was little effect on arterial blood pressure (e.g. Roddie and Shepherd 1958; Zoller et al. 1972; Johnson et al. 1974; Abboud and Mark 1979). However, the evidence for this is by no means conclusive. A small change in blood pressure, when it influences all arterial baroreceptors at a sensitive part of the baroreceptor response curve, is a very potent stimulus. Also, distension of the low-pressure receptors, particularly those in the great veins and atria, does not normally result in depressor responses (see Linden and Kappagoda 1982). Furthermore, the concept of "cardiopulmonary receptors" is one to be avoided, since it implies a homogeneous group of nerve receptors with similar reflex effects and this is far from the truth.

2.3.2.6 Exercise

There have been several reports that both phasic and isometric exercise result in constriction of veins in a non-exercising limb (Merritt and Weissler 1959; Bevegård and Shepherd 1965a; Robinson and Wilson 1968; Seaman et al. 1973; Detry et al. 1974). The venoconstriction appears only to be transient during mild exercise but becomes more sustained with more severe exercise (Hanke et al. 1969). However, if body temperature rises, the venoconstrictor response to exercise is reduced or abolished (Zitnik et al. 1971; Wenger and Roberts 1980). The mechanisms which may be responsible for any venous responses which may occur during exercise are not known.

2.4 Importance of Capacitance in Muscle and Cutaneous Circulations

2.4.1 Magnitude of Responses

Most of the earlier misconceptions about the extent of active capacitance responses in musculo-cutaneous circulations arose from inadequacies of the methods used to study them. Studies in which flow was allowed to changes (e.g. Mellander 1960) resulted in consistent changes in volume. However, the changes in volume are almost entirely secondary to the changes in flow (Hainsworth et al. 1983c). If flow is held constant, changes in volume, even in response to direct electrical stimulation of the sympathetic nerve supply, are trivial and no measurable changes occur at all in response to baroreceptor stimulation. The conclusions concerning the absence of significant capacitance responses in the limbs were obtained mainly from carefully controlled studies using experimental animals (usually dogs). However, it seems highly likely that the same applies to humans. There is no evidence for important reflex capacitance responses in the human limbs, although cutaneous veins do seem to constrict in response to emotional stimuli.

Cutaneous veins may respond weakly to some reflex stimuli. They dilate weakly in response to stimulation of baroreceptors and chemoreceptors. This is in marked contrast to the large changes which these stimuli induce in resistance vessels. There thus appears to be an effective dissociation of the sympathetic nerve supply to arteries and veins. However, an alternative explanation may be that there is a dissociation between the sympathetic innervation to cutaneous vessels (arteries and veins) and that to muscle vessels. This possibility is supported to some extent by recordings in humans from nerves presumed to be innervating skin and muscle blood-vessels (Wallin et al. 1975).

The absence of responses of veins in muscle tissue either to reflex stimulation or to direct stimulation of the efferent nerves is likely to be due to lack of smooth muscle in these vessels, poor innervation, or both. This view is supported by the findings of Lesh and Rothe (1969) and Hebert and Marshall (1985). Also, histological studies have shown a paucity of innervation to deep muscle veins (Fuxe and Sedvall 1965; Bevan et al. 1974). The conclusion to be drawn is that muscle veins do not contribute in any measurable way to capacitance responses.

2.4.2 Physiological Significance of Lack of Capacitance Responses in Musculo-cutaneous Circulation

Evidence has been presented that there are no important reflex capacitance responses in the limb circulation, at least in the dog and almost certainly in man also. It is likely that this lack of response also applies to muscular and cutaneous circulations generally. It is of interest to consider the physiological significance of this lack of response. At first thought, it would seem that, owing to large hydrostatic pressures which can occur in vessels in the limbs, particularly in upright man, it would be of extreme importance to limit capacitance vessels distension by powerful and active contraction of the vascular smooth muscle. However, the consequences of a powerful venous constriction would not necessarily be wholly beneficial. The volume of blood in both legs of an adult man in the horizontal position is about 250 ml and this doubles to about 500 ml during passive upright tilting (Gauer and Thron 1965). During active standing or walking the volume of blood accumulated in the limbs would be likely to be less owing to the pressure on the blood-vessels from contracting muscle. The total volume of blood, therefore, which is contained within this large mass of tissue is very small. This contrasts with the abdominal circulation, in which the mass of tissue is relatively small but the volume of blood relatively large. Powerful active constriction of veins in the musculo-cutaneous circulations, therefore, would have only a relatively small direct effect on the distribution of blood volume.

The circulations to muscle and skin are different from those to most other regions of the body in that the flow can change over a very wide range. Blood flow to skeletal muscle ranges from $20-50$ ml$\cdot$kg^{-1} tissue at rest to $500-750$ ml$\cdot$kg^{-1} during severe exercise. In a 70-kg man, total muscle blood flow can increase from only about 1.0 l$\cdot$min^{-1} to as much as 30 l$\cdot$min^{-1} during maximal exercise in an athlete (Folkow and Neil 1971). Similarly, in skin, total blood flow can range from about 20 ml$\cdot$min^{-1} to 3.0 l$\cdot$min^{-1}. The importance of this is that if the veins draining these regions were to constrict powerfully, the effects on the body would be detrimental. Constriction of veins draining muscle tissue would have only a small effect on the regional blood volume but, particularly at high flow rates, it would markedly increase the resistance in the postcapillary part of the circulation and this would result in a large loss of fluid from the capillaries to the tissue (Oberg 1964; Hadjiminas and Oberg 1968; Mellander and Johansson 1968; Mitzner and Goldberg 1975). The consequence of this would be that prolonged constriction of veins draining muscle and skin would result in a decrease rather than an increase in effective circulating blood volume. The lack of such a response is therefore likely to be of considerable benefit.

3 Abdominal Vascular Bed

Contrary to the case in the limb circulation, in the abdomen studies of vascular capacitance have been confined mainly to experimental animals.

Some of the studies have examined responses in the whole or most of the abdominal circulation, whereas others have examined the major components: intestine, liver and spleen. The few studies in humans of volume changes in the splanchnic circulation provide little information on active capacitance responses because neither blood flow nor venous pressure could be controlled adequately.

3.1 Studies in Whole Abdominal Vascular Bed

3.1.1 Methods of Study

The two main approaches have been to study changes in venous pressure at constant vascular volume and changes in volume when venous pressure is controlled.

Bartelstone (1960), by a combination of surgical ligation and clamping of major vessels, isolated the circulation to the animal below the level of the diaphragm from that above it. Once the arterial and venous pressures had equilibrated, changes in venous pressure provided an index of capacitance vessel responses. This principle has been applied frequently to the study of the limb circulation and to the whole body but it has not been used recently for the abdomen. The disadvantages are its complexity, the very short period for which recordings can be taken before ischaemic changes take place, the likelihood of secondary reflex changes occurring, and the problem of stress relaxation or creep in blood-vessels (Alexander 1963).

An entirely different approach to the study of capacitance in the splanchnic circulation was devised by Brooksby and Donald (1971). This involved the determination of inflow and outflow to the splanchnic region and, calculating a change in volume by integration of the difference between these flows. The major problem is that, because neither flows nor venous pressures were controlled and the region was incompletely isolated, the difference between inflow and outflow does not necessarily indicate an active capacitance change. Indeed, in a later paper the same authors (Brooksby and Donald 1972) estimated that changes in volume which were obtained when flow was changed mechanically were only one-third less than those occurring in response to sympathetic nervous stimulation.

A preparation to solve the problems of the above technique was devised by Hainsworth and Karim (1976), who made a much more complete surgical isolation of the region by tying or cutting all structures immediately above the diaphragm, perfusing the abdominal aorta at constant flow and draining the abdominal inferior vena cava at constant pressure. Because the region was isolated, inflow was constant and outflow pressure

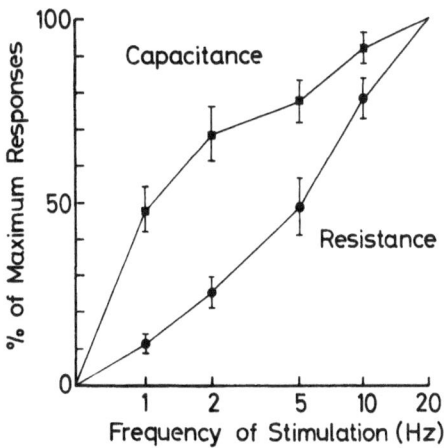

Fig. 5. Capacitance and resistance responses of the abdominal circulation to stimulation of both splanchnic nerves. Values are means ± SE of values obtained in 14 dogs and are expressed relative to values at 20 Hz. Capacitance responses at 1−5 Hz were significantly ($p < 0.001-0.01$) greater relative to the maximum than the corresponding resistance responses. (Modified from Karim and Hainsworth 1976)

was constant, changes in capacitance could be calculated from changes in outflow. This method addresses the major problems, although there are some disadvantages. Even though total flow to the region is constant, there is likely to be a change in the distribution between the various components which could affect the volume. Also, there may be a change in the pressure distending small vessels due to constriction of the veins downstream and this could have passive effects. Nevertheless, these effects would be likely to be small and, if anything, would probably slightly underestimate capacitance changes.

Another approach to the determination of capacitance in the abdominal circulation is by the estimation of volume by indicator dilution and estimation of pressure by direct recording (Rothe 1983c). This method can only be applied reliably when blood flow and venous pressures are constant, conditions which would be unlikely to occur unless the region was isolated and perfused.

3.1.2 Responses to Direct Stimulation of Sympathetic Nerves

Capacitance vessels in the abdominal circulation are particularly sensitive to sympathetic nerve stimulation. When the splanchnic nerves are stimulated at 1 Hz the decrease in capacitance in the abdomen is about half the response obtained at 20 Hz (Karim and Hainsworth 1976). The differences in the relative responses of resistance and capacitance vessels are seen in Fig. 5. It has been argued (Stark 1968) that the difference in the shapes of the resistance and the capacitance response curves is merely a geometrical consequence of the Posseuille equation and the fact that resistance is a function of $1/r^4$, whereas capacitance is a function of r^2. From the data of Karim and Hainsworth (1976), in which the resistance response to stimulation of 20 Hz was +135%, it can be calculated that the radius of a simple model represented by a single tube would be reduced by 20%. This would have resulted in a reduction in volume of 36%. At 1 Hz stimulation, the resistance change was only 14% and this would correspond to a volume change of 6%. Thus, if capacitance vessels were constricted by the same amount as resistance vessels, the response at 1 Hz stimulation would be only 17% of that at 20 Hz, whereas the measured value was nearly 50% of that at 20 Hz. Therefore, it can be concluded that capacitance vessels in the abdominal circulation shorten relatively more than resistance vessels at low stimulus frequencies.

3.1.3 Reflex Responses

3.1.3.1 Baroreceptor Reflexes

Bartelstone (1960) reported changes in pressure in the occluded lower half of an animal using the major vessel occlusion technique. However, apart from the methodological problems outlined above, it is not possible to quantitate responses using this preparation.

Using the vascularly isolated perfused preparation of the abdomen, Hainsworth and Karim (1976) reported relatively large capacitance responses to changes in carotid sinus pressure (Fig. 6). Samoilenko and Tkachenko (1971) also reported a change in volume in the cat's abdominal vascular bed in response to a change in carotid pressure. Hainsworth and Karim (1976) found that the maximal response to a change in carotid pressure was equivalent to electrical stimulation of both splanchnic nerves with a frequency of about 5 Hz. Graded decreases in carotid pressure from above saturation levels resulted in relatively greater capacitance responses at high pressures compared with the responses of resistance. This was shown to be due to the greater sensitivity of capacitance vessels to sympathetic activity because when the reflex responses of both resistance and capacitance vessels were related to responses to nerve stimulation it appeared that the impulse activities to the two types of vessel were not different (Fig. 7).

Stimulation of aortic baroreceptors resulted in qualitatively similar responses in the abdominal circulation (Karim et al. 1978). Responses of both vascular capacitance and resistance to changes in aortic pressure were influenced by the level of carotid sinus pressure, both being inhibited by elevating carotid pressure. The smallness of responses from aortic receptors compared with those from carotid receptors may have been due partly to the fact that both carotid sinuses were distended, whereas only the aortic arch receptors, innervated by the left aortic nerve, would have been distended and the subclavian baroreceptors, innervated by the right aortic nerve, were excluded from the stimulus.

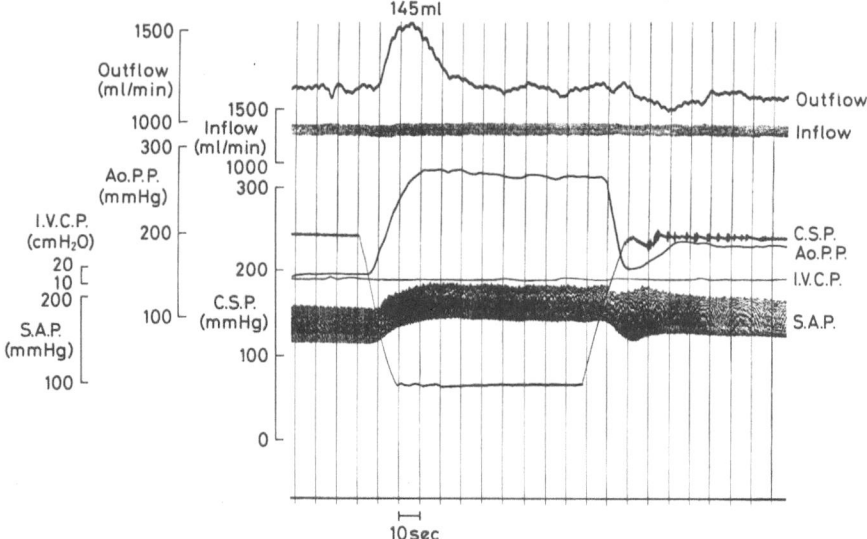

Fig. 6. Responses in the vascularly isolated abdominal circulation to a large step decrease in carotid sinus pressure. Traces of outflow from inferior vena cava, inflow, carotid sinus pressure (*C.S.P.*), abdominal aortic perfusion pressure (*Ao.P.P.*), inferior vena caval pressure (*I.V.C.P.*), systemic arterial blood pressure (*S.A.P.*). Inflow and inferior vena caval pressure were held constant. A decrease in carotid sinus pressure resulted in an increase in arterial perfusion pressure and a transient increase in outflow which, on integration, denoted the expulsion of 145 ml blood from the region. Note that the changes were reversed when carotid pressure was increased but that the change in outflow was slower. (Hainsworth and Karim 1976)

3.1.3.2 Other Reflexes

Stimulation of both carotid and aortic chemoreceptors with either venous blood or injections of sodium cyanide results in a decrease in capacitance in the abdominal circulation (Hainsworth et al. 1983a,b). The similarity of the vascular responses to stimulation of the two chemoreceptor areas contrasts with the different cardiac responses: stimulation of carotid chemoreceptors results in bradycardia and a negative inotropic response, whereas stimulation of aortic chemoreceptors results in tachycardia and a positive inotropic response (Daly and Scott 1958; Hainsworth et al. 1979; Karim et al. 1980b). The relative magnitudes of the response of resistance and capacitance vessels to chemoreceptor stimulation depends on the baroreceptor input. At high carotid sinus pressures (> 17 kPa) the responses of resistance were significantly greater and those of capacitance significantly smaller than the responses at lower carotid pressures (Hainsworth et al. 1983a). This difference can be explained in terms of the difference in sensitivities of resistance and capacitance to sympathetic nervous activity. At high carotid pressures the efferent sympathetic nervous activity would be low and so a small increase due to chemoreceptor activa-

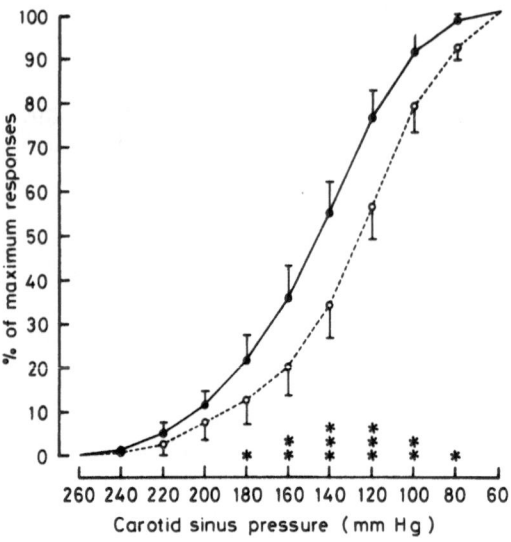

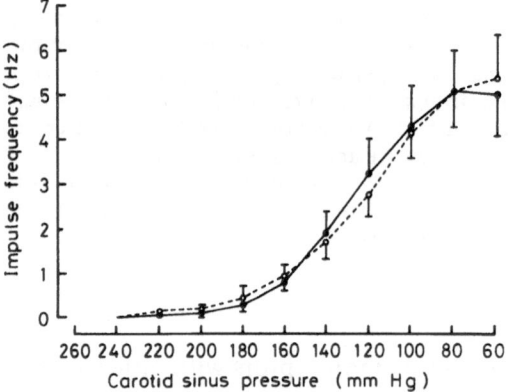

Fig. 7a,b. Responses of resistance (*open circles*) and capacitance (*closed circles*) in the abdominal circulation to changes in carotid sinus pressure. **a** Resistance and capacitance responses at each step of carotid pressure expressed as percentages of the changes when carotid pressure was reduced to 8 kPa (60 mmHg). Relative capacitance responses were significantly greater than the corresponding resistance responses (* $p < 0.05$, ** $p < 0.01$, *** $p < 0.005$; $n = 10$). **b** Relates frequencies of electrical stimulation of splanchnic nerves required to produce resistance and capacitance responses identical to those obtained at each value of carotid pressure. This shows that the relatively more complete responses of capacitance at the intermediate values of carotid pressure were not due to differences in the sympathetic nervous activities to the two types of vessel but were due to the greater sensitivity of capacitance to the sympathetic activity. Note also that the responses of resistance and capacitance on decreasing carotid pressure to 20 kPa (150 mmHg) were similar to those obtained in response to stimulation at 1 Hz, the responses at 16 kPa (120 mmHg) were equivalent to those at about 3 Hz, and those at 8 kPa were equivalent to stimulation at about 5 Hz. (Hainsworth and Karim 1976)

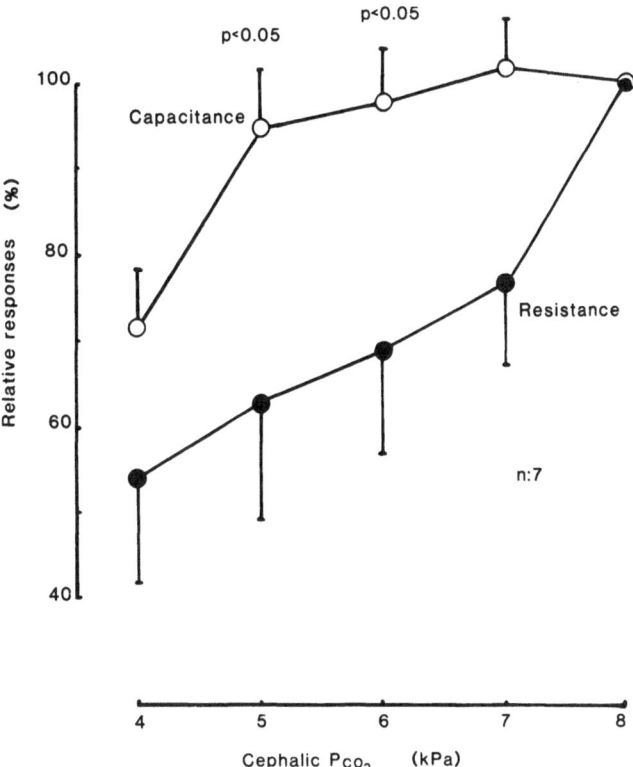

Fig. 8. Responses of resistance and capacitance in the abdominal circulation to large step decreases in carotid sinus pressure at various values of cephalic P_{CO_2}. Responses at cephalic P_{CO_2} of 8 kPa have been expressed as 100%. Results show that at P_{CO_2} of 5 and 6 kPa the responses of capacitance, relative to that at 8 kPa, were significantly greater than the corresponding resistance responses. Results can be explained by capacitance vessels, unlike resistance vessels, being almost maximally constricted at low carotid sinus pressure and values of cephalic P_{CO_2} of 5 kPa and above. (Ford et al. 1985b)

tion would lead to a relatively large capacitance response. At lower carotid pressures, when there would be a higher sympathetic efferent discharge, most of the capacitance response would already have been taken up, so resistance responses would predominate.

Reflex cardiovascular responses are importantly influenced by the level of carbon dioxide perfusing the cephalic circulation (Hainsworth et al. 1984, 1985; Soladoye et al. 1985; Ford et al. 1985a). An increase in cephalic P_{CO_2} results in constriction of both resistance and capacitance vessels in the abdomen and enhances their responses to baroreceptor stimulation (Ford et al. 1985a). It is of interest to note that the effect on capacitance vessels of changing cephalic P_{CO_2} occurs at values below normal arterial levels, whereas the effect on resistance occurs mainly when P_{CO_2} is increased above normal (Fig. 8). This can also be explained in terms of the relative sensitivities of the two types of vessel to sympathetic activation. At normal values of cephalic P_{CO_2} and low carotid sinus pressure, capacitance vessels, but not resistance vessels, are maximally constricted. Therefore, increases in cephalic P_{CO_2} can only affect the constriction of resistance vessels.

There is little information on the responses of the whole abdominal circulation to stimulation of other reflexogenic areas. Ford et al. (1985b) described the responses of the abdominal circulation to stimulation of cardiac receptors with veratridine injected into the aortic root. At low carotid sinus pressures this resulted in dilatation of both resistance and capacitance vessels. However, when carotid pressure was increased to dilate the abdominal vessels, intracoronary veratridine resulted in no further response of resistance, but capacitance vessels then constricted. The significance of this observation is unknown and it will be necessary to examine the responses to more physiological stimuli.

3.2 Studies in the Intestinal Vascular Bed

3.2.1 Methods of Study

Most of the techniques which have been applied to other regions have been used in the intestine. Venous responses have been assessed by direct observation of veins (Altura and Zweifach 1965; Johnson 1967; Gaehtgens and Uekermann 1971; Furness and Marshall 1974). Responses have also been assessed, when the circulation to part of the intestine was occluded, from changes in venous pressure (Hooker 1918) or from volume-pressure curves (Alexander 1954). Iizuka et al. (1970) perfused a mesenteric vein at constant flow and assessed responses from changes in perfusion pressure. Changes in volume of loops of intestine have been determined by plethysmography (Folkow et al. 1963), weighing (Selkurt and Johnson 1958; Donald and Aarbus 1974) and determination of transit time of injected indicators (Rothe et al. 1978; Johns and Rothe 1978). The changes in calculated intestinal volume can only be regarded as a change in capacitance provided that flow and venous pressure do not change (see page 106).

3.2.2 Responses to Stimulation of Sympathetic Nerves

Stimulation of sympathetic nerves results in constriction of intestinal veins (Alexander 1954; Eckstein et al. 1969; Iizuka et al. 1970). Donald and Aarhus (1974) reported that an iliac segment decreased its weight by 10% on maximal sympathetic stimulation. However, flow was not controlled and a decrease in flow induced by constriction of the arterial inflow resulted in a change in volume which was only one-third smaller than the response to stimulation. Rothe et al. (1978) found that the change in intestinal blood volume which occurred during noradrenaline infusion was similar to that resulting from a similar decrease in flow without noradrenaline. However, they did show that at constant flow perfusion there was a decrease in volume of 24% in response to high rates of infusion of noradrenaline.

3.2.3 Reflex Responses

3.2.3.1 Baroreceptor Reflexes

The evidence for a major contribution by capacitance vessels in the intestinal vascular bed to baroreceptor reflexes is not overwhelming and most previous studies are open to some degree of criticism. Eckstein et al. (1969) and Iizuka et al. (1970) reported that there was a small constriction of a perfused colic vein in response to carotid occlusion. However, carotid occlusion is an unsatisfactory method of eliciting the baroreceptor reflex because the intensity of the stimulus is unknown and uncontrolled. It is possible that responses might have been greater if carotid pressure had changed over a wider range. Responses of intestinal capacitance vesvels to carotid occlusion have also been reported by Hadjiminas and Oberg (1968), Samoilenko and Tkachenko (1971) and Tkachenko et al. (1978). However, Hadjiminas and Oberg (1968) did not control intestinal blood flow and the other two papers do not give any quantitative data.

3.2.3.2 Other Reflexes

Injection of sodium cyanide or nicotine into a carotid artery failed to cause any response of a perfused colic vein but injection of the same drugs in the aortic root resulted in a constriction (Iizuka et al. 1970). However, it is uncertain whether this response was due to stimulation of aortic chemoreceptors. The responses of intestinal capacitance vessels to stimulation of chemoreceptors therefore remains uncertain. There have been no adequate studies of the responses of intestinal vascular capacitance in other cardiovascular reflexes.

3.3 The Liver

3.3.1 Methods of Study

Studies of capacitance in the liver are complicated by the vascular anatomy. Blood enters the liver through both the hepatic artery and the portal vein. The drainage of blood from the liver may be influenced by various hepatic sphincter mechanisms, although the importance of such mechanisms is uncertain and may vary in different species (Knisely et al. 1957; Greenway and Oshiro 1973; Andrews et al. 1973).

The techniques used for determining hepatic capacitance are broadly similar to those used for other regions. Volume changes have been estimated from changes in diameter, assessed by a variety of transducers (Guntheroth and Mullins 1963) and by plethysmography (Greenway et al. 1969; Carneiro and Donald 1977a). Bennett and Rothe (1981) determined changes in hepatic blood volume from the difference between inflow in the portal vein and hepatic artery and outflow in the hepatic vein, and Lutz et al. (1967) perfused the hepatic artery and/or the portal vein at constant flow and assessed capacitance responses from the changes in hepatic venous outflow. More recently, Cousineau et al. (1985) assessed hepatic volume in dogs by use of an indicator dilution technique.

Because of the dual circulation and the high degree of compliance of the liver, it is difficult to determine active capacitance responses as distinct from passive volume changes. In the cat, hepatic compliance varies with the venous pressure from about 100 ml·kPa^{-1} per litre liver tissue at low venous pressures (about 0.3 kPa) to about 25 ml·kPa^{-1} at venous pressures of about 1–3 kPa (Lautt and Greenway 1976). In addition, at moderately raised venous pressures there is continual filtration of fluid, the rate of which is directly proportional to the hepatic venous pressure (Greenway and Lautt 1970). Bennett and Rothe (1981) studied hepatic compliance in dogs with control of both the inflow and the hepatic venous pressure and found a relatively constant compliance over a wide range of venous pressures. They also noted that changes in flow resulted in important changes in hepatic volume even when venous pressure was unchanged.

3.3.2 Responses to Direct Stimulation of Sympathetic Nerves

Direct stimulation of the hepatic nerves results in a reduction of liver blood volume in the cat as measured by plethysmography (Greenway et al. 1968; Greenway and Oshiro 1972). These responses are similar to those resulting from infusion of catecholamines and other vasoactive drugs (Greenway and Lautt 1972a,b; Greenway et al. 1985). Similar responses occur in the livers of dogs in response to sympathetic nerve stimulation (Carneiro and Donald 1977a; Guntheroth and Mullins 1963). However, in some of these studies it was not possible to distinguish responses due to active constriction of capacitance vessels from those due to the decreased passive distension due to the reduction in flow.

The only adequate study of active capacitance responses in the liver circulation was made by Bennett et al. (1982). They found that, during constant flow perfusion of the hepatic artery and portal vein, stimulation of the hepatic nerves at 5 Hz reduced hepatic blood volume by an average of 76 ml/kg tissue. Responses during constant pressure perfusion were larger. The results of Bennett et al. (1982) did not show the same degree of sensitivity to low frequencies of stimulation as was reported by Karim and Hainsworth (1976) in the whole abdominal circulation.

Infusion of catecholamines seems to change hepatic and splanchnic blood volume by a beta-adrenergic dilator effect on hepatic outflow resistance (see page 150). The importance of this is uncertain.

3.3.3 Reflex Responses

Lautt and Greenway (1972) obtained little or no change in liver volume of cats in response to carotid occlusion. However, carotid occlusion is an unsatisfactory way of eliciting a baroreceptor reflex, and furthermore hepatic blood flow was not occluded. An increase in carotid sinus pressure resulted in an increase in liver blood volume in the dog (Carneiro and Donald 1977a). Although in these experiments hepatic arterial and portal venous flows were not controlled, they were measured and the changes were relatively small. Very similar results were also obtained using a different technique by Cousineau et al. (1985). These results would seem to suggest that hepatic capacitance does participate in the baroreceptor reflex.

Several other less specific interventions result in decreases in liver blood volume. These interventions include haemorrhage (Carneiro and Donald 1977b). However, in all those studies not only were the methods of assessing capacitance inadequate, but the procedures did not examine specific reflex mechanisms.

3.4 The Spleen

In the dog and cat, in contrast to the case in man, the spleen is a relatively large blood-filled organ and has the potential to function as a large blood reservoir (Barcroft and Barcroft 1923; Barcroft and Stephens 1927). The potential importance of the spleen in the control of total abdominal vascular capacitance was demonstrated by Karim and Hainsworth (1976), who noted that the change in capacitance to splanchnic nerve stimulation was reduced by about one-third when the spleen was excluded from the response.

3.4.1 Methods of Study

Direct observation of the change in size of the spleen in response to splanchnic nerve stimulation can leave one in no doubt that it can have a capacitance role. Guntheroth and Mullins (1963) used techniques similar to those which they used for the liver to determine changes in splenic diameter in conscious dogs. Responses of the spleen can be assessed from the changes in its weight (Greenway and Stark 1969; Opdyke and Ward 1973) or changes in volume measured by plethysmography (Carneiro and Donald 1977b), or by determining the reduction in response of the whole body or abdomen after its removal (Karim and Hainsworth 1976; Shoukas et al. 1981). Webb-Peploe (1969) determined splenic responses from the changes in venous pressure in a spleen which was totally vascularly occluded (isovolumic).

3.4.2 Responses to Direct Stimulation of Sympathetic Nerves

In dogs and cats at least, the spleen contracts powerfully to sympathetic nerve stimulation. The maximal response of the spleen is a decrease in flow of about 75% and a decrease in volume or weight estimated in different studies at between 25% and 70% (Greenway et al. 1968; Opdyke and Ward 1973; Donald and Aarhus 1974). Since, in these studies, the spleens were normally perfused and there was always a large increase in resistance, part of the volume change would have been due to the change in flow. However, Donald and Aarhus (1974) showed that only small responses resulted from large changes in flow, so it is reasonable to assume that most of the change in volume which occurs when the splenic nerves are stimulated is due to active contraction of the spleen.

The spleen appears to be particularly responsive to low frequencies of stimulation. In the cat, Greenway et al. (1968) reported that responses were maximal at 3 Hz, and in the dog Donald and Aarhus (1974) reported that at 2 Hz, 80% of the maximal response was obtained.

Contraction of the spleen, at least in cats and dogs, results in the release into the circulation of blood having a high haematocrit. Donald and Aarhus (1974) found that in dogs the haematocrit in the splenic venous blood increased to a peak of 80% during splenic nerve stimulation, and Opdyke (1970) also found a large increase in haematocrit in response to infusions of adrenaline or noradrenaline. The blood with the high haematocrit from the spleen is rapidly diluted by the blood from the rest of the splanchnic circulation, and in the blood in the inferior vena cava a change of only 1% occurred in response to splanchnic nerve stimulation (Karim and Hainsworth 1976).

3.4.3 Reflex Responses

There seems to be little doubt that the spleen participates in the baroreceptor reflex. Brender and Webb-Peploe (1969) obtained an increase in the pressure in an isovolumic spleen in response to carotid sinus hypotension. The spleen contracts in response to occlusion of the carotid arteries in dogs (Carneiro and Donald 1977b) and cats (Tkachenko et al. 1976). Shoukas et al. (1981) evaluated the role of the spleen in the carotid sinus reflex by determining the changes in total body capacitance before and after removing or occluding the spleen. The results indicated that the spleen is responsible for about a quarter of the total response of the systemic capacitance.

There have been several other studies which have found a powerful splenic contraction in response to other, non-specific stimuli, including haemorrhage and hypoxia (Guntheroth and Mullins 1963; Greenway and Stark 1969; Opdyke and Ward 1973; Greenway and Lister 1974; Carneiro and Donald 1977b).

3.5 Capacitance in the Human Splanchnic Circulation

3.5.1 Methods of Study

In order to determine changes in capacitance in any region it is essential to define the region under study, to control the blood flow, to keep venous pressure constant and then to determine changes in volume. Clearly, in the human splanchnic circulation it is difficult to measure the relevant variables, let alone to control them. It is possible to estimate the volume of the splanchnic circulation by the use of indicators and the determination of the mean transit time. The theory and experimental details have been given by Rothe et al. (1978) and Rothe (1983c). Pressures are determined by arterial and venous cannulation. However, it should be noted that any calculated changes in volume using these techniques do not necessarily represent capacitance changes because the region is highly distensible and volume may change consequent upon changes in pressure or flow.

3.5.2 Responses in the Human Splanchnic Circulation

There have been no studies of the volume responses in the splanchnic vascular bed to stimulation of specific sensory areas. Stimulation of carotid baroreceptors by application of a subatmospheric pressure to the neck results in splanchnic vasoconstriction (Abboud et al. 1979), but the effect on capacitance is not known. Effective circulating blood volume has been reduced by haemorrhage (Price et al. 1966) or by application of lower body negative pressure (Rowell et al. 1972; Johnson et al. 1974; Abboud et al. 1979), which reduces splanchnic blood flow and splanchnic or central blood volume. However, because both flow and venous pressure

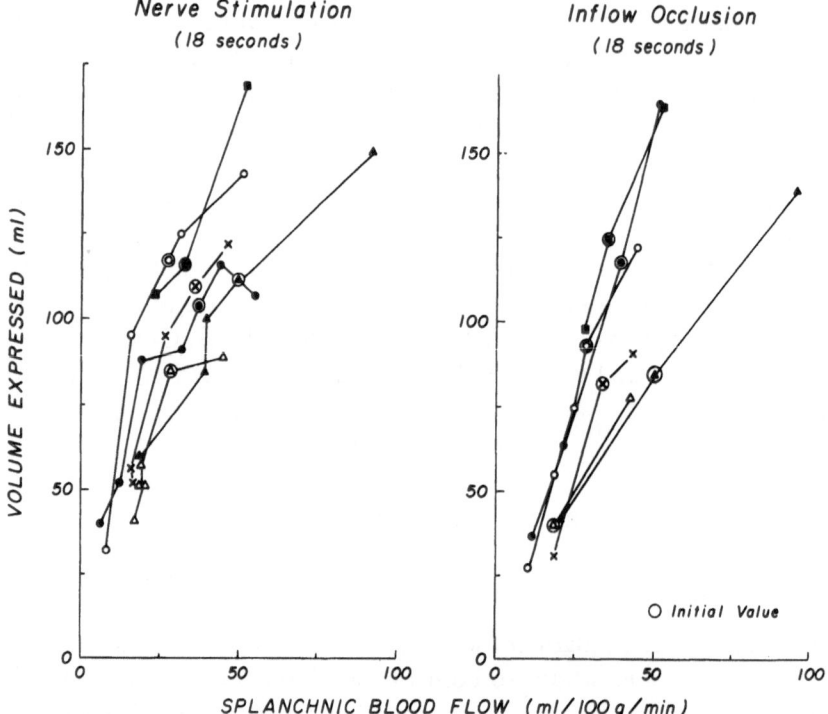

Fig. 9. Changes in splanchnic blood volume occurring in response to splanchnic nerve stimulation and arterial inflow occlusion. Initial splanchnic blood flow was varied by infusion of vasoactive drugs. Results show that the changes in volume in response to stimulation or occlusion were proportional to the initial splanchnic flow. Also, the changes in volume in response to sympathetic nerve stimulation were not greatly in excess of those induced passively by inflow occlusion. (Brooksby and Donald 1972)

decrease, it is uncertain to what extent volume changes are due to active constriction of capacitance vessels.

There has been shown to be a marked decrease in splanchnic blood volume during exercise (Wade et al. 1956). However, again splanchnic flow decreased. Venous pressure was not reported and, due to the cardiac response, it may have decreased.

3.6 Importance of Capacitance in the Abdominal Circulation

3.6.1 Magnitude of Responses

There seems to be no doubt that many physiological stimuli result in changes in the volume of blood in the abdomen. However, in many of the studies it is uncertain to what extent these changes are due to active constriction of capacitance vessels, rather than being passive changes secondary to the changes in flow resulting from constriction of resistance

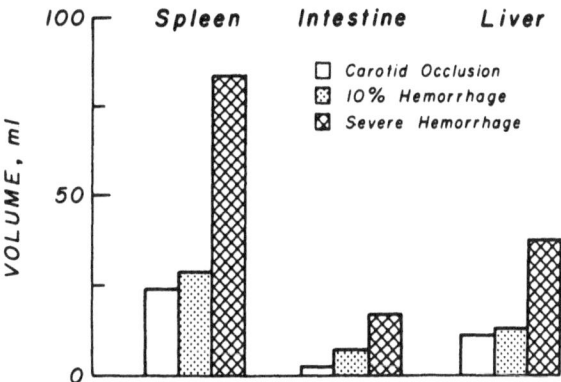

Fig. 10. Changes in blood volume in spleen, intestine and liver, expressed relative to a 10-kg dog, in response to carotid occlusion and to moderate and severe haemorrhage. Note that blood flow was not controlled, so that volume changes would include both active and passive components. The responses of the spleen were greater than the sum of those of the liver and intestine. (Carneiro and Donald 1977b)

vessels. The relative importance of active versus passive changes can be assessed either by determining the changes in volume in response to a stimulus during constant flow and constant pressure perfusion, or by comparing the change in volume which occurs in response to a stimulus during constant pressure perfusion with the change occurring when the flow is changed artificially by the same amount. Brooksby and Donald (1972) found that a decrease in flow caused by arterial occlusion resulted in a change in volume which was only about one-third less than that which occurred in response to stimulation of the sympathetic nerves (Fig. 9). Thus, although sympathetic stimulation undoubtedly results in an active constriction of capacitance vessels in the abdomen, during normal perfusion there would be a decrease in flow which would result in passive changes in volume which are probably of nearly as much importance.

The magnitude of the active capacitance change in the abdominal circulation is not greatly less than that obtained in response to the same stimuli in the whole body. For example, Shoukas and Sagawa (1973) reported that the change in whole body capacitance in response to a large change in carotid sinus pressure averaged $7.5 \text{ ml} \cdot \text{kg}^{-1}$ and Hainsworth and Karim (1976) obtained $5 \text{ ml} \cdot \text{kg}^{-1}$ from the abdominal circulation alone. This evidence, together with the finding that the musculo-cutaneous circulation contributes little if anything to the capacitance response, strongly suggests that the responses in the abdomen are by far the most important in the body.

The major components of the abdominal circulation, gut, liver and spleen, all make some contribution to the overall response. The relative magnitude of these contributions is uncertain. Carneiro and Donald (1977b) reported that the spleen made a greater contribution than the gut and the liver together (Fig. 10). However, other investigators have concluded that the response of the spleen is about one-third or one-quarter of the total (Karim and Hainsworth 1976; Shoukas et al. 1981). The difference in responses may be related to different venous pressures, different flows or different stimuli. The role of the liver remains to be established. There is some evidence that, apart from the splenic response, much of the response of the total abdominal capacitance, at least to infusion of catecholamines, may be due to changes in hepatic outflow resistance (e.g. Rutlen et al. 1981).

3.6.2 Sensitivity of Abdominal Capacitance Vessels to Sympathetic Nerve Activity

Low frequencies of discharge in the sympathetic efferent nerves result in near maximal responses of capacitance vessels. This has been shown for the whole abdominal circulation (Karim and Hainsworth 1976; Hainsworth and Karim 1976) and the spleen (Greenway et al. 1968; Donald and Aarhus 1974), but the liver may have a more gradual stimulus-response relationship (Bennett et al. 1982).

The high degree of sensitivity of capacitance vessels to sympathetic activity is able to explain many of the observations of the reflex responses in the region. It can explain why a decrease in carotid sinus pressure, from a level which effectively inhibits sympathetic outflow to the abdomen, results in constriction of capacitance vessels at higher pressures than those at which resistance vessels are powerfully constricted (Hainsworth and Karim 1976). It explains why, when carotid pressure is high and sympathetic outflow is inhibited, an increase in sympathetic discharge due to stimulation of chemoreceptors results in larger capacitance and smaller resistance responses than those occurring at low carotid pressures (Hainsworth et al. 1983a). Also, it explains why an increase in cephalic P_{CO_2}, which increases sympathetic outflow at low carotid pressures, enhances the effects of the baroreceptor reflex on capacitance at lower levels of cephalic CO_2 than those which enhance resistance responses (Ford et al. 1985a).

3.6.3 Role of the Abdominal Circulation as a Blood Reservoir

The capacitance function of the abdominal circulation arises not only from the effects of active constriction of vessels but also from the changes in the volume that can occur in response to changes in the venous distending pressure. Such a change can result from a change in the blood flow (e.g. Brooksby and Donald 1972), which would influence mainly the pressure distending the venules and very small veins, and from changes in large vein pressure (e.g. Karim and Hainsworth 1976), which affect all veins. The compliance of the abdominal circulation was found to be about 7 ml·kPa^{-1} per kilogram body weight (Karim and Hainsworth 1976). Thus a change in abdominal venous transmural pressure of 1 kPa would result in almost exactly the same change in blood volume as that resulting from maximal stimulation of the sympathetic nerves. The effects of changes in venous distending pressure are likely to be of considerable significance during changes in posture or changes in central venous pressure from any other cause, including cardiac failure, and during changes in blood volume.

The abdominal vascular bed is much better designed to serve a capacitance function than muscle and skin vasculatures. In the limb, for example, the volume of blood relative to that of the tissue is very small. Also, under conditions of stress, muscle blood flow is very high. Most of the increase in cardiac output on exercise passes through the limbs. Therefore, as pointed out in the discussion of the musculo-cutaneous circulation, an intense constriction of limb veins would have little effect on vascular volume, but would be likely to lead to an accelerated loss of fluid from the capillaries to the tissues. The situation in the abdominal circulation is very different. The volume of blood is very large relative to the tissue volume. Estimates of abdominal blood volume range from about 20% to 50% of total blood volume (Delorme et al. 1951; Johnstone 1956; Chien 1963). The magnitude depends largely on whether the spleen is included and on the degree of distension at the time the measurement is made. Contrary to the case in the limb, abdominal visceral blood flow is unlikely to increase to any great extent due to metabolic requirements. Under conditions of stress, such as exercise or haemorrhage, when sympathetic efferent discharge would be greatly increased, abdominal flow would be reduced and volume would decrease both as the result of active constriction of capacitance vessels and as the result of the decreased flow. Under such conditions, abdominal blood volume could be reduced by as much as half (Wade et al. 1956; Price et al. 1966; Rowell et al. 1972; Cousineau et al. 1985).

Not only does the abdominal vascular bed make the major contribution to whole body capacitance response, therefore, but because of its large blood volume and high compliance, its volume can also be changed passively. It must thus be regarded as the body's major blood reservoir, which can take up or release, actively and passively, the major part of any change in circulating blood volume.

4 The Pulmonary Circulation

In most regions of the body blood flow is regulated to suit either the local metabolic requirements of the tissue or other functional needs. Constriction of resistance vessels decreases the flow of blood to the region and thus results in a greater proportion of cardiac output being diverted elsewhere. In the lung, the situation is totally different. Pulmonary blood flow is normally entirely dependent on systemic blood flow and not on vasomotor responses in the lung. Also, the volume of blood within the lung is determined largely by factors outside the pulmonary circulation. A change in pulmonary blood volume would result from a transient difference between the outputs of the right and left hearts. This difference may result from the direct effects of one or both of the ventricles, for example by a change in rate or inotropic state, or it may result from a change in venous return due to postural changes or exercise, or from changes in intrapulmonary airways pressure or volume.

The distribution of resistance to flow through the lung differs from that in systemic vascular beds in that, instead of small arteries and arterioles providing nearly all the resistance to flow, resistance is much more evenly distributed along the various blood-vessels. Over half the total vascular resistance in the lung is estimated to be in the pulmonary capillaries and veins (Brody et al. 1968). Also, less than half of the pulmonary blood volume is normally contained within veins and, furthermore, the volume in the various pulmonary vessels varies with lung inflation (Howell et al. 1961; Permutt et al. 1961). Therefore, contrary to the case in the systemic circulation, it is not justified to consider arterioles as resistance vessels and veins as capacitance vessels.

4.1 Methods of Study

A change in pulmonary vascular capacitance can only be determined in preparations in which pulmonary blood flow and pulmonary venous pressures are held constant. Consideration of pulmonary vascular capacitance is complicated by the fact that major contributions arise from precapillary, capillary and postcapillary vessels. Thus if during constant flow perfusion there is an increase in pulmonary arterial pressure, the net volume change will be influenced by a change in the distension of the precapillary vessels. In the systemic circulation this is of minimal importance because so little blood is contained within the active resistance vessels, but in the lung this could result in a significant effect on the volume.

4.1.1 Constant Flow Perfusion: Measurement of Pressure Changes

Essentially, this technique determines pulmonary vascular resistance and was used by Daly and Daly (1959a) in their "vasosensory perfused living animal preparation". Because a significant proportion of the pulmonary blood volume is contained in precapillary vessels and the distribution of resistance is more even along the series vessels, a change in resistance at constant flow perfusion must be accompanied by a change in volume.

4.1.2 Constant Flow Perfusion: Measurement of Volume Changes

A technique is to bypass both sides of the heart and replace with two pumps of exactly equal outflow (Shoukas 1982). The pulmonary veins drain into a reservoir and, since the flows into the pulmonary artery and out of the pulmonary venous reservoir are constant and equal, a change in the reservoir volume provides a measure of the change in pulmonary blood volume. By changing the pulmonary venous pressure by altering the height of the venous reservoir, information can be obtained about the pulmonary vascular compliance.

The relative effects on (mainly) the pulmonary arteries and (mainly) the pulmonary veins can be assessed by determining the relationship between a change in lung volume and either arterial or venous pressure with the other pressure held constant (Shoukas 1975). Changes in arterial pressure are effected by changing the rate of perfusion, with adjustments being made to the level of the venous reservoir. Changes in venous pressure are effected by changing the reservoir level while controlling arterial pressure by making adjustments to the flow rate.

4.1.3 Indicator Dilution Methods

There have been several studies which have used the mean pulmonary transit time of indicators to determine pulmonary blood volume in man (Milner et al. 1960) and dogs (Feeley et al. 1963). Pulmonary blood volume is given as cardiac output multiplied by the difference between the mean transit time from pulmonary artery to brachial artery and that from left atrium to pulmonary artery. This method gives values of about 6 ml/kg in man and 11 ml/kg in the dog (Milner 1980). However, as already mentioned, when flow and arterial and venous pressures are uncontrolled, pulmonary blood volume is likely to have little relationship to any control mechanisms within the lung.

4.2 Responses to Adrenergic Stimulation

Infusions of adrenaline and noradrenaline result in pulmonary vasoconstriction (Feeley et al. 1963). This seems to affect mainly the "downstream" vessels, presumably veins, whereas serotonin has a greater effect
on "upstream" vessels (Maron and Dawson 1980). Using the constant
perfusion and reservoir technique, Shoukas (1982) obtained small but
significant reductions in pulmonary blood volume and compliance in
response to infusions of adrenaline.

4.3 Responses to Reflex Stimulation

Daly and Daly (1959b) reported that an increase in carotid sinus pressure
resulted in a decrease in pulmonary arterial perfusion pressure, which, at
a constant blood flow and constant venous pressure, implies that there
was a decrease in resistance and that there would also have been an increase
in pulmonary blood volume (capacitance). Shoukas (1982) determined the
changes in pulmonary blood volume during constant flow perfusion in
response to large changes in carotid sinus pressure. The change in pulmonary capacitance was about $1 \text{ ml} \cdot \text{kg}^{-1}$, which was much less than the
change in systemic capacitance ($7.5 \text{ ml} \cdot \text{kg}^{-1}$). Constriction of the pulmonary vascular bed occurs in response to stimulation of carotid chemoreceptors (Daly and Daly 1959a). However, that response was frequently
not obtained unless the bronchial circulation was obstructed.

4.4 Importance of Pulmonary Vascular Capacitance

The distensibility of the pulmonary circulation is undoubtedly of importance in that it permits transient imbalance between the outputs of the
left and right hearts. However, it seems unlikely that the lung is an important controllable reservoir of blood. Only about 9% of blood volume is
contained within the lungs (Dock et al. 1961; Milner et al. 1960) and, in
man, this volume would rarely change by more than 200 ml (Milner
1980), most of which would occur directly as the result of events outside
the lungs. Changes in pulmonary blood volume would result from mechanical factors, change in cardiac output or posture, and changes due to
active constriction of pulmonary blood-vessels are likely to be small.

It seems likely, therefore, that although it has a role in the balance
between left and right heart outputs, the pulmonary vascular bed plays a
relatively unimportant part as a blood reservoir and its active capacitance
role is trivial in comparison with that of the systemic circulation.

5 Whole Body Capacitance

5.1 Methods of Study

Estimates of capacitance in the whole body necessarily involve relatively complex perfusion techniques. Any values obtained represent a lumped value and a change may be due to the effect of increases in capacitance in some regions and decreases in others. The principles of measurement of changes in capacitance are essentially the same as for regions of the body. Changes in volume can be determined at constant flow and constant venous pressure or changes in pressure can be determined in the arrested circulation.

5.1.1 Changes in Volume at Constant Flow
and Constant Venous Pressure

The technique requires either a left heart bypass or a complete cardio-pulmonary bypass and a constant flow perfusion of the systemic circulation. A change in the total systemic vascular capacitance can be assessed from a change in the total venous outflow provided the venous pressure did not change. Kahler et al. (1962) and Shoukas and Sagawa (1971) bypassed both the heart and the lungs and perfused the systemic circulation through an oxygenator from a venous reservoir. A change in the systemic blood volume resulted in an opposite change in the reservoir volume. Brunner et al. (1981) described a modified approach which avoided the necessity for an oxygenator. The systemic veins drained into one reservoir from which blood was pumped at constant flow to the pulmonary artery and the pulmonary veins drained into a second reservoir from which blood was pumped at constant flow into the systemic arteries. Under stable conditions the output of the two pumps would be exactly equal. A change in total systemic capacitance would result in a change in the volume of the reservoir draining the systemic veins.

An attempt to assess changes in total body capacitance in conscious dogs has recently been reported by Bennett et al. (1984). The technique involved a preliminary operation during which heart block was created, ventricular stimulating electrodes were implanted, an electromagnetic flow transducer was placed round the aorta and catheters were inserted in the right atrium and aorta. At the subsequent investigation, cardiac output, as assessed by the aortic flow transducer, was held constant by ventricular pacing and changes in capacitance were assessed from the changes in central venous pressure. This is an ingenious method and can provide useful information. However, the magnitude of a change in central venous pressure would be influenced to some extent by any

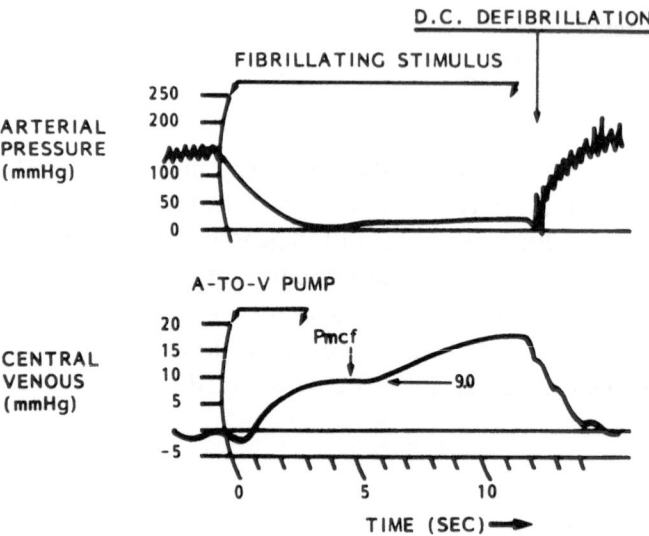

Fig. 11. Technique for estimating mean circulatory filling pressure (*Pmcf*). Heart was fibrillated and blood transferred rapidly, at about 2 l/min, from arteries to veins so that arterial and venous pressures equilibrate at about 5 s. Note that there was an increase in venous pressure after about 6 s, which was attributed to a reflex increase in sympathetic nerve activity. (Rothe 1983c)

change in right heart output, which was not controlled, and lung blood volume. Furthermore, there may be reflex changes due to changes in cardiac rate or volume of blood in the great veins, heart or lungs. However, these secondary changes may be reduced or prevented by vagal blockade or vagotomy.

5.1.2 Mean Circulatory Filling Pressure

Mean circulatory filling pressure is the pressure equilibrated throughout all systemic blood-vessels immediately after stopping the heart (Guyton et al. 1955). Since the bulk of the blood volume is contained within the veins, the degree of constriction of the veins and the magnitude of the blood volume are factors which are mainly responsible for determining the value of the mean circulatory pressure. Usually, the heart is stopped by inducing ventricular fibrillation and the arterial and venous pressures are rapidly equated by use of a pump (Fig. 11); Rothe and Dress (1976) showed that repeated estimates could be made provided that the period of circulatory arrest did not exceed 10 s and at least 2 min was allowed between estimates. After about 7 s of arrest arterial and venous constriction occur, which by 1 min become intense (Rothe 1976). An alternative approach which has been used in rats is to arrest the circulation by use of right atrial or pulmonary arterial occluders (Samar and Coleman 1978; Yamamoto et al. 1980).

A detailed description of the method has been given by Rothe (1983c). Several assumptions are made. Mean circulating filling pressure is assumed to be representative of the pressure distending the small veins and is independent of the pre-existing cardiac output. It also assumes that a reliable value can be obtained within 7 s, because after this time reflex changes will occur which will change the pressure (Fig. 11). It may be difficult to equilibrate the vascular pressures in this time and the presence of venous valves interferes with the rapid equilibration. Also, the effect of stress relaxation of veins (Alexander 1963) may lead to instability. Another disadvantage of the technique is that it is not possible to make continuous measurements of mean circulating filling pressure and, if there has been a change in the circulating blood volume between successive estimates, a change in the recorded pressure will not accurately represent an active change in capacitance.

5.2 Responses to Infusions of Catecholamines

Infusions of adrenaline and noradrenaline have been used as methods of bringing about a generalized constriction of capacitance vessels. Most studies have employed the constant perfusion and reservoir technique. The largest responses were reported in the early studies of Ross et al. (1961a) and Braunwald et al. (1963), in which it was reported that infusion of both adrenaline and noradrenaline resulted in an astonishingly large change in reservoir volume averaging 19 ml/kg with a maximum of 37 ml/kg. These volumes, which are far in excess of those reported by other investigators, may partly be due to changes in tissue volume, because responses were assessed over as long as 8 min, and partly due to inadequate control of venous pressures. Most other investigations give maximum volume changes in response to infusions of adrenaline and noradrenaline of $5-10$ $ml \cdot kg^{-1}$ (Caldini et al. 1974; Muller-Ruchholtz et al. 1977a,b; Shoukas and Brunner 1980; Shoukas 1982). The large responses of total body capacitance following infusion of adrenaline and noradrenaline contrast with the absence of a change in response to infusions of angiotensin II in doses which cause similar pressor responses (Rose et al. 1962). Infusion of noradrenaline results in an increase in mean circulatory filling pressure and a decrease in whole body compliance (Drees and Rothe 1974; Fig. 12).

The mechanism of action of the catecholamines in reducing systemic capacitance is not absolutely clear but seems to be complex. Imai et al. (1978) reported that the decrease in capacitance in response to isoprenaline was abolished and that to noradrenaline reversed following beta-adrenoceptor blockade. Rutlen et al. (1981) found that isoprenaline

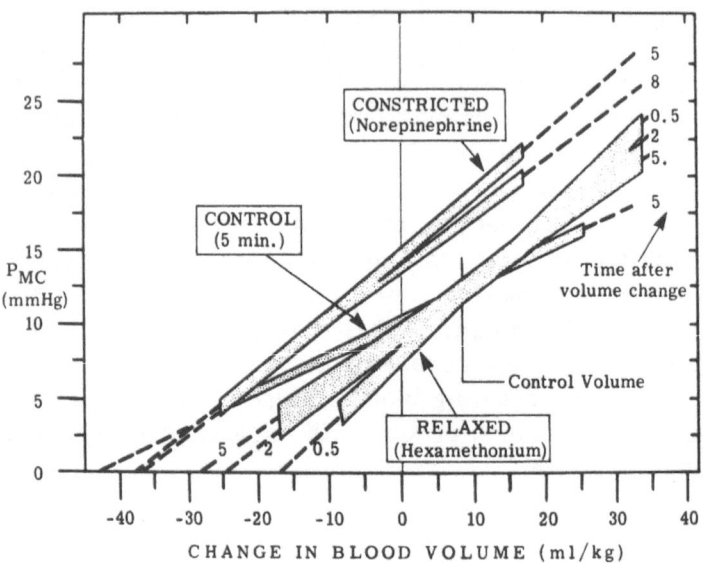

Fig. 12. Estimated values of mean circulatory filling pressure following changes in blood volume and showing the effects of vasoconstriction from infusion of noradrenaline and vasodilatation from administration of hexamethonium. Results show 95% confidence bands. (Drees and Rothe 1974)

resulted in a small decrease in vascular volume and noradrenaline in a much larger change. However, after splenectomy the response to noradrenaline was similar to that during isoprenaline infusion, suggesting that the large alpha-adrenergic effect was predominantly due to splenic contraction. The isoprenaline-induced response was prevented by beta-adrenoreceptor blockade. The whole of the capacitance response was shown to arise from the splanchnic circulation, and evidence was presented suggesting that the mechanism was mainly due to a reduction in postsinusoidal hepatic vascular resistance and that the changes were much reduced by portal venting which effectively bypassed the liver. Thus the suggestion is that, apart from the spleen, the control of total systemic vascular capacitance is almost entirely the result of changes in outflow resistance from the splanchnic circulation.

Muller-Ruchholtz et al. (1977b) also concluded that both alpha- and beta-adrenoceptors were responsible for the change in capacitance. Part of the effect may have been a reflex change because denervation of baroreceptors reduced or abolished the capacitance response to isoprenaline. However, this is probably not the whole explanation because Rutlen et al. (1981) still obtained responses to isoprenaline after ganglion blockade.

In conscious dogs, Bennett et al. (1984) determined changes in central venous pressure during constant cardiac output. They found that infusion of the alpha-adrenoceptor agonist phenylephrine caused large changes in

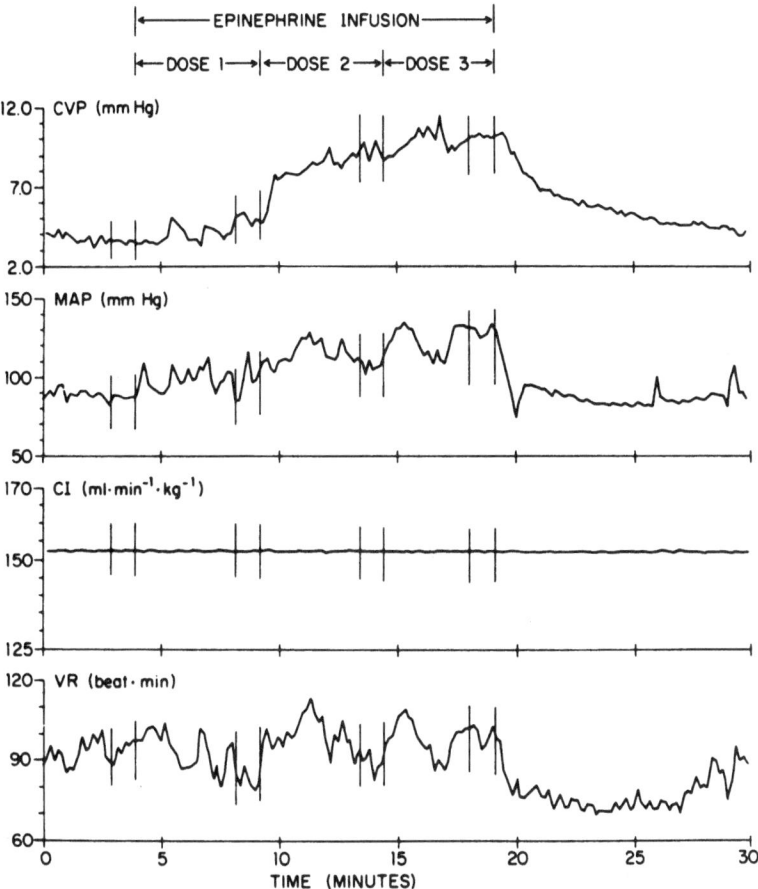

Fig. 13. Results from conscious dog in which cardiac output was held constant by electrical pacing of heart and changes in total body capacitance were assessed from changes in total body capacitance were assessed from changes in central venous pressure. Traces are of central venous pressure (*CVP*), mean arterial blood pressure (*MAP*), cardiac index (*CI*) (from implanted flowmeter) and ventricular rate (*VR*). Adrenaline in increasing doses resulted in increases in arterial blood pressure and in central venous pressure. (Bennett et al. 1984)

arterial pressure but little effect on venous pressure. Adrenaline resulted in little change in arterial pressure but an increase in venous pressure of 0.7 kPa (Fig. 13), and isoprenaline decreased arterial pressure and increased venous pressure. The changes in venous pressure were attenuated following beta-adrenoceptor blockade with propranolol. Although the possible effects of reflexes have not been excluded, these experiments do support the view that stimulation of beta-adrenoceptors can be an important mechanism for decreasing vascular capacitance.

Finally, studies have also been carried out in the rat using the technique of circulatory arrest by inflation of a right atrial balloon (Trippodo 1981), in which it was shown that an infusion of adrenaline resulted in an increase in mean circulatory filling pressure.

5.3 Reflex Responses

5.3.1 Baroreceptors

There have been several studies of the effects of changes in carotid sinus pressure on the reservoir volume during cardiopulmonary bypass and constant flow perfusion of the systemic circulation. Ross et al. (1961a) and Braunwald et al. (1963) reported changes in volume as high as 13 ml·kg^{-1}. However, these responses may have been exaggerated because of the long time over which assessments were made and because venous pressure was not precisely controlled. Much smaller responses were obtained by Muller-Ruchholtz et al. (1979) using a similar preparation. Shoukas and Sagawa (1973) changed carotid pressure between 10 and 27 kPa and reported an average change in reservoir volume, and hence in capacitance, of 7.5 ml·kg^{-1}. Much larger responses were obtained when the region was perfused at constant pressure instead of at constant flow. Similar responses were obtained in subsequent studies (Brunner et al. 1981; Shoukas 1982). Shoukas and Brunner (1980) obtained somewhat larger responses, but in that study there was poor control of temperature. Brunner et al. (1981) perfused the whole systemic circulation at constant flow and collected the blood from the splanchnic and extrasplanchnic circulations separately to test the hypothesis that a change in total body volume in response to a change in carotid pressure might have resulted from a differential effect on different resistance vessels, perhaps causing, a smaller distension to a highly compliant region. They found, however, that there was only a small (5%) distribution of blood away from the splanchnic circulation when carotid pressure was decreased, and this was not considered sufficient to explain the response. They also noted that almost all the change in volume was from the splanchnic circulation.

   Shoukas and Sagawa (1973) reported no difference between the sinus pressures at which the "gains" were maximal for resistance and capacitance responses. This contrasts with the results of Hainsworth and Karim (1976), who found that, for the abdominal circulation at least, resistance responses occurred at significantly lower carotid sinus pressures. This difference may be related to the method of assessing the responses. Shoukas and Sagawa (1973) determined the gains, defined as the response to a small step change in carotid pressure, whereas Hainsworth and Karim (1976) determined the cumulative responses and observed that the resistance curves were displaced to correspond with lower carotid values than the capacitance curves. Replotting the data of Shoukas and Sagawa (1973) does show a small displacement of the curves at least qualitatively similar to the results of Hainsworth and Karim (1976).

Bennett et al. (1984) occluded the carotid arteries in conscious dogs in which cardiac output was controlled (see page 147) and obtained no change in central venous pressure unless the vagi had been blocked, in which case there was a small increase in central venous pressure (+ 0.1 kPa). This response seems to be small in comparison with the results from the anaesthetized animal preparations, but the stimulus, carotid occlusion, is likely to have been smaller than in the other studies.

5.3.2 Responses to Other Interventions

Haemorrhage resulted in a decrease of total body compliance as estimated by determining mean circulation pressure and its responses to infusions and withdrawals of volumes of blood (Rothe and Drees 1976). Ventricular distension has been reported to result in a decrease in total body capacitance (Salisbury et al. 1960; Ross et al. 1961b). However, in these studies ventricular systolic and, particularly, diastolic pressures were raised well beyond normal ranges. Furthermore, venous pressure was not reported and the slow time course of the responses casts doubt on the extent to which changes were actually due to capacitance vessel constriction. The role of cardiac receptors in the control of capacitance, therefore, remains uncertain.

There has been only one study which aimed to determine the capacitance responses to stimulation of atrial receptors (Rutlen 1981). In this, distension of the left atrium by a large balloon resulted in a small decrease in vascular volume which was reversed when the atrium was distended following vagotomy or administration of the beta-adrenoceptor antagonist propranolol. However, in this study there would have been changes in cardiac and pulmonary blood volumes, which makes interpretation of the results difficult. Also, the absence of any response until 4 min after atrial obstruction casts doubt on the responses being due to active capacitance vessel constriction.

A decrease in blood volume occurs in response to hypoxia (Kahler et al. 1962) and this does not occur after chemoreceptor denervation. Also, the mean circulatory filling pressure is increased during hypoxia (Smith and Crowell 1967). However, there have been no studies of the responses of total body capacitance to a discrete stimulus applied to chemoreceptors.

Increase in intracerebral pressure results in an intense vasoconstriction: the Cushing response. There is also evidence of venous participation in this response (Rodbard and Stone 1956; Brown 1956; Brashear and Ross 1970). Richardson and Fermoso (1964) reported an increase in mean circulating filling pressure. Stein et al. (1983) determined total body capacitance from changes in reservoir volume during constant right atrial perfusion and reported that cerebral compression resulted in a decrease in total body capacitance of about 10 ml·kg^{-1}, which probably represents the maximal response obtainable.

6 Importance of Vascular Capacitance

In the introduction to this review I noted that Folkow and Mellander (1964) claimed that veins were "at least as reactive and well controlled as other elements in the circulation and had an important role to play in cardiovascular integration". Earlier, Folkow (1959) had stated that changes in vascular capacitance form "the basis of a rapidly mobilisable, centrally controlled intravascular fluid reserve, which is so necessary for good pump performance in a closed flow circuit". Mark and Eckstein (1968) considered that "the venous system functions as a dynamic reservoir from which blood is pumped by the heart". Clearly, then, these prominent research workers, as well as many others (e.g. Gauer and Thron 1962; Weil and Shubin 1970; Shepherd and Vanhoutte 1975; Rothe 1979, 1983a,b), regard veins and vascular capacitance as playing a major role in cardiovascular control. In this section I will examine whether the claims made for the importance of vascular capacitance are supported by the evidence.

6.1 Extent and Magnitude of Capacitance Changes

One important functional difference between vascular resistance and capacitance is that control of resistance is concerned with the regulation of flow in the tissue in which the particular resistance vessels are situated, whereas the control of capacitance vessels are concerned with the whole body and cardiac filling. Constriction of capacitance vessels in a particular region has little or no direct functional effect on that region.

The evidence now seems to be overwhelming, from experiments on both animals and man, that blood-vessels in the muscle and cutaneous circulations make little or no active contribution to the body's capacitance control. The lung may make a small contribution, but the pulmonary circulation is entirely dependent on the systemic circulation and, due to its distensibility, any active capacitance response in the lung is totally overshadowed by secondary effects due to changes in cardiac output and posture. Thus the great bulk of the tissue in the body is of no importance in capacitance control. This leaves the 20% or so of the blood volume which is contained in the splanchnic circulation. There is evidence, at least in experimental animals if not in man, that a reduction in blood volume in the splanchnic vascular bed can occur which is not dependent on changes in blood flow. However, in dog it has been shown that about 40% of the maximal response is due to contraction of the spleen. This is a very large contactile organ in the dog but in man it is relatively insignificant.

Most reports of the magnitude of the response of total body capacitance to powerful reflex stimuli, for example a large change in carotid sinus pressure, give responses of the order of 7.5 ml·kg^{-1}. Rather larger changes may occur in response to extreme interventions, such as cerebral ischaemia. However, such interventions do not occur as a normal part of daily life. If the normal maximal response in the dog is assessed to be 7.5 ml·kg^{-1} and 40% of this is due to splenic contraction, the response of the rest of the vasculature becomes 4.5 ml·kg^{-1}. Since man does not have a large contractile spleen like the dog, and if it is assumed that the rest of the human circulation behaves like that of the dog, the maximal capacitance response will be 4.5 ml·kg^{-1}, which for a 70-kg man represents a volume change of 315 ml.

The mechanisms responsible for the change in capacitance in the splanchnic circulation excluding the spleen are not clear. There is undoubtedly adequate evidence that intestinal veins can constrict in response to nervous stimuli (e.g. Iizuka et al. 1970; Donald and Aarhus 1974), but the extent to which this is responsible for the overall response is not known. The liver may have a more important role that was hitherto not recognized. In addition to the likelihood that hepatic volume can actively decrease, there is evidence that at least when there are increased levels of catecholamines, regulation of hepatic venous outflow resistance may make an important contribution to the total volume change.

6.2 Active Versus Passive Volume Changes

In the hind limb circulation the active capacitance responses resulting from electrical stimulation of the sympathetic nerves are very small indeed and no significant responses are obtained to reflex stimuli (Lesh and Rothe 1969; Hainsworth et al. 1983c). Nevertheless, when the region is perfused at a constant pressure, stimulation of sympathetic nerves either directly or reflexly does significantly reduce the volume of blood in the region (Mellander 1960; Hainsworth et al. 1983c). Hainsworth et al. (1983c) found the magnitude of this active plus passive response in the limb to be about ten times as large as the active response determined under conditions of constant flow, and Brooksby and Donald (1972) found that even in the splanchnic circulation, which is the one region which can contribute significantly to capacitance responses, when flow to the region was reduced artificially by the same amount as that occurring during stimulation of sympathetic nerves, the change in volume was only one-third less than that which occurred during sympathetic stimulation. The flow-related changes in vascular volume can be attributed to changes in pressure gradient along the veins. If the downstream, or large vein, pressure is constant, then the pressure in the smaller upstream vein will change by an amount which is such that the pressure gradient is proportional to flow (Poiseuille's equation). Since most of the blood is contained in the small veins and these veins are distensible, changes in flow, even at constant pressure in the large veins, result in changes in volume.

Large changes in regional vascular volumes occur when there are generalized changes in venous pressure due, for example, to changes in posture. Estimates of total vascular compliance in both man and dog give values of 15–20 $ml \cdot kg^{-1} \cdot kPa^{-1}$ (Echt et al. 1974; Drees and Rothe 1974; Larochelle and Ogilvie 1976; Shoukas and Sagawa 1973). During upright tilting or quiet standing in man the greatest pressures occur in the veins of the lower parts of the legs and feet. However, relatively little blood is pooled in these vessels because, even when they are fully distended, only a small additional amount of blood can be stored in them due to the limit to their distension. It has been estimated that in recumbent man the legs contain about 250 ml blood and that on upright tilting this increases to about 500 ml, a quite small increase of about 250 ml. Also, to some extent an increase in the volume of blood in the legs on standing would be offset by a gravitational decrease in the volume of blood contained in the veins in the upper part of the body. The abdominal vessels, however, are much more distensible. In the dog, abdominal vascular compliance is about 10 $ml \cdot kg^{-1} \cdot kPa^{-1}$ (Karim and Hainsworth 1976). Thus although only about 20% of blood volume is contained within this region, it is responsible for the major portion of the total body compliance. To some extent, the accumulation of blood in the abdominal capacitance vessels during postural stress may be limited by the external pressure on the vessels exerted by the weight of the viscera, which are effectively contained within a relatively non-elastic bag. Nevertheless, it is not difficult to see how, in man, stresses including gravitational and respiratory manoeuvres, including positive pressure ventilation, can result in the retention in the abdominal circulation of large volumes of blood. The passive changes in volume in the abdominal capacitance vessels which occur with changes in venous pressure resulting from changes in posture are likely to be approximately equal and opposite to thsoe which occur actively when the sympathetic nerves are maximally excited. It should also be noted that changes in venous pressure will result also in changes in capillary pressure, and hence changes in blood volume will occur due to exchange of fluid between the capillaries and the tissues.

6.3 Potential Importance of Capacitance Vessels

The main function of the control of vascular capacitance is said to be the regulation of cardiac output. Even a modest constriction of veins, because they contain so much blood, is said to boost venous return, increase central venous pressure and increase cardiac output. Guyton (1955, 1963) described the relationships between venous return, central venous pressure and cardiac output. There is an equilibrium value of right atrial pressure which is determined by venous return and the contractile properties of the heart. Guyton regards the mean circulatory filling pressure (the pressure which would exist when the heart is stopped and the pressure in all blood-vessels is rapidly equated) as the primary determinant of cardiac filling, and therefore of cardiac output. Since most of the blood is in the veins, the two factors which are of greatest importance in determining the mean circulatory pressure are the volume of blood in the circulation and the state of constriction of the capacitance vessels. Since various nervous and reflex stimuli cause changes in the mean circulatory pressure and the time course of these responses is such that they could not be due to changes in vascular volume, the implication is that capacitance responses must be of prime importance in the control of cardiac output.

There seem to be a number of difficulties with this line of reasoning, however. The use of the concept of mean circulatory pressure ignores any dynamic consideration in the circulation. At high outputs there would be likely to be significant inertial forces (momentum) which would propel blood back to the heart. The mean circulatory pressure is unaffected by blood flow, whereas the pressure gradient along the blood-vessels is directly proportional to flow. Also, it ignores effects of pressure generated by extravascular tissues and any pumping action caused by phasic compression of veins.

Some idea as to the likely effect of a change in total body capacitance on cardiac output can be obtained by studying the effects of haemorrhage and transfusion. However, the effects of changes in blood volume on cardiac output are variable. Bishop et al. (1964) found that, in conscious dogs, an increase in right atrial pressure of about 0.5 kPa was sufficient to more than double cardiac output. More recently, Barnes et al. (1979) reported that much larger increases in left atrial pressure (+2.7 kPa) were required to result in the doubling of cardiac output. Infusions to conscious dogs result in increases in cardiac output of 4 ml·kg^{-1}·min^{-1} for each

1 ml·kg^{-1} infused (Vatner and Boettcher 1978). The sensitivity of cardiac output to right atrial pressure is much greater when the influence of reflexes is prevented (Herndon and Sagawa 1969). Frye and Braunwald (1960) studied normal humans and found that infusion of 1500 ml blood had no significant effect on cardiac output unless reflexes were prevented by ganglionic blockade, in which case cardiac output increased by about 50%.

The studies of the effects of haemorrhage and transfusion show that a mechanism can exist whereby constriction of capcitance vessels can result in an increase in the flow of blood to the heart and a consequent increase in cardiac output. Shoukas and Sagawa (1973) reported that the maximal capacitance response to change in carotid sinus pressure was 7.5 ml·kg^{-1}. They calculated that this would result in an increase in mean circulatory pressure of 0.5 kPa, which according to Guyton's curves would result in an increase in cardiac output of about 60%.

6.4 Vascular Capacitance in Perspective

So far in this review, I have examined the changes in capacitance in various regions of the body and in the whole body in response to a variety of stimuli. Many of the earlier studies can be seriously faulted because they failed to distinguish changes in volume which occurred as a direct result of constriction of blood-vessels from responses due to changes in the vascular distending pressures consequent upon changes in venous pressure or changes in blood flow. When only the studies are considered which have satisfactorily met the criteria for assessing active capacitance changes, it is apparent that significant active capacitance changes occur only in the abdominal circulation. The bulk of the body mass, muscle and skin, has no active role to play in the control of vascular capacitance.

When responses to resistance and capacitance are compared, it is apparent that in tissues in which there is a change in capacitance in response to a stimulus there is no evidence that the efferent discharge rate to capacitance vessels is any different from that to resistance vessels; the observations that stepwise decreases in baroreceptor distending pressure result in an initial preponderance of capacitance, and the interaction of the responses to the stimulation of different reflexes, can be explained by the greater sensitivity of capacitance vessels to sympathetic activity.

In the dog, the maximum capacitance response of the whole body to a reflex stimulus is only about 7.5 ml·kg^{-1}, and of this at least one-third can be attributed to contraction of the spleen. Unlike the dog, man does not have a powerfully contractile spleen so, if the rest of his vasculature behaves like that of the dog, the total response in man would be only

4.5 ml·kg^{-1}. This would represent a total actively mobilizable blood reserve of 315 ml in a 70-kg man. A transfusion of this volume of blood would have a negligible effect on cardiac output, at least in presence of active reflexes. If reflexes did not modify the response the change in cardiac output would be somewhat larger. If the calculations are based on the effects of a change in whole body capacitance (excluding spleen) and if the mean circulatory filling pressure is determined and the data are applied to Guyton's curves (Shoukas and Sagawa 1973), the predicted maximal increase in cardiac output will be about 40%.

The next question to consider is under what circumstances would a capacitance response be expected to occur. Capacitance vessels in the abdomen, which is the only region to make a significant contribution, are very sensitive to low frequencies of sympathetic nerve activity. Responses of 50% of maximal occur during stimulation at only 1 Hz. Comparison of the responses to direct and reflex stimuli indicate that at a carotid sinus pressure of about 20 kPa the degree of constriction of abdominal capacitance vessels is the same as that during electrical stimulation at 1 Hz; i.e. responses are half maximal. At 17 kPa carotid pressure, responses are 70% maximal. A number of assumptions are required to relate these studies of responses to non-pulsatile stimulation of the carotid sinus in the anaesthetized dog with those of responses to pulsatile stimulation of all arterial baroreceptors in conscious man. Nevertheless, it is not unreasonable to assume that when man is standing, the efferent sympathetic activity is equivalent to a regular stimulus of at least 1 Hz. Firm evidence on this is lacking, but experiments from relatively intact anaesthetized animals give values for the resting sympathetic discharge rate of about 2 Hz (Folkow 1959; Koizumi and Brooks 1972). Some recordings have been made in man from presumed cutaneous sympathetic efferent nerves and, although these may not be related to the frequency of discharge in the abdominal circulation and the activity tends to occur in bursts, the discharge frequency does increase substantially with the assumption of an upright posture (Burke et al. 1977; Hallin and Torebjork 1974) and it is unlikely to be less than the equivalent of stimulation at less than 1 Hz. If it is assumed that in the recumbent position sympathetic efferent discharge is near zero, then on upright tilting or standing the change in capacitance would be about 60% of maximal, i.e. about 200 ml. This would leave a further capacitance reserve of only about 150 ml when the sympathetic system is strongly stimulated.

The rather small active capacitance responses can be contrasted with the much larger changes which occur in response to a change in the venous distending pressure. If it is assumed that total body compliance (mainly venous) is about 20 ml·kPa^{-1}·kg^{-1}, then in a 70-kg man a change in venous pressure of 1 kPa would result in a change in volume of nearly

1.5 1. The pressure change required to produce the same volume change as the maximum possible capacitance change is of the order of 0.2 kPa (2 cmH$_2$O). Changes in venous distending pressure arising from changes in flow can result in larger changes in volume than those which occur as the result of constriction of capcitance vessels. Indeed, in most of the body only passive changes in volume occur. Thus it is not an exaggeration to state that arterioles are probably of greater importance in the control of venous pressure and hence their volume than are the veins.

In addition to these effects of blood flow and body position, the transmural pressure of veins is importantly affected by external pressure from surrounding tissues. Contraction of limb muscles compresses and empties the deep veins. The weight of the abdominal viscera, as well as abdominal wall and diaphragmatic contractions, compresses the abdominal veins. Veins in the upper part of the body normally remain almost empty by the influence of gravity. Thus events external to the veins can result in larger changes in venous volume than those resulting from contraction of the vein wall itself.

6.5 Conclusion

In this review I have presented evidence that vascular capacitance can be controlled by a number of cardiovascular reflexes. In particular, changes in the stimulus to baroreceptors can result in near maximal responses of capacitance. The vessels in the abdomen are almost certainly the only ones which make a significant contribution to the overall response and these vessels are particularly sensitive to low frequencies of sympathetic discharge. The overall maximal reflex change in capacitance, particularly if the spleen is excluded, is quite small in relation to changes in volume which occur secondarily to changes in tissue blood flow or in venous pressure from other causes. There are likely to be few manoeuvres which would result in an active capacitance response in man greater than about 200 ml and the maximal response is unlikely to be much more than about 300 ml. It seems likely that most of the capacitance "reserve" would be taken up when man assumes an upright posture and it is unlikely that any large further mobilizable reserve is available. However, although the total response seems to be small, it may not be entirely unimportant. Standing reduces cardiac output by about one-third under normal circumstances and, presumably, despite constriction of the capacitance vessels. Without this constriction, it would probably be equivalent to a small haemorrhage, and following even quite small haemorrhage postural hypotension is much more common.

Thus capacitance control is probably of some significance during postural changes, but it is hard to envisage many other circumstances when it would be likely to be of any great importance.

References

Abboud FM, Eckstein JW (1966) Comparative changes in segmental vascular resistance in response to nerve stimulation and to norepinephrine. Circ Res 18:263–277

Abboud FM, Mark AL (1979) Cardiac baroreceptors in circulatory control in humans. In: Hainsworth R, Kidd C, Linden RJ (eds) Cardiac receptors. Cambridge University Press, Cambridge, pp 437–462

Abboud FM, Schmid PG, Eckstein JW (1968) Vascular responses after alpha-adrenergic blockade. I. Responses of capacitance and resistance vessels to norepinephrine in man. J Clin Invest 47:1–9

Abboud FM, Eckberg DL, Johannsen UJ, Mark AL (1979) Carotid and cardiopulmonary baroreceptor control of splanchnic and forearm vascular resistance during venous pooling in man. J Physiol (Lond) 286:173–184

Alexander RS (1954) The participation of the venomotor system in pressor reflexes. Circ Res 2:405–409

Alexander RS (1963) The peripheral venous system. In: Hamilton WF (ed) Circulation. American Physiological Society, Washington DC, pp 1084–1098 (Handbook of physiology, sect 2, vol 2)

Al-Shamma YMH, Hainsworth R (1985) The cardiovascular responses to upright tilting and lower body negative pressure in man. J Physiol (in press)

Altura BM, Zweifach BW (1965) Antihistamines and vascular reactivity. Am J Physiol 209:545–549

Andrews WHH, Ritchie HD, Maegraith BG (1973) An assessment of the physiological importance of the large hepatic venous sluices in the dog. Q J Exp Physiol 58: 325–333

Anrep GV, Pascual W, Rossler R (1936) Respiratory variations of the heart rate. Proc R Soc Lond (Biol) 119:191–217

Ardill BL, Bhatnagar VM, Fentem P (1968) Observation of changes in volume of a congested limb as a means of studying the behaviour of capacity vessels. J Physiol (Lond) 194:627–644

Baker CH (1966) Vascular volume changes following histamine release in the dog forelimb. Am J Physiol 211:661–666

Baker DG, Coleridge HM, Coleridge JCG (1979) Vagal afferent "C" fibres from the ventricle. In: Hainsworth R, Kidd C, Linden RJ (eds) Cardiac receptors. Cambridge University Press, Cambridge, pp 117–139

Barcroft H, Barcroft J (1923) Observations on the taking up of carbon monoxide by the haemoglobin in the spleen. J Physiol (Lond) 58:138–144

Barcroft J, Stephens JG (1927) Observations on the size of the spleen. J Physiol 64: 1–22

Barnes GE, Chevis BC, Granger HJ (1979) Regulation of cardiac output during rapid volume loading. Am J Physiol 237:R197–R202

Bartelstone HJ (1960) Role of the veins in venous return. Circ Res 8:1059–1076

Baum T, Hosko MJ (1965) Responses of resistance and capacitance vessels to central nervous system stimulation. Am J Physiol 209:236–242

Bennett TD, Rothe CF (1981) Hepatic capacitance responses to changes in flow and hepatic venous pressure in dogs. Am J Physiol 240:H18–H28

Bennett TD, MacAnespie CL, Rothe CF (1982) Active hepatic capacitance responses to neural and humoral stimuli in dogs. Am J Physiol 242:H1000–H1009

Bennett TD, Wyss CR, Scher AM (1984) Changes in vascular capacity in awake dogs in response to carotid sinus occlusion and administration of catecholamines. Circ Res 55:440–453

Bevan JA, Hosmer DW, Ljung B, Pegram BL, Su C (1974) Innervation pattern and neurogenic response of rabbit veins. Blood Vessels 11:172–182

Bevegård BS, Shepherd JT (1965a) Changes in tone of limb veins during supine exercise. J Appl Physiol 20:1–8

Bevegård BS, Shepherd JT (1965b) Effect of local exercise of forearm muscles on forearm capacitance vessels. J Appl Physiol 20:968–974

Bevegård BS, Shepherd JT (1966) Circulatory effects of stimulating the carotid arterial receptors in man at rest and during exercise. J Clin Invest 45:132–142

Biscoe TJ, Purves MJ, Sampson SR (1970) The frequency of nerve impulses in single carotid body chemoreceptor afferent fibres recorded in vivo. J Physiol (Lond) 208: 121–131

Bishop VS, Stone HL, Guyton AC (1964) Cardiac function curves in conscious dogs. Am J Physiol 207:677–682

Boreus LO, Hollenberg NK (1972) Venous constriction in the response to head-up tilt in man. Can J Physiol Pharmacol 50:317–320

Brashear RE, Ross JC (1970) Hemodynamic effects of elevated cerebrospinal fluid pressure: alterations with adrenergic blockade. J Clin Invest 49:1324–1333

Braunwald E, Ross J Jr, Kahler RL, Gaffney TE, Goldblatt A, Mason DT (1963) Reflex control of the systemic venous bed. Effects on venous tone of vasoactive drugs, and of baroreceptor and chemoreceptor stimulation. Circ Res 12:539–550

Brender D, Webb-Peploe MM (1969) Influence of carotid baroreceptors on different components of the vascular system. J Physiol (Lond) 205:257–274

Britton SL, Donald DE (1982) Response of large hindlimb veins of dog to aortic arch stimulation. Am J Physiol 242:H1050–H1055

Brod J, Cachovan M, Bahlman J, Bauer GE, Celsen B, Sippel R, Hundeshagen H, Feldman U, Rienhoff O (1976) Haemodynamic responses to an acute emotional stress (mental arithmetic) with special reference to the venous side. Aust NZ J Med 6 (Suppl 2):19–25

Brody JS, Stemmler EJ, Dubois AB (1968) Longitudinal distribution of vascular resistance in the pulmonary arteries, capillaries and veins. J Clin Invest 47:783–799

Brooksby GA, Donald DE (1971) Measurement of changes in blood flow and blood vlume in the splanchnic circulation. J Appl Physiol 31:930–933

Brooksby GA, Donald DE (1972) Release of blood from the splanchnic circulation in dogs. Circ Res 31:105–118

Brown E, Greenfield ADM, Goei JS, Plassaras G (1966) Filling and emptying the low-pressure blood vessels of the human forearm. J Appl Pysiol 21:573–582

Brown FK (1956) Cardiovascular effects of acutely raised intracranial pressure. Am J Physiol 185:510–514

Browse NL, Hardwick PJ (1969) The deep-breath-venoconstriction reflex. Clin Sci 37: 125–135

Browse NL, Shepherd JT (1966) Response of veins of canine limb to aortic and carotid chemoreceptor stimulation. Am J Physiol 210:1435–1441

Browse NL, Donald DE, Shepherd JT (1966a) Role of the veins in the carotid sinus reflex. Am J Physiol 210:1424–1434

Browse NL, Lorenz RR, Shepherd JT (1966b) Response of capacity and resistance vessels of dog's limb to sympathetic nerve stimulation. Am J Physiol 210:95–102

Browse NL, Shepherd JT, Donald DE (1966c) Differences in response of veins and resistance vessels in limbs to same stimulus. Am J Physiol 211:1241–1247

Brunner MH, Shoukas AA, MacAnespie CL (1981) The effect of the carotid sinus baroreceptor reflex on blood flow and volume redistribution in the total systemic vascular bed of the dog. Circ Res 48:274–285

Burch GE (1975) Psychogenic and neurogenic effects on the intact forearm vein of man. Palovian J Biol Sci 10:130–141

Burch GW, Murtadha M (1956) Study of venomotor tone in short intact venous segment of forearm of man. Am Heart J 51:807–828

Burke D, Sundlof G, Wallin BG (1977) Postural effects of muscle nerve sympathetic activity in man. J Physiol (Lond) 272:399–414

Burki N, Guz A (1970) The distensibility characteristics of the capacitance vessels of the forearm in normal subjects. Cardiovasc Res 4:93–98

Burton AC (1954) Relation of structure to function of the tissues of the wall of blood vessels. Physiol Rev 34:619–642

Caldini P, Permutt S, Waddell JA, Riley RL (1974) The effect of epinephrine on pressure, flow and volume relationships in the systemic circulation of dogs. Circ Res 34: 606–623

Calvelo MG, Abboud FM, Ballard DR, Abdel-Sayed W (1970) Reflex vascular responses to stimulation of chemoreceptors with nicotine and cyanide. Circ Res 27:259–276

Carlsten A, Folkow B, Grimby G, Homberger CA, Thulesius O (1958) Cardiovascular effects of direct stimulation of carotid sinus nerve in man. Acta Physiol Scand 44: 138–145

Carneiro JJ, Donald DE (1977a) Change in liver blood flow and blood content in dogs during direct and reflex alteration of hepatic sympathetic nerve activity. Circ Res 40:150–158

Carneiro JJ, Donald DE (1977b) Blood reservoir function of dog spleen, liver, and intestine. Am J Physiol 232:H67–H72

Carswell F, Hainsworth R, Ledsome JR (1970) The effects of distension of the pulmonary vein-atrial junctions upon peripheral vascular resistance. J Physiol (Lond) 207:1–14

Chien S (1963) Cell volume, plasma volume and cell percentage in splanchnic circulation of splenectomized dogs. Circ Res 12:22–28

Coleridge JCG, Coleridge HM (1984) Afferent vagal C fibre innervation of the lungs and airways and its functional significance. Rev Physiol Biochem Pharmacol 99: 1–110

Coles DR, Patterson GC (1957) The capacity and distensibility of the blood vessels in the human hand. J Physiol (Lond) 135:163–170

Coles DR, Kidd BSL, Moffat W (1957) Distensibility of blood vessels of the human calf determined by local application of subatmospheric pressures. J Appl Physiol 10:461–468

Comroe JH Jr, Mortimer L (1964) The respiratory and cardiovascular responses of temporarily separated aortic and carotid bodies to cyanide, nicotine, phenyldiguanide, and serotonin. J Pharmacol Exp Ther 146:33–41

Cousineau D, Goresby DA, Rose CP, Lee S (1985) Reflex sympathetic effects on liver vascular space and liver perfusion in dogs. Am J Physiol 248:H186–H192

Cruz JC, Grover RF, Reeves JT, Maher JT, Cymerman A, Denniston JC (1976) Sustained venoconstriction in man supplemented with CO_2 at high altitude. J Appl Physiol 40:96–100

Daly I de B, Daly M de B (1959a) The effects of stimulation of the carotid body chemoreceptors on the pulmonary vascular bed in the dog; the "vasosensory controlled perfused living animal" preparation. J Physiol (Lond) 148:201–219

Daly I de B, Daly M de B (1959b) The effects of stimulation of the carotid sinus baroreceptors on the pulmonary vascular bed in the dog. J Physiol (Lond) 148:220–226

Daly M de B, Robinson BH (1968) An analysis of the reflex systemic vasodilator response elicited by lung inflation in the dog. J Physiol (Lond) 195:387–406

Daly M de B, Scott MJ (1958) The effects of stimulation of the carotid body chemoreceptors on heart rate in the dog. J Physiol (Lond) 144:148–166

Daly M de B, Scott MJ (1962) An analysis of the primary cardiovascular reflex effects of stimulation of the carotid body chemoreceptors in the dog. J Physiol (Lond) 162:555–573

Davis DL (1963) Effect of sympathetic stimulation on dog paw volume. Am J Physiol 205:989–994

DeGeest H, Levy MN, Zieske H (1965) Reflex effects of cephalic hypoxia, hypercapnia and ischemia upon ventricular contractility. Circ Res 17:349–358

Delius W, Kellerova E (1971) Reactions of arterial and venous vessels in the human forearm and hand to deep breath or mental strain. Clin Sci 40:271–282

Delorme EJ, Macpherson AIS, Mukherjee SR, Rowlands S (1951) Measurement of the visceral blood volume in dogs. Q J Exp Physiol 36:219–231

De Pasquale NP, Burch GE (1963) Effect of angiotensin II on the intact forearm veins of man. Circ Res 8:239–245

Detry JMR, Wyss CR, Rowell LB (1974) Increased forearm vascular compliance during exercise with nitroglycerin. Acta Cardiol (Brux) 29:31–43

Diana JN, Schwinghamer J, Young S (1968) Direct effect of histamine on arterial and venous resistance in isolated dog hindlimb. Am J Physiol 214:494–505

Di Salvo J, Parker PE, Scott JB, Haddy FJ (1971) Carotid baroreceptor influence on total and segmental resistances in skin and muscle vasculatures. Am J Physiol 220: 1970–1978

Dock DS, Kraus WL, McGuire LB, Hyland JW, Haynes FW, Dexter L (1961) The pulmonary blood volume in man. J Clin Invest 40:317–328

Donald DE, Aarhus LL (1974) Active and passive release of blood from canine spleen and small intestine. Am J Physiol 227:1166–1172

Donald DE, Edis AJ (1971) Comparison of aortic and carotid baroreflexes in the dog. J Physiol (Lond) 215:521–538

Donegan JF (1921) The physiology of the veins. J Physiol (Lond) 55:226–245

Downing SE, Mitchell JH, Wallace AG (1963) Cardiovascular responses to ischemia, hypoxia and hypercapnia of the central nervous system. Am J Physiol 204:881–887

Drees JA, Rothe CF (1974) Reflex venoconstriction and capacity vessels pressure-volume relationships in dogs. Circ Res 34:360–373

Duggan JJ, Love VL, Lyons RH (1953) A study of reflex venomotor reactions in man. Circulation 7:869–872

Echt M, Lange L, Gauer OH (1974) Changes of peripheral venous tone and central transmural venous pressure during immersion in a thermo-neutral bath. Pfluegers Arch 352:211–217

Eckberg DL, Cavanaugh MS, Mark AL, Abboud FL (1975) A simpliefied neck suction device for activation of carotid baroreceptors. J Lab Clin Med 85:167–173

Eckstein JW, Horsley AW (1960) Effects of hypoxia on peripheral venous tone in man. J Lab Clin Med 56:847–853

Eckstein JW, Wendling MG, Abboud FM (1965) Forearm venous responses to stimulation of adrenergic receptors. J Clin Invest 44:1151–1159

Eckstein JW, Mark AL, Schmid PG, Iizuka T, Wendling MG (1969) Responses of capacitance vessels to physiologic stimuli. Trans Am Clin Climatol Assoc 81:57–64

Epstein SE, Beiser GD, Stampfer M, Braunwald E (1968) Role of the venous system in baroreceptor-mediated reflexes in man. J Clin Invest 47:139–152

Ernsting J, Parry DJ (1957) Some observations on the effects of stimulating the stretch receptors in the carotid artery of man. J Physiol (Lond) 137:45P

Feeley JW, Lee TD, Milnar WR (1963) Active and passive components of pulmonary vascular response to vasoactive durgs in the dog. Am J Physiol 205:1193–1199

Folkow B (1959) The efferent innervation of the cardiovascular system. Verh Dtsch Ges Herz Kreislaufforsch 25:84–96

Folkow B, Mellander S (1964) Veins and venous tone. Am Heart J 68:397–408

Folkow B, Neil E (1971) Circulation. Oxford University Press, London

Folkow B, Lundgren O, Wallentin I (1963) Studies on the relationship between flow resistance, capillary filtration coefficient and regional blood volume in the intestine of the cat. Acta Physiol Scand 57:270–283

Ford R, McGregor KH, Hainsworth R (1984) Reflex vascular responses from intracoconary veratridine in dogs. Clin Sci 67:1–2P

Ford R, Hainsworth R, McGregor KH (1985a) Abdominal vascular responses to intracoronary injections of veratridine. J Physiol (Lond) 362:39P

Ford R, Hainsworth R, Rankin AJ, Soladoye AO (1985b) Abdominal vascular responses
to changes in carbon dioxide tension in the sephalic circulation of anaesthetized
dogs. J Physiol (Lond) 358:417–431

Frye RL, Braunwald E (1960) Studies on Starling's law of the heart. I. The circulatory
response to acute hypervolemia and its modification by ganglionic blockade. J Clin
Invest 39:1043–1050

Furness JB, Marshall JM (1974) Correlation of the directly observed responses of
mesenteric vessels of the rat to nerve stimulation and noradrenaline with the distribution of adrenergic nerves. J Physiol (Lond) 239:75–88

Fuxe K, Sedvall G (1965) The distribution of adrenergic nerve fibres to the blood
vessels in skeletal muscle. Acta Physiol Scand 64:75–86

Gaehtgens P, Uekermann U (1971) The distensibility of mesenteriv microvessels.
Pfluegers Arch 330:206–216

Gauer OH, Thron HL (1962) Properties of veins in vivo: integrated effects of their
smooth muscle. Physiol Rev 42 (Suppl 5):283–303

Gauer OH, Thron HL (1965) Postural changes in the circulation. In: Hamilton WF
(ed) Circulation. American Physiological Society, Washington DC, pp 2409–
2439 (Handbook of physiology, sect 2, vol 3)

Gero J, Gerova M (1968) Sympathetic regulation of collecting vein. Experientia 24:
811–812

Gerova M, Gero J (1975) The ability of collecting veins to sustain sympathetic constriction. Physiol Bohemoslov 24:193–198

Gilbert CA, Stevens PM (1966) Forearm vascular response to lower body negative pressure and orthostasis. J Appl Physiol 21:1265–1272

Glick G, Wechsler AS, Epstein SE (1969) Reflex cardiovascular depression produced
by stimulation of pulmonary stretch receptors in the dog. J Clin Invest 48:467–
473

Green HD (1950) In: Glasser O (ed) Medical physics, vol 2. Year Book, Circulation:
Physical Principles, Chicago, p 231

Green HD, Rapela CE, Conrad MC (1963) Resistance (conductance) and capacitance
phenomena in terminal vascular beds. In: Hamilton WF (ed) Circulation. American
Physiological Society, Washington DC (Handbook of physiology, sect 2, vol 2)

Greenfield ADM, Patterson GC (1956) On capacity and distensibility of blood vessels
of the human forearm. J Physiol (Lond) 131:290–306

Greenway CV, Lautt WW (1970) Effects of hepatic venous pressure on transsinusoidal
fluid transfer in the liver of the anesthetized cat. Circ Res 26:697–703

Greenway CV, Lautt WW (1972a) Effects of adrenaline, isoprenaline and histamine
on transsinusoidal fluid filtration in the cat liver. Br J Pharmacol 44:185–191

Greenway CV, Lautt WW (1972b) Effects of infusions of catecholamines, angiotensin,
vasopressin and histamine on hepatic blood volume in the anaesthetized cat. Br J
Pharmacol 44:177–184

Greenway CV, Lister GE (1974) Capacitance effects and blood reservoir function in
the splanchnic vascular bed during non-hypotensive haemorrhage and blood volume
expansion in anaesthetized cats. J Physiol (Lond) 237:279–294

Greenway CV, Oshiro G (1972) Comparison of the effects of hepatic nerve stimulation
on arterial flow, distribution of arterial and portal flows and blood content in the
livers of anaesthetized cats and dogs. J Physiol (Lond) 227:487–501

Greenway CV, Oshiro G (1973) Effects of histamine on hepatic volume (outflow
block(in anaesthetized dogs. Br J Pharmacol 47:282–290

Greenway CV, Stark RD (1969) Vascular responses of the spleen to rapid haemorrhage
in the anaesthetized cat. J Physiol (Lond) 204:169–179

Greenway CV, Lawson AC, Stark RD (1968) Vascular responses of the spleen to nerve
stimulation during normal and reduced blood flow. J Physiol (Lond) 194:421–433

Greenway CV, Stark RD, Lautt WW (1969) Capacitance responses and fluid exchange in the cat liver during stimulation of the hepatic nerves. Circ Res 25:227–284

Greenway CV, Seaman KL, Innes IR (1985) Norepinephrine on venous compliance and unstressed volume in cat liver. Am J Physiol 248:H468–H476

Greenwood PV, Hainsworth R, Karim F, Morrison GW, Sofola OA (1977) Peripheral vascular responses from lung inflation. J Physiol 273:55–56P

Greenwood PV, Hainsworth R, Karim F, Morrison GW, Sofola OA (1980) Reflex inotropic responses of the heart from lung inflation in anaesthetized dogs. Pfluegers Arch 386:199–204

Guntheroth WG, Mullins GL (1963) Liver and spleen as venous reservoir. Am J Physiol 204:35–41

Guyton AC (1955) Determination of cardiac output by equating venous return curves with cardiac response curves. Physiol Rev 35:123–129

Guyton AC (1963) Venous return. In: Hamilton WF (ed) Circulation. American Physiological Society, Washington DC, pp 1099–1133 (Handbook of physiology, vol 2, sect 2, chap 32)

Haddy FJ, Molnar JI, Campbell RW (1961) Effects of denervation and vasoactive agents on vascular pressures and weight of dog forelimb. Am J Physiol 201:631–638

Hadjiminas J, Oberg B (1968) Effecsts of carotid baroreceptor reflexes on venous tone in skeletal muscle and intestine of the cat. Acta Physiol Scand 72:518–532

Hainsworth R (1974) Circulatory responses from lung inflation in anesthetized dogs. Am J Physiol 226:247–255

Hainsworth R, Karim F (1973) Left ventricular inotropic and peripheral vasomotor responses from independent changes in pressure in the carotid sinuses and cerebral arteries in anaesthetized dogs. J Physiol (Lond) 228:139–155

Hainsworth R, Karim F (1976) Responses of abdominal vascular capacitance in the anaesthetized dog to changes in carotid sinus pressure. J Physiol (Lond) 262:659–677

Hainsworth R, Linden RJ (1979) Reflex control of vascular capacitance. In: Guyton AC, Young DB (eds) Cardiovascular physiology III, vol 18. University Park Press, Baltimore, MD, pp 67–124

Hainsworth R, Ledsome JR, Carswell F (1970) Reflex responses from aortic baroreceptors. Am J Physiol 218:423–429

Hainsworth R, Karim F, Stoker JB (1975) The influence of aortic baroreceptors on venous tone in the perfused hind limb of the dog. J Physiol 244:337–351

Hainsworth R, Karim F, Sofola OA (1979) Left ventricular inotropic responses to stimulation of carotid body chemoreceptors in anaesthetized dogs. J Physiol (Lond) 287:455–466

Hainsworth R, Karim F, McGregor KH, Wood LM (1983a) Responses of abdominal vascular resistance and capacitance to stimulation of carotid chemoreceptors in anaesthetized dogs. J Physiol (Lond) 334:409–419

Hainsworth R, Karim F, McGregor KH, Rankin AJ (1983b) Effects of stimulation of aortic chemoreceptors on abdominal vascular resistance and capacitance in anaesthetized dogs. J Physiol (Lond) 334:421–431

Hainsworth R, Karim F, McGregor KH, Wood LM (1983c) Hind-limb vascular-capacitance responses in anaesthetized dogs. J Physiol (Lond) 337:417–428

Hainsworth R, McGregor KH, Rankin RJ, Soladoye AO (1984) Cardiac inotropic responses from changes in carbon dioxide tension in the cephalic circulation of anaesthetized dogs. J Physiol (Lond) 357:23–35

Hainsworth R, Rankin AJ, Soladoye AO (1985) Effect of sephalic carbon dioxide tension on the cardiac inotropic response to carotid chemoreceptor stimulation in dogs. J Physiol (Lond) 358:405–416

Hall JE, Schwinghamer JM, Lalone B (1976) Mechanisms of blood vessel constriction during hemorrhage. Am J Physiol 230:569–578

Hallin RG, Torebjork HE (1974) Single unit sympathetic activity in human skin nerves during rest and various manoeuvres. Acta Physiol Scand 92;303–317

Hanke D, Schlepper M, Westermann K, Witzleb E (1969) Venentonus, Haut- und Muskeldurchbluting an Unterarm und Hand bei Beinarbeit. Pfluegers Arch 309: 115–127

Hardy JD, Hellon RF, Sutherland K (1964) Temperature-sensitive neurones in the dog's hypothalamus. J Physiol 175:242–253

Hassan AAM, Hicks MN, Linden RJ, Mary DASG, Walters GE (1985) Effect of distension of the urinary bladder on efferent cardiac sympathetic nerves in the dog. J Physiol (Lond) 367:48P

Hebert MT, Marshall JM (1985) Responses of skeletal muscle microcirculation of the rat to baroreceptor stimulation. J Physiol (Lond) 364:70P

Herndon CW, Sagawa K (1969) Combined effects of aortic and right atrial pressures on aortic flow. Am J Physiol 217:65–72

Hintze A, Thron HL (1961) Das Verhalten der menschlichen Handvenen bei akuter arterieller Hypoxie. Pfluegers Arch 274:227–251

Hollenberg NK, Boreus LO (1972) The influence of the rate of filling on apparent venous distensibility in man. Can J Physiol Pharmacol 50:310–316

Hooker DR (1918) The veno-pressor mechanism. Am J Physiol 46:591–598

Howell JBL, Permutt S, Proctor DF, Riley RL (1961) Effect of inflation of the lung on different parts of the pulmonary vascular bed. J Appl Physiol 16:71–76

Iizuka T, Mark AL, Wendling MG, Schmid PG, Eckstein JW (1970) Differences in responses of saphenous and mesenteric veins to reflex stimuli. Am J Physiol 219: 1066–1070

Imai Y, Satoh K, Taira N (1978) Role of the peripheral vasculature in changes in venous return caused by isoproterenol, norepenephrine, and methoxamine in anesthetized dogs. Circ Res 43:553–561

Janssens WJ, Vanhoutte PM (1978) Instantaneous changes in alpha-adrenoceptor affinity caused by moderate cooling in canine cutaneous veins. Am J Physiol 234: H330–H337

Johns BL, Rothe CF (1978) Delayed vascular compliance and fluid exchange in the canine intestine. Am J Physiol 234:H660–H669

Johnson PC (1967) Measurement of microvascular dimensions in vivo. J Appl Physiol 23:593–596

Johnson JM, Rowell LB, Niederberger M, Eisman MM (1974) Human splanchnic and forearm vasoconstrictor responses to reductions of right atrial pressure in dogs. Circ Res 34:515–524

Johnstone FRC (1956) Measurement of splanchnic blood volume in dogs. Am J Physiol 185:450–452

Kahler RL, Goldblatt A, Braunwald E (1962) The effects of acute hypxoia on the systemic venous and arterial systems and on myocardial contractile force. J Clin Invest 41:1553–1562

Karim F, Hainsworth R (1976) Responses of abdominal vascular capacitance to stimulation of splanchnic nerves. Am J Physiol 231:434–440

Karim F, Kidd C, Malpus CM, Penna PE (1972) The effects of stimulation of the left atrial receptors on sympathetic efferent nerve activity. J Physiol (Lond) 227: 243–260

Karim F, Hainsworth R, Handey RP (1978) Reflex responses of abdominal vascular capacitance from aortic baroreceptors in dogs. Am J Physiol 235:H488–H493

Karim F, Araneda G, Hainsworth R (1980a) The influence of perfusate temperature on the responses of a superfucial vein in the carotid baroreceptor reflex in dogs. Pfluegers Arch 383:79–85

Karim F, Hainsworth R, Sofola OA, Wood LM (1980b) Responses of the heart to stimulation of aortic body chemoreceptors in dogs. Circ Res 46:77–83

Knisely MH, Harding F, Debacker H (1957) Hepatic sphincters. Science 125:1023–1026

Koizumi K, Brooks CMcC (1972) The integration of autonomic system reactions. Ergebn Physiol 67:1–68

Larochellé P, Ogilvie RI (1976) Effective vascular compliance and venous diameter in dogs. Can J Physiol Pharmacol 54:154–159

Lautt WW, Greenway CV (1972) Hepatic capacitance vessel responses to bilateral carotid occlusion in anaesthetized cats. Can J Physiol Pharmacol 50:244–247

Lautt WW, Greenway CV (1976) Hepatic venous compliance and role of liver as a blood reservoir. Am J Physiol 231:292--295

Ledsome JR, Linden RJ (1964) A reflex increase in heart rate from distension of the pulmonary vein-atrial junctions. J Physiol (Lond) 170:456–473

Ledsome JR, Linden RJ (1967) The effect of distending a pouch of the left atrium on heart rate. J Physiol 193:121–129

Lee KD, Mayou RA, Torrance RW (1964) The effect of blood pressure upon chemoreceptor discharge to hypoxia, and the modification of this effect by the sympathetic adrenal system. Q J Exp Physiol 49:171–183

Lesh TA, Rothe CF (1969) Sympahtetic and hemodynamic effects on capacitance vessels in dog skeletal muscle. Am J Physiol 217:819–827

Linden RJ, Kappagoda CT (1982) Atrial receptors. Cambridge University Press, Cambridge

Linden RJ, Mary DASG, Weatherill D (1982) The response in efferent cardiac sympathetic nerves to stimulation of atrial receptors, carotid sinus baroreceptors and carotid chemoreceptors. Q J Exp Physiol 67:151–163

Lutz J, Peiper U, Segarra-Domerech J, Bauereisen E (1967) Das Druck-Volumendiagramm und Elastizitätswerte des gesamten Lebergefäß-Systems der Katze in situ. Pfluegers Arch 295:315–327

Malliani A (1979) Afferent cardiovascular sympathetic nerve fibres and their function in the neural regulation of the circulation. In: Hainsworth R, Kidd C, Linden RJ (eds) Cardiac receptors. Cambridge University Press, Cambridge, pp 319–354

Manchanda SK, Bhattarai R (1974) Central nervous control of venous tone. I. Effect of sympathetic chain stimulation on cutaneous capacitance and resistance vessels. Indian J Physiol Pharmacol 18:3–13

Manchanda SK, Bhattarai R, Nayar U (1975) Central nervous control of venous tone. III. Responses of capacitance and resistance vessels of skin to bulbar and hypothalamic stimulation. Indian J Physiol Pharmacol 19:105–120

Mancia G (1975) Influence of carotid baroreceptors on vascular responses to carotid chemoreceptor stimulation in the dog. Circ Res 36:270–276

Mark AL, Eckstein JW (1968) Venomotor tone and central venous pressure. Med Clin North Am 52:1077–1090

Maron MB, Dawson CA (1980) Pulmonary venoconstriction caused by elevated cerebrospinal fluid pressure in the dog. J Appl Physiol 49:73–78

Martin DA, White KL, Vernon CR (1959) Influence of emotional and physical stimuli on pressure in the isolated vein segment. Circ Res 7:580–587

Mason DR, Braunwald E (1964) Studies on digitalis. Effects of ouabain on forearm vascular resistance and venous tone in normal subjects and in patients with heart failure. J Clin Invest 43:532–543

Mellander S (1960) Comparative studies on the adrenergic neuro-hormonal control of resistance and capacitance blood vessels in the cat. Acta Physiol Scand 50 (Suppl 176):1–86

Mellander S, Johansson B (1968) Control of resistance, exchange and capacitance functions in the peripheral circulation. Pharmacol Rev 20:117–196

Menninger RP, Baker CH (1976) Hypothalamic and brachial nerve effects on circulation of isolated canine fore-limb. Am J Physiol 230:797–803

Merritt FL, Weissler AM (1959) Reflex venomotor alterations during exercise and hyperventilation. Am Heart J 58:382–387

Milner WR (1980) Pulmonary circulation. In: Mountcastle VB (ed) Medical physiology, vol 2. Mosby, St Louis, pp 1108–1117

Mitzner W, Goldberg H (1975) Effects of epinephrine on resistive and compliant properties of the canine vasculature. J Appl Physiol 39:272–280

Morris TW, Abbrecht PH, Leverett SD Jr (1974) Diameter-pressure relationships in the unexposed femoral vein. Am J Physiol 227:782–788

Mowassaghi A, Westermann KW, Witzler E (1969) Venomotorische Reaktionen bei Veränderungen des intrapulmonalen Druckes durch Über- und Unterdruckatmung. Pfluegers Arch 305:340–350

Muller-Ruchholtz ER, Losch HM, Gund E, Lochner W (1977a) Effect of alpha adrenergic receptor stimulation on integrated systemic venous bed. Pfluegers Arch 370: 241–246

Muller-Ruchholtz ER, Losch HM, Gund E Lochner W (1977b) Effect of beta adrenergic receptor stimulation on integrated systemic venous bed. Pfluegers Arch 370: 247–251

Muller-Ruchholtz ER, Grund E, Hauer F, Lapp ER (1979) Effect of carotid pressoreceptor stimulation on integrated systemic venous bed. Basic Res Cardiol 74: 467–476

Nachev CH, Collier J, Robinson B (1971) Simplified method for measuring compliance of superficial veins. Cardiovasc Res 5:147–156

Oberg B (1964) Effects of cardiovascular reflexes on net capillary fluid transfer. Acta Physiol Scand 62 (Suppl 229):1–98

Oberg B (1967) The relationship between active constriction and passive recoil of the veins at various distending pressures. Acta Physiol Scand 71:233–247

Opdyke DF (1970) Hemodynamics of blood flow through the spleen. Am J Physiol 219:102–106

Opdyke DF, Ward CJ (1973) Spleen as an experimental model for the study of vascular capacitance. Am J Physiol 225:1416–1420

Paessler H, Schlepper M, Westermann KW, Witzleb E (1968) Venentonusreaktionen in kapazitiven Hautgefäßen bei passiver und aktiver Orthostase. Pfluegers Arch 302: 315–332

Page EG, Hickson JB, Sieker HO, McIntosh HE, Pryor WW (1955) Reflex venomotor activity in normal persons and in patients with postural hypotension. Circulation 11:262–270

Paintal AS (1973) Vagal sensory receptors and their reflex effects. Physiol Rev 53: 159:227

Pelletier CL (1972) Circulatory responses to graded stimulation of the carotid chemoreceptors in the dog. Circ Res 31:431–443

Pelletier CL, Shepherd JT (1972) Venous responses to stimulation of carotid chemoreceptors by hypoxia and hypercapnia. Am J Physiol 223:97–103

Pelletier CL, Clement DL, Shepherd JT (1972) Comparison of afferent activity in canine aortic and sinus nerves. Circ Res 31:557–568

Permutt S, Howell JBL, Proctor DF, Riley RL (1961) Effect of lung inflation on static volume characteristics of pulmonary vessels. J Appl Physiol 16:64–70

Price HL, Deutsch S, Marshall BE, Stephen GW, Behar MG, Neufeld GR (1966) Hemodynamic and metabolic effects of hemorrhage in man, with particular reference to the splanchnic circulation. Circ Res 18:469–474

Quiroz AC, Burch GE, Malaret GE (1960) Spontaneous variations in venous tone in man. J Appl Physiol 15:255–257

Rice AJ, Leeson CR, Long JP (1966) Localization of venoconstrictor responses. J Pharmacol Exp Ther 154:539–545

Richardson TQ, Fermoso JD (1964) Elevation of mean circulatory pressure in dogs with cerebral ischemia-induced hypertension. J Appl Physiol 19:1133–1134

Robinson BF, Wilson AG (1968) Effect on forearm arteries and veins of attenuation of the cardiac response to leg exercise. Clin Sci 35:143–152

Robinson BF, Epstein SE, Kahler RL, Braunwald E (1966) Circulatory effects of acute expansion of blood volume. Circ Res 19:26–32

Rodbard S, Stone W (1956) Pressor mechanisms induced by intracranial compression. Circulation 12:883–890

Roddie IC, Shepherd JT (1958) Receptors in the high pressure and low pressure vascular systems. Their role in the reflex control of the human circulation. Lancet 1: 493–496

Rose JC, Kot PA, Coh JN, Freis ED, Eckert GE (1962) Comparison of effects of angiotensin and norepinephrine on pulmonary circulation, systemic arteries and veins, and systemic vascular capacity in the dog. Circulation 25:247–252

Ross J Jr, Frahm CJ, Braunwald W (1961a) Influence of carotid baroreceptors and vasoactive drugs on systemic vascular volume and venous distensibility. Circ Res 9: 75–82

Ross J Jr, Frahm CJ, Braunwald E (1961b) The influence of intracardiac baroceptors on venous return, systemic vascular volume and peripheral resistance. J Clin Invest 40:563–572

Rothe CF (1976) Reflex vascular capacity reduction in the dog. Circ Res 39:705–710

Rothe CF (1979) Reflex control of the veins in cardiovascular function. Physiologist 22(1):28–35

Rothe CF (1983a) The venous system. The physiology of the capacitance vessels. In: Shepherd JT, Abboud FM (eds) The cardiovascular system. Peripheral circulation and organ blood flow. American Physiological Society, Bethesda, pp 397–452 (Handbook of physiology, sect 2, vol 3)

Rothe CF (1983b) Reflex control of veins and vascular capacitance. Physiol Rev 63:1281–1342

Rothe CF (1983c) Measurement of circulatory capacitance and resistance. In: Linden RJ (ed) Techniques in cardiovascular physiology. Techniques in life sciences. Elsevier, Ireland Shannon, pp P306/1-P306/28

Rothe CF, Drees JA (1976) Vascular capacitance and fluid shifts in dogs during prolonged hemorrhagic hypotension. Circ Res 38:347–356

Rothe CF, Johns BL, Bennett TD (1978) Vascular capacitance of dog intestine using mean transit time of indicator. Am J Physiol 234:H7–H13

Rowell LB, Detry JMR, Blackmon JR, Wyss C (1972) Importance of the splanchnic vascular bed in human blood pressure regulation. J Appl Physiol 32:213–220

Rutlen DL, Supple EN, Powell WJ Jr (1981) Beta-adrenergic regulation of total systemic intravascular volume in the dog. Circ Res 48:112–120

Sagawa K, Ross JM, Guyton AC (1961) Quantitation of cerebral ischemic pressor response in dogs. Am J Physiol 200:1164–1168

Salisbury PF, Galletti PM, Lewin RJ, Rieben PA (1959) Stretch reflexes from the dog's lung to the systemic circulation. Circ Res 7:62–67

Salisbury PF, Cross CE, Rieben PA (1960) Reflex effects of left ventricular distension. Circ Res 8:530–534

Saltzman EW (1957) Reflex peripheral venoconstriction induced by carotid occlusion. Circ Res 5:149–152

Samar RE, Coleman TG (1978) Measurement of mean circulatory filling pressure and vascular capacitance in the rat. Am J Physiol 234:H94–H100

Samoilenko AV, Tkachenko BI (1971) Changes in systemic vascular resistance and capacitance elicited by some systemic reflexes. Cor Vasa 13:136–146

Sampson SR, Hainsworth R (1972) Responses of aortic body chemoreceptors of the cat to physiological stimuli. Am J Physiol 222:953–958

Samueloff SL, Chinyanga (1968) Effect of venous distension on neurogenic venoconstriction in dog's limb. Cardiovasc Res 2:187–192

Samueloff SL, Bevegård BS, Shepherd JT (1966) Temporary arrest of the circulation to a limb for the study of venomotor reactions in man. J Appl Physiol 21:341–346

Seaman RG, Wiley RL, Zechman FW, Goldey JA (1973) Venous reactivity during static exercise (handgrip) in man. J Appl Physiol 35:858–860

Selkurt EE, Johnson PC (1958) Effect of acute elevation of portal venous pressure on mesenteric blood volumes, interstitial fluid volumes and hemodynamics. Circ Res 6:592–599

Shadle OW, Zukof M, Diana J (1958) Translocation of blood from the isolated dog's hindlimb during levarterenol infusion and sciatic nerve stimulation. Circ Res 6: 326–333

Sharpey-Schafer EP (1961) Venous tone. Br Med J 2:1589–1595

Sharpey-Schafer EP (1963) Venous tone: effects of reflex changes, humoral agents and exercise. Br Med Bull 19:145–148

Shepherd JT, Vanhoutte P (1975) Veins and their control. Saunders, Philadelphia

Shoukas AA (1975) Pressure-flow and pressure-volume relations in the active pulmonary vascular bed of the dog determined by two-port analysis. Circ Res 37: 809–818

Shoukas AA (1982) Carotid sinus baroreceptor reflex control and epinephrine. Circ Res 51:95–101

Shoukas AA, Brunner MC (1980) Epinephrine and the carotid sinus baroreceptor reflex: influence on capacitive and resistive properties of the total systemic vascular bed of the dog. Circ Res 47:249–257

Shoukas AA, Sagawa K (1971) Total systemic compliance measured as incremental volume-pressure ratio. Circ Res 28:277–289

Shoukas AA, Sagawa K (1973) Control of total systemic vascular capacity by the carotid sinus baroreceptor reflex. Circ Res 33:22–23

Shoukas AA, MacAnespie CL, Brunner MH, Watermeier L (1981) The importance of the spleen in blood volume shifts of the systemic vascular bed caused by the carotid sinus baroreceptor reflex in the dog. Circ Res 49:759–766

Smith EE, Crowell JW (1967) Influence of hypoxia on mean circulatory pressure and cardiac output. Am J Physiol 212:1067–1069

Smyth HS, Sleight P, Pickering GW (1969) Reflex regulation of arterial pressure during sleep in man. Circ Res 24:109–121

Soladoye AO, Rankin AJ, Hainsworth R (1985) Influence of carbon dioxide tension in the cephalic circulation on hind-limb vascular resistance in anaesthetized dogs. Q J Exp Physiol 70:

Stark RD (1968) Conductance or resistance. Nature 217:513–531

Stein PM, MacAnespie CL, Rothe CF (1983) Total body vascular capacitance changes during high intracranial pressure in dogs. Am J Physiol 245:H947–H956

Thoren P (1979) Reflex effects of left ventricular mechanoreceptors with afferent fibres in the vagal nerve. In: Hainsworth R, Kidd C, Linden RJ (eds) Cardiac receptors. Cambridge University Press, London, pp 259–278

Tkachenko BI, Vinogradova MI, Makovskaja VA (1976) Relation of efferent impulse activity in splenic nerve to reflexly induced reactions of resistance and capacitance vessels of spleen. Experientia 32:1012–1014

Tkachenko BI, Vinogradova MI, Makovskaja VA (1978) A correlation of responses of the resistance and capacitance vessels of the intestine and kindey to changes of impulse in postganglionic nerves under pressor reflexes. Experientia 34:1298–1299

Tripathi A, Shi X, Wenger CB, Nadel ER (1984) Effect of temperature and baroreceptor stimulation on reflex venomotor responses. J Appl Physiol 57:1384–1392

Trippodo NC (1981) Total circulatory capacity in the rat. Effects of epinephrine and vasopressin on compliance and unstressed volume. Circ Res 49:923–931

Vanhoutte P, Leusen I (1969) The reactivity of isolated venous preparations to electrical stimulation. Pfluegers Arch 306:341–353

Vanhoutte PM, Lorenz RR (1970) Effect of temperature on reactivity of saphenous, mesenteric, and femoral veins of the dog. Am J Physiol 218:1746–1750

Vanhoutte PM, Shepherd JT (1970) Effect of temperature on reactivity of isolated cutaneous veins of the dog. Am J Physiol 218:187–190

Vatner SF, Boettcher DH (1978) Regulation of carciac output by stroke volume and heart rate in conscious dogs. Circ Res 42:557–561

Verrier RC, O'Neil TJ, Lefer AM (1969) Functional capacity of resistance and capacitance vessels in adrenal insufficiency. Am J Physiol 217:341–347

Wade OL, Combes B, Childs AW, Wheeler HO, Cownard A, Bradley SE (1956) The effect of exercise on the splanchnic blood flow and splanchnic blood volume in normal man. Clin Sci 15:451–463

Wallin BG, Sundlof G, Delius W (1975) The effect of carotid sinus nerve stimulation on muscle and skin nerve sympathetic activity in man. Pfluegers Arch 358:101–110

Watson WE (1962) Distensibility of the capacity blood vessels of the human hand during sleep. J Physiol (Lond) 161:392–398

Webb-Peploe MM (1969) The isovolumetric spleen: an index of reflex changes in splanchnic vascular capacity. Am J Physiol 216:407–413

Webb-Peploe MM, Shepherd JT (1968a) Response of large hindlimb veins of the dog to sympathetic nerve stimulation. Am J Physiol 215:299–307

Webb-Peploe MM, Shepherd JT (1968b) Responses of the supervicial limb veins of the dog to changes in temperature. Circ Res 22:737–746

Webb-Peploe MM, Shepherd JT (1968c) Responses of dogs' cutaneous veins to local and central temperature changes. Circ Res 23:693–699

Webb-Peploe MM, Shepherd JT (1968d) Peripheral mechanism involved in response of dogs' cutaneous veins to local temperature change. Circ Res 23:701–708

Webb-Peploe MM, Shepherd JT (1969) Beta-receptor mechanisms in the superficial limb veins of the dog. J Clin Invest 48:1328–1335

Wiedeman MP (1959) Responses of subcutaneous vessels to venous distension. Circ Res 7:238–242

Weil JV, Byrne-Quinn E, Battock DJ, Grover RF, Chidsey CA (1971) Forearm circulation in man at high altitude. Clin Sci 40:235–246

Weil MH, Shubin H (1970) Change in venous capacitance during cardiogenic shock – a search for the third dimension. Am J Cardiol 26:613–614

Wenger CB, Roberts MF (1980) Control of forearm venous volume during exercise and body heating. J Appl Phyysiol 48:114–119

Whitney RJ (1953) The measurement of volume changes in human limbs. J Physiol (Lond) 121:1–27

Widdicombe JG (1973) Reflex control of breathing. In: Widdicombe JG (ed) Respiratory physiology. University Park Press, Baltimore, pp 273:302 (MTP Internatiol review of science, vol 2)

Wood JE, Eckstein JW (1958) A tandem forearm plethysmograph for study of acute responses to the peripheral veins of men: the effect of environmental and local temperature change and the effect of pooling blood in the extremities. J Clin Invest 37:41–50

Wood JE, Roy SB (1970) The relationship of peripheral venomotor responses to high altitude pulmonary edema in man. Am J Med Sci 259:56–65

Wood LM, Hainsworth R, McGregor KH (1985) Effects of lung inflation on abdominal vascular resistance in anaesthetized dogs. Q J Exp Physiol 70:

Yamamoto J, Trippodo NC, Ishise S, Frohlich ED (1980) Total vascular pressure-volume relationship in the conscious rat. Am J Physiol 238:H823–H828

Zellis R, Mason DR (1969) Comparison of the reflex reactivity of skin and muscle veins in the human forearm. J Clin Invest 48:1870–1877

Zingher D, Grodins FS (1964) Effect of carotid baroreceptor stimulation upon the forelimb vascular bed of the dog. Circ Res 14:392–399

Zitnik RS, Lorenz RR (1969) Sensitivity of methods for detection of active changes in venous wall tension. Am J Cardiol 24:220–223

Zitnik RS, Ambrosioni E, Shepherd JT (1971) Effect of temperature on cutaneous venomotor reflexes in man. J Appl Physiol 31:507–512

Zoller RP, Mark AL, Abboud FM, Schmid PG, Heistad DD (1972) The role of low pressure baroreceptors in reflex vasoconstrictor responses in man. J Clin Invest 51:2967–2972

Zoster R, Tom H (1967) Adrenoceptive sites in the veins. Br J Pharmacol Chemother 31:407–419

Rev. Physiol. Biochem. Pharmacol., Vol. 105
© by Springer-Verlag 1986

Electrical Breakdown, Electropermeabilization and Electrofusion

ULRICH ZIMMERMANN

Contents

1 Introduction. 176

2 Electropermeabilization 177
    2.1   Overcoming the Radius Dependence of Membrane Potential 188
    2.2   Electroinjection of Foreign Molecules into Cells 190

3 Fusion with Dielectrophoresis 201

4 Fusion without Dielectrophoresis 215

5 Electrofusion 219
    5.1   Planar Lipid Bilayer Membranes 220
    5.2   Liposomes 221
    5.3   Bacteria. 222
    5.4   Fungi 224
    5.5   Yeast 224
    5.6   Plant Protoplasts 226
    5.7   Animal Cells 231
    5.7.1 Hybridoma. 234
    5.7.2 Blastomeres and Eggs 236

6 Electrofusion as a Tool in Membrane Research 237

7 The Possible Role of Electrofusion in Evolution 239

8 Conclusions 242

Appendix: Mechanism of Electrical Breakdown. 243

References 250

Lehrstuhl für Biotechnologie der Universität Würzburg, Röntgenring 11, 8700 Würzburg, Federal Republic of Germany

This work was supported by grants from the BMFT, the DFVLR, and the Deutsche Forschungsgemeinschaft (SFB 176)

1 Introduction

In vitro cell-to-cell fusion has proved to be a valuable tool in membrane research, genetic mapping and, in particular, genetic engineering (1–6). Fusion mediated by chemicals and inactivated virus has led to new strains of bacteria (7, 8), yeasts and fungi (9, 10), hybrid plants (11, 12), and hybrid mammalian cell lines (13, 14). Although cell fusion was introduced as early as 1909 by Küster (15), the standard conventional techniques still have many, and sometimes severe, limitations. Chemically induced and virus-induced fusion is therefore still more of an art than a real science.

To overcome the shortcomings of the standard techniques, Zimmermann and co-workers developed an electrical fusion technique (16–28). In electrofusion the cells are brought into tight membrane contact in an inhomogeneous field, and the cell pairs or multiples are then fused by short field pulses of high intensity. This results in a controlled increase of the permeability in localized zones of the membrane (so-called reversible electrical breakdown, electropermeabilization or electroporation) (18, 29, 30). Electrofusion provides selectivity for desirable fusion products, efficient control of the fusion process, prediction of the fusion conditions, and high yields of viable hybrids.

The results obtained so far with electrofusion have been reviewed a number of times from different aspects (19–28). However, the author believes that rapid development in this field in the last two years justifies a new critical review of this method and its potential applications. The need for a critical discussion of recently published results was recognised when it became clear that a certain lack of understanding of the physical processes underlying electrofusion was creating apparent confusion rather than new insights.

An understanding of these processes is both an important prerequisite for the application of this universal fusion method and a guide to the kind of electronic components required for electrofusion. These of course make a significant contribution to the successful electrical production of viable hybrids.

This review also aims to point out that electropermeabilization is a valuable tool for sequestering such foreign substances as drugs, proteins, and DNA into cells without causing damage or deterioration of cellular functions (electroinjection). Although this technique has been known for more than a decade (29–34) it has only recently aroused increasing interest among biologists and medical researchers.

2 Electropermeabilization

The key step in the standard and modified electrofusion techniques of attached cells, or in DNA-transfection of free suspended cells is the reversible electrical breakdown of the cell membrane (18, 19, 29, 30, 35–37). Electrical breakdown of the cell membrane is achieved when the cell, cell pairs, or cell multiples are exposed to a field pulse of high intensity (kV/cm range) and short duration (ns to μs range) (Fig. 1). The membrane capacitor is charged by the external electrical field due to charge movement, so that a corresponding membrane potential is induced and this is superimposed on the intrinsic membrane potential. The transformation of external fields into in situ fields and membrane potentials is described by Schwan (38); a rigorous mathematical treatment has been published by Jeltsch and Zimmermann (39).

If the total membrane potential exceeds a critical value of about 1 V, the phenomenon of reversible electrical breakdown – which is known from solid state physics – occurs. The electrical breakdown of the membrane capacitor causes greater or lesser local perturbations of the membrane structure, which in turn increase the permeability of the membrane.

For the biological applications considered in this review, it is of particular significance that these field-induced perturbations of the membrane structure are reversible – particularly at higher temperatures (37°C) – (analogous to the self-healing technical capacitors), provided both that the field strength was not too high and that the duration of the external field pulse was short enough (nanoseconds to a couple of microseconds). Otherwise permeabilization of the membrane in response to the field pulse is irreversible (18, 37) (Fig. 1). Zimmermann et al. (30) therefore termed this effect "reversible electrical breakdown," in order to distinguish it from irreversible damage to the membrane in response to field pulses of longer duration. For a more detailed discussion the reader is referred to the extensive literature on this particular subject matter (18–20, 23, 24).

The term "electrical breakdown" refers to the primary event which then leads to secondary changes in the membrane properties and subsequently – owing to osmotic processes – to changes in the cell interior (18). This term reveals nothing about the molecular mechanism leading to electrical breakdown, which still has to be elucidated. A number of models have been proposed (35, 37, 40–65), among them one which postulates that electrical breakdown leads to local instabilities in the membrane by way of electromechanical compression and electrical field-induced tension, so that pores are able to form (35, 37, 40–50, 64, 65). These fill up with conducting intracellular and extracellular solutions, thus

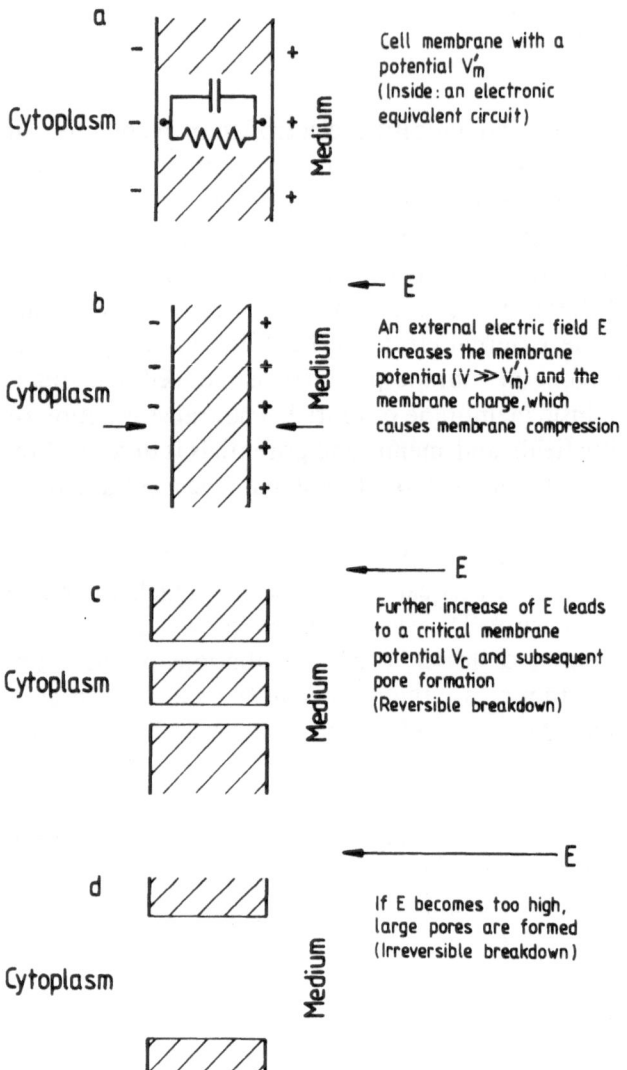

Fig. 1a–d. Schematic diagram of reversible and irreversible breakdown.

a The membrane can be considered as a capacitor filled with a dielectric (*hatched area*). The normal resting potential difference across the membrane, V'_m, is about 10 mV.

b Exposure of the membrane (cell) to an external electrical field (pulse) leads to the build-up of a membrane potential difference V due to charge separation across the membrane. V is proportional to the field strength E and to the radius of the cell. The increase in the membrane potential leads to a reduction in the membrane thickness.

c Breakdown of the membrane occurs if the critical breakdown voltage V_c (of the order of 1 V) is reached by a further increase in the external field strength. It is assumed that breakdown causes the formation of transmembrane pores (filled with conductive solution), which leads to an immediate discharge of the membrane and, in turn, to a decompression of the membrane. Breakdown is reversible if the product of number and size of the pores is small in relation to the total membrane surface.

d At supracritical field strengths (but also at longer exposure times) larger and larger areas of the membrane are subjected to breakdown. If the size and number of pores become large in relation to the total membrane surface, reversible breakdown turns into irreversible breakdown associated with mechanical destruction of the cell

Fig. 2. Schematic diagram of a cell exposed to high electrical field strengths. The membrane potential V_m which is built up across the membrane in response to the external field, is at its highest value at membrane sites oriented in field direction (poles) and progressively decreases towards the equator (perpendicular to the field direction). For membrane sites in perpendicular orientation to the field direction, the potential V_m is zero. The breakdown voltage V_c is therefore first reached in field direction, $E = E_c$. It is only reached in membrane sites orientated at a certain angle ϑ to the field direction if supracritical field strengths $(E \gg E_c)$ are applied. Breakdown of the membrane is indicated by the formation of transmembrane pores (see Fig. 1)

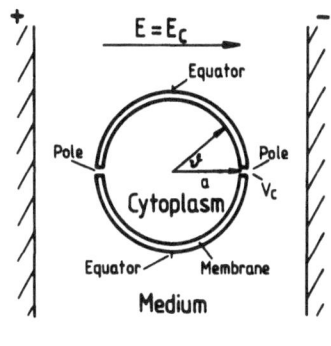

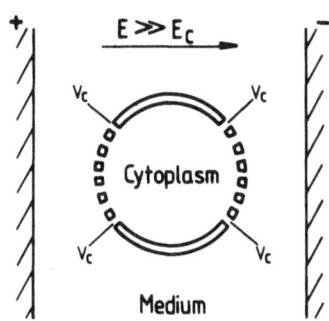

$$V_c = 1.5 \cdot a \cdot E_c \cdot \cos \vartheta$$

causing the electrical conductivity and permeability of the membrane to increase (Figs. 1 and 2). Although the pore model is certainly a very lucid one, we have to point out that the molecular interpretation of electrical breakdown in terms of pores is not proven. On the contrary, evidence provided by Zimmermann and co-workers (22, 66, 67), that organelles (Fig. 3) and whole cells (Figs. 4 and 5) can be sequestered by electrical means across the membrane of the host cell under appropriate field conditions, seems to rule out the formation at least of stable pores (see below). Likewise, the observation by Vienken et al. (68) of electrofusion of two myeloma cells on pronase/polylysine-treated microslides cannot be reconciled with the idea of simple stable pores. These authors observed that after the cells had been aligned on the surface of the microslide with the aid of an alternating field and subsequent application of field pulses, they first moved away from each other before approaching each other again after about one minute and fusing. On the basis of his experiments where he changed the sequence in which the square pulse and the alternating field were applied, Sowers (69) also came to the conclusion that pore formation was not involved. He concluded from

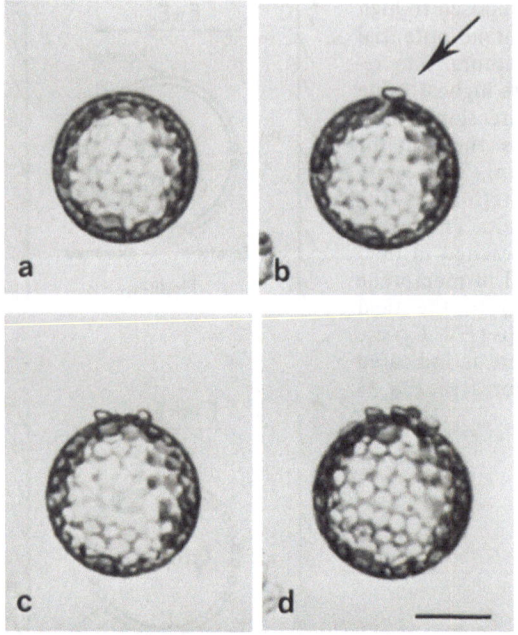

Fig. 3a—d. Field-induced release of chloroplasts (*arrow*) from a mesophyll cell protoplast of *Avena sativa*. **a** Before field application; **b** after injection of a field pulse of 4 kV/cm and 40-μs duration; **c, d** after the injection of a second and a third pulse, respectively, of the same strength and duration. The time interval between subsequent pulses was about 5 s. *Bar* in **d**, 10 μm. From (66)

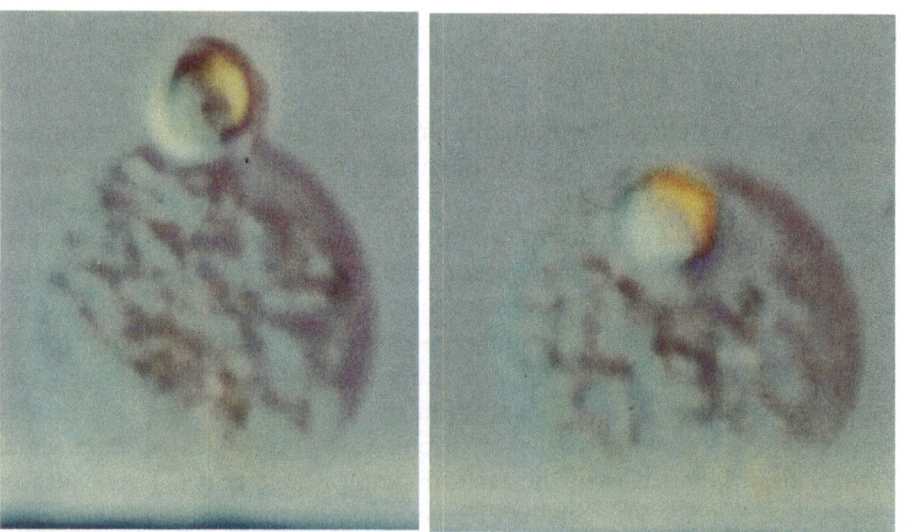

Fig. 4a,b. Field-induced uptake of a mouse lymphocyte cell through the membrane of a Friend cell (virus-induced leukaemic cell). The lymphocyte cell was placed on top of the Friend cell (**a**) using the arrangement for dielectrophoresis (see Fig. 19) and an alternating field of 1 MHz and 200 V/cm. Uptake and incorporation of the lymphocyte cell occur after application of a field pulse of 3.5 kV/cm and 20 μs duration (**b**). The cells were suspended in an isotonic mannitol solution. Photograph **b** was taken 30 s after field application. Magnification × 1000. From (67)

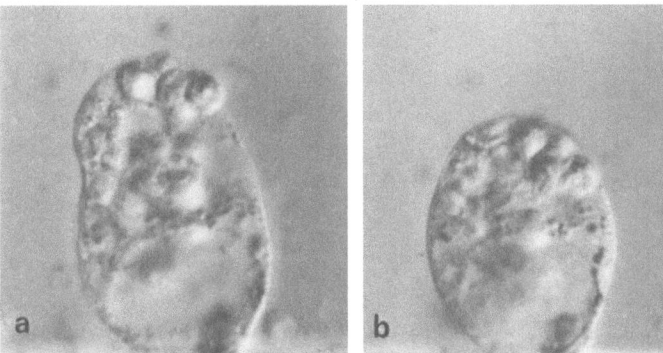

Fig. 5a,b. Electrical field induced uptake ("Endocytosis") of 11 Friend cells into a *Petunia* protoplast **a** 2 min and **b** 5 min after application of a field pulse of 2 kV/cm strength and 20 μs duration. Interference phase contrast micrographs; ⊢——⊣ 20 μm. From (200)

studies of different pulse forms that more complicated mechanisms in the membrane must come into play as a result of the field.

The fusion experiments of Lucy and co-workers [(70) see also (71, 72)] and electrical breakdown studies on red blood cells (73, 74) suggest that – at least for red blood cells – secondary osmotic processes are involved in fusion and uptake (see below). Osmotic processes may also well account for the observations of Vienken et al. (68) and Sowers (69).

The term "electroporation", introduced by Neumann et al. (75) to the term "reversible electrical breakdown", is very appealing, but may not be totally accurate under the circumstances. In fact, it may well be misleading since other permeability changes in the membrane structure elicited by the breakdown pulse are not taken into account. This could lead to completely erroneous interpretations of the events involved in electrofusion and electrotransfection and hence to erroneous experimental concepts. In particular, studies by Deuticke and co-workers (76, 77) have shown that electrical breakdown reduces the asymmetry of the biological membrane by an increased transbilayer mobility rate of the lipid molecules. Reduced asymmetry leads to increased permeability. The input of electrical energy from the field pulse corresponds roughly with the activation energy for the flip-flop (Zimmermann, unpublished data). However, as pointed out in the Appendix the flip-flop process does not necessarily contradict the pore model. It is also conceivable that permeability changes are brought about by phase transitions in the lipid phase of the membrane (62).

The particular problem of the mechanism of electrical breakdown is considered in more detail in the Appendix. The information provided there should serve to further optimise electrofusion and electrotransfec-

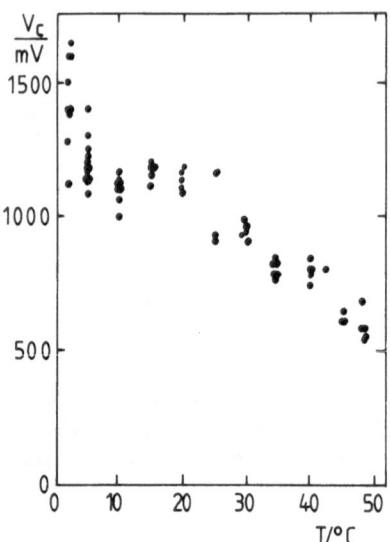

Fig. 6. Breakdown voltage V_C as a function of the temperature T. V_0 is defined as the maximum voltage to which a planar lipid membrane made up of oxidized cholesterol/n-decane can be charged by a charge pulse of 400 ns duration. For further details see (37)

tion , and the interest reader is referred to this section for detailed information. In any case, on the basis of these considerations, I would regard the use of the term electroporation as inappropriate – at least for the time being. The terms electropermeabilization or electroinjection, on the other hand, are quite acceptable if we place more emphasis on the consequence of reversible electrical breakdown. The increased permeability of the membrane is, after all, the significant event as far as biological applications are concerned.

Like reversible electrical breakdown, these terms give no indication of the type of molecular changes induced in the membrane by the critical field pulse.

The breakdown voltage is in the order of 1 V, corresponding to a field pulse of kilovolt per centimetre intensity. The breakdown voltage is not dependent on the pH of the solution, but is influenced by some other membrane and system parameters [e.g., (19, 35, 37, 78)], particularly temperature (37, 40, 79) (Fig. 6). Direct breakdown measurements on planar lipid bilayer membranes (37) and on giant algal cells of *Valonia utricularis* (40, 79) have shown that the breakdown voltage of a unit lipid-protein membrane is about 2 V at 4°C, 1 V at 20°C (room temperature) and about 0.5 V at 30°C–40°C. Fusion experiments are usually carried out at room temperature, so that a value of about 1 V can be assumed when calculating the critical field strength with the integrated Laplace equation (Eq. 1). DNA transfection by electrical field pulses, on the other hand, has to be performed at 4°C because otherwise the life span of the field-induced perturbations (pores) in the membrane would be too short to sequester sufficient DNA or other macromolecules across the membrane [e.g. (18, 30, 33, 80, 81)]. At low temperatures it is therefore necessary to apply pulses of higher field strength in order to induce electrical breakdown. Potter et al. (82) applied field strengths of 4–8 kV/ cm at 4°C for DNA transfection and were therefore apparently unable to achieve the breakdown voltage for most cell volumes. This accounts for the relatively low yield quoted by these authors. We will be returning to this point later on.

The critical field strength for electrofusion and DNA transfection is generally calculated for spherical cells according to Eq. 1 (Fig. 2):

$$V_c = 1.5 \ E_c \cdot a \cdot \cos\vartheta \tag{1}$$

where V_c = breakdown voltage; a = cell radius; E_c = critical field strength; and ϑ = angle between a given membrane site and field direction.

Equation 1 is derived for certain boundary conditions and is only valid for the stationary case, i.e., the external field has to be applied until the corresponding potential has been established across the membrane. The time required for this depends on the relaxation time of the membrane (20):

$$1/\tau = \frac{1}{R_m \cdot C_m} + \frac{1}{a \cdot C_m \ (\rho_i + 0.5 \ \rho_e)} \tag{2}$$

where R_m = specific membrane resistance ($\Omega \ cm^2$); C_m = specific membrane capacitance ($\mu F \ cm^{-2}$); ρ_i = internal specific resistance ($\Omega \ cm$); and ρ_e = external specific resistance ($\Omega \ cm$).

Equation 2 demonstrates that the relaxation time of a spherical cell consists of two terms, an RC term and a term determined by the intracellular and extracellular conductivity.

When cells are field-treated in conductive solutions (e.g., loading of cells with such substances as proteins and DNA), the RC term is usually negligible compared with the second term. In nonconductive or weakly conductive solutions, on the other hand, such as those used in standard electrofusion, the RC term is in the same order of magnitude as the conductivity term and must therefore be taken into account in any calculations. For a cell diameter of a few micrometres the relaxation time for the charging process of the membrane is about 500 ns or less in conductive solutions, so that we can assume that the equilibrium potential is reached with pulse durations of a few microseconds. In weakly conductive solutions, on the other hand, the relaxation time is in the order of a few microseconds. On the basis of our experience we know that electropermeabilization and electrofusion generally require pulse durations of less than 20 μs. Since the equilibrium potential is reached after roughly five times the relaxation time, the final value for the membrane potential is not always achieved for a given field strength under these conditions, so that the actual value of the membrane voltage is lower than the one calculated with Eq. 1. Given the different relaxation times in conductive and nonconductive solutions, it is also possible to explain why pulses up to 40 μs in duration can be applied in standard electrofusion in weakly conductive solutions, whereas similar exposure times in DNA transfection in conductive solutions lead to irreversible changes. Erythrocytes represent an exception here since they have no nucleus [e.g. (17)]

A closer inspection of Eq. 2 shows that the following approximation can be used to calculate the relaxation time for electrofusion in weakly conductive solutions with a conductivity in say the millisiemens per centimeter range (because the term RC can be neglected under these conditions). This is because the internal specific resistance is generally in the order of 100 Ω cm and thus about 2 orders of magnitude lower than the external specific resistance:

$$\tau = a \cdot C_m (\rho_i + 0.5\, \rho_e) \tag{3a}$$

and for $\rho_i \ll \rho_e$

$$\tau = 0.5\, a \cdot C_m\, \rho_e \tag{3b}$$

Since it is easy to measure the external specific resistance (or conductivity) of media, Eq. 3b can be used to estimate which fraction of the induced membrane potential calculated with Eq. 1 is present for a given pulse duration of the electrical field. In fusion experiments, it should be borne in mind that as a rule only unbuffered or weakly buffered solutions can be used (for reasons of conductivity). This means that considerable changes in the absolute conductivity (and hence in the specific resistance) of the solution may occur when cells become leaky in these solutions during prolonged incubation. If such changes are ignored, the electrical boundary conditions of electrofusion may well alter to the extent that fusion no longer takes place.

Recently Pilwat and Zimmermann (83) showed that electrical breakdown studies also allow direct measurement of the internal conductivity so that the exact relaxation time can be calculated according to Eq. 3a for weakly conductive solutions. In this method the signal of cells generated in the orifice of a particle analyser (Coulter Counter) is measured before and after the breakdown of the membrane in dependence on the external conductivity (Figs. 7a,b). The signal of the particle, compared with that of the intact cell, is reduced once breakdown has occurred. The reduction of the signal is due to the breakdown of the membrane and the high

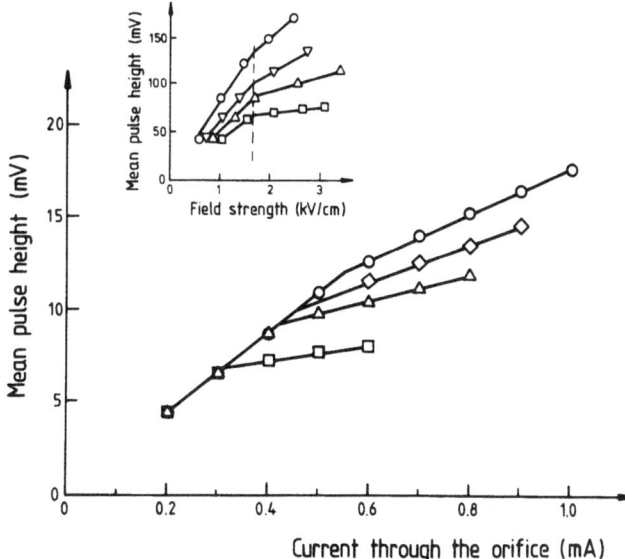

Fig. 7a. The effect of increasing resistivity of the suspension medium on the pulse height of size distributions of human red blood cells measured as a function of the current through the orifice of a particle volume analyser. The different resistivities were achieved by adding isotonic sucrose solution to an isotonic NaCl solution. Resistivities and final sucrose concentrations: o––o 69.4 Ω cm, 0 mmol; $\diamond$––$\diamond$ and $\triangledown$––$\triangledown$, 98.6 Ω cm, 90 mmol; $\triangle$––$\triangle$ 120 Ω cm, 120 mmol; $\square$––$\square$ 145.5 Ω cm, 150 mmol. Before electrical breakdown for a given current the mean pulse heights of the size distributions are independent of the resistivity of the suspension medium. This is because the analyser uses a "current-sensing" amplifier. After electrical breakdown the mean pulse heights of the size distributions depend on the resistivity of the medium. With increasing resistivities of the medium and therefore higher resistances in the orifice the electrical breakdown occurs at smaller currents. Nevertheless, the electrical field strength for breakdown is constant for all resistivities of the incubation medium as demonstrated in the inset. From (83)

internal conductivity, which influences the height of the signal after breakdown.

The method of Pilwat and Zimmermann (83) has so far been applied to the determination of the internal conductivity of human red blood cells. Preliminary experiments with myeloma cells show, however, that this procedure can also be used for measurements on nucleated cells (Broda, Pilwat and Zimmermann, unpublished data).

Since the relaxation time and the breakdown voltage depend on the radius (second term), it may be necessary, under certain conditions, to make corresponding adjustments to the strength and the duration of the critical field pulse. For example, when experimenting with eggs (e.g., *Xenopus* eggs with a radius of 500 μm to 1 mm), the pulse duration has to be extended to 5 ms in order to achieve reversible electrical breakdown (84). On the other hand, the application of field pulses of such long

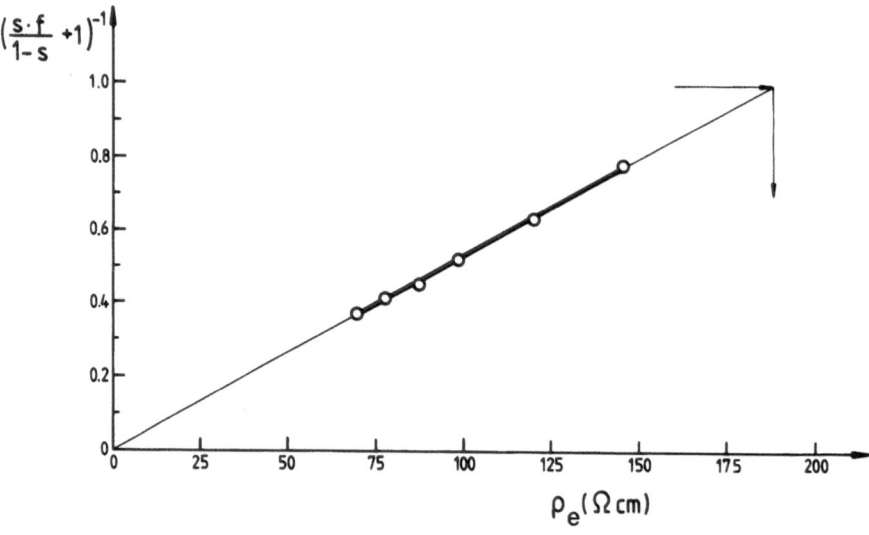

Fig. 7b. The relationship between the reciprocal value of the term $[s \cdot f/(1 - s)] + 1$ and the resistivity of the medium, ρ_e, for the data presented in **a**. The resulting straight line can be extrapolated for $([s \cdot f/(1 - s)] + 1)^{-1} = 1$ leading to a value for the resistivity of the cell interior. The result indicates that the measurement of the intracellular resistance can be conducted at any extracellular resistance. s, slope ratio of post-breakdown to pre-breakdown signal; f, shape factor. For further details see (83)

duration to the much smaller plant protoplasts, mammalian cells, and bacteria undoubtedly leads to irreversible deterioration of the membrane, since the final value for the membrane voltage is reached much more rapidly because of the shorter relaxation times mentioned above. Under these conditions the cells are exposed to an intense field for some time after electrical breakdown, so that secondary processes such as local heating, electrophoresis in the membrane, and pronounced osmotic processes which are ultimately lethal for the cells come into play (18, 85, 86). Since plant protoplasts have a somewhat larger diameter than animal, yeast, or bacterial cells, it is possible to use slightly longer pulses (up to about 40 μs) in these biological systems without causing irreversible damage.

The dependence of the relaxation time and the breakdown voltage on cell radius can also lead to considerable problems if cells with substantially different radii are to be fused, e.g., myeloma cells with unstimulated lymphocytes. The latter have a mean radius of 3 μm, while myeloma cells have a radius of 7 μm. At a given field strength the induced membrane potential is greater for a larger cell than for a smaller cell. However, it has to be noted that, if the exposure times in the electrical field are not long enough to allow the stationary value of the membrane potential to be established, the differences in the actual membrane potential between a large and a small cell become smaller. In any case, because of this dependence on the radius, the breakdown voltage for smaller cells is only reached at external field strengths which are higher than those required for larger cells. The critical field strengths which bring about the breakdown voltage for smaller cells may however be supercritical for the large cells, so that reversible electrical breakdown turns into irreversible destruction of the membranes of the cells. A related event is shown very clearly in Fig. 4a. A Friend cell is aligned with a lymphocyte cell by dielectrophoresis (see below), and field pulses are applied which are not high enough to break through the membrane of the small lymphocyte, but are sufficient to break through the membrane of the Friend cell. Under these conditions the small cells penetrate the membrane of the larger cell without fusion (Fig. 4b). Uptake of many Friend cells into a plant protoplast is shown in Fig. 5 using similar field conditions. Generally similar considerations also apply to electropermeabilization experiments for DNA transfection. If the cell suspension exhibits a large volume of distribution, electrical breakdown at a given critical field strength will only occur in those cells with a large volume, and not in those cells with a smaller volume. At very much higher field strengths, on the other hand, electropermeabilization will also occur in the smaller cells, but the larger cells may be irreversibly destroyed in the process. There are various ways of overcoming these difficulties in fusing cells of different sizes.

As far as electropermeabilization of free suspended cells is concerned, the simplest method is to remove the largest or smallest cells in order to narrow down the size distribution. Various methods for doing this have been described (19).

Since the total electrical field in the membrane or the total membrane voltage determine the electric breakdown behaviour of the membrane (87, 88), it is possible in principle to achieve a reasonably good agreement between the actual membrane potentials of large and small cells at a given external field strength by changing the intrinsic field, e.g., by reducing the intrinsic field in the membranes of the larger cells. Both types of cell then exhibit roughly the same electrical breakdown behaviour. The intrinsic membrane potential is made up of the diffusion potential, the surface potential and those potentials resulting from the presence of electrogenic pumps in the membrane (89). The surface potentials in particular can be specifically influenced by changes in the charge density and by the ionic strength in the solutions (89). This is possibly the reason why field conditions can be found in which myeloma cells and unstimulated lymphocytes can be fused with a good yield (90, 91). Changes in the surface charges and hence the surface potentials could also be responsible for the fusion-promoting effect of pretreatment of cells with digestive enzymes. Zimmermann et al. (92, 93) found that pretreatment of cells with enzymes like pronase and dispase resulted in a field stability of the cell membrane, i.e., cells suffered no adverse side effects when exposed to field strengths which would normally lead to irreversible changes in the membrane and cell. By combining this field stability with the electrofusion method, it was possible for the first time to produce mouse and human hybridoma cells [(94), see also (95)]. Pretreatment with dispase in particular seems to be suitable for a number of applications, since it does not influence the viability and ability of the cells to divide, even if a few molecules are taken up into the cell interior during electropermeabilization or electrofusion (see below). However, with the current commercially available fusion media (90, 91) pretreatment with enzymes is in many cases unnecessary, since the presence of Ca^{2+} and Mg^{2+} ions and albumin in the fusion media changes the surface charge in the appropriate manner. However, since these substances are also capable of inducing other effects in biological membranes, we cannot rule out the possibility that other causes are responsible for the increased fusion of myeloma cells and lymphocytes in such solutions.

The dependence of the induced membrane potential on the cell radius can also be largely eliminated by physical means, as will be described in detail in the following section.

2.1 Overcoming the Radius Dependence of Membrane Potential

Fourier analysis shows that each square pulse consists or can be thought to consist of a sum of sine waves with different frequencies. For a pulse length of 5–15 μs, which is typical in electrofusion, most of the frequencies of these waves are at 100 kHz, although there are also some higher frequencies. The assumption that cells are poor conductors is only justified for lower frequencies. At low frequencies the cell capacitor prevents the interior from participating in the conduction process. When the frequency, f, increases significantly, the membrane capacity become short-circuited, and the membrane potential built up across the membrane is lower than predicted by Eq. 1. The frequency dependence of the membrane potential is described by Eq. 4:

$$V_m = \frac{1.5 \cdot E \cdot a \cdot \cos \vartheta}{\sqrt{1 + (2\,\pi f \cdot \tau)^2}} \tag{4}$$

where τ is the relaxation time given by Eq. 3. For high frequencies, a combination of Eq. 3 and 4 yields:

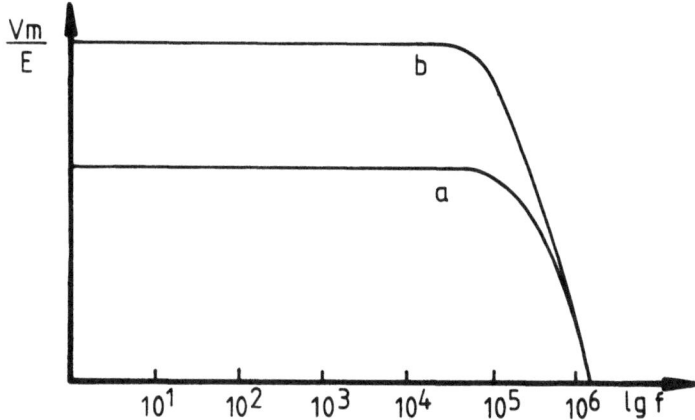

Fig. 8. A plot of the membrane potential V_m per unit E as a function of the frequency f according to Eq. 4. Curve *a* corresponds to lymphocytes and curve *b* to myeloma cells. The following data were used: curve *a*, radius = 3 μm, C = 1 μF/cm^2, ρ_i = 65 mS/cm and ρ_e = 320 μs/cm; curve *b*, radius = 7.5 μm, C = 1 μF/cm^2, ρ_i = 25 mS/cm, ρ_e = 320 μS/cm. It is evident that the two curves merge into each other above a frequency of about 1 MHz. (Büchner and Zimmermann, unpublished data)

$$V_m = \frac{1.5\, E \cdot \cos \vartheta}{2\,\pi f \cdot C_m\,(\rho_i + 0.5\,\rho_e)} \tag{5}$$

i.e., the induced membrane potential becomes independent of the radius and drops linearly with the frequency [see (20)]. The cut-off frequency, f_c, is given by:

$$f_c = 1/2\,\pi\tau \tag{6}$$

Figure 8 shows typical curves for the dependence of the membrane potential on frequency for myeloma cells and lymphocytes. For the plots values were used which fit with the standard electrofusion conditions. It is evident that above a frequency of 1 MHz the induced membrane potential is the same for both the small and the large cells, although it is considerably reduced compared with the membrane potential at low frequencies or under steady state conditions. The external field strength must therefore be correspondingly increased in order to achieve membrane potentials compared to the situation at low frequencies. The problem of the radius dependence of the membrane potential in electrical breakdown experiments can thus be avoided if pulses are used which consist mainly of since waves with frequencies above 1 MHz. This cannot be achieved with a single pulse, but is possible with a sequence of consecutive pulses, each lasting 5 μs. Fourier analysis shows that this block of pulses arises from the superimposition of sine waves whose frequency lies mainly around 2 MHz.

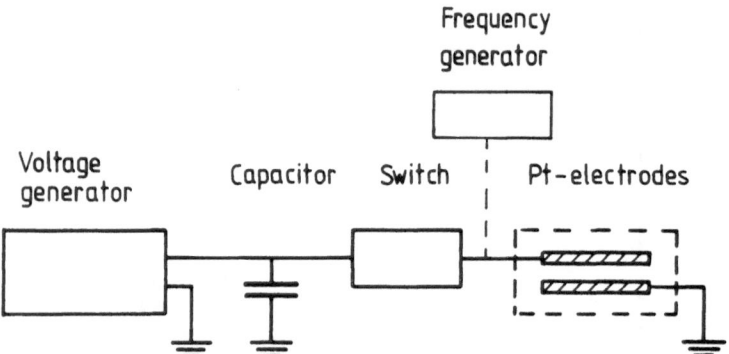

Fig. 9. Experimental arrangement for electrical breakdown of cell membranes. Suspended cells are pipetted into the gap between two cylindrical or flat Pt-electrodes arranged in parallel (say 500 µm to 1 cm apart). The electrodes are connected to a capacitor via a switch. The capacitor is charged by a voltage generator to a certain voltage (up to 20 kV/cm). When the switch is closed, the capacitor is exponentially discharged through the cell suspension. For uptake of foreign substances (e.g., genes) into the cell by means of electrical breakdown, the substances to be incorporated are added to the solution prior to the application of the field pulse. Field application is usually performed at 4°C. The subsequent resealing process of the field-induced perturbations in the membrane is achieved by raising the temperature to 37°C. For electrofusion a frequency generator is additionally connected to the electrodes (*dashed line*). The frequency generator is used for dielectrophoretic collection of the cells (see Fig. 19)

Preliminary studies in this direction in our laboratory have demonstrated that a sequence of three pulses at intervals of 5 µs results in high yields in the fusion of myeloma cells and lymphocytes.

This method of overcoming the radius dependence of the membrane potential could have interesting implications, since the possibility exists of developing a universally applicable protocol for the electrofusion and electropermeabilization of completely different cells.

2.2 Electroinjection of Foreign Molecules into Cells

Zimmermann and co-workers were able to demonstrate more than 10 years ago (18, 30–34, 80, 81, 96–98) that electropermeabilization of the cell membrane would allow substances to be incorporated in the cell without destroying its integrity. Most of these experiments were made by the high voltage discharge technique (Fig. 9).

In the early experiments radioactively labelled albumin was entrapped in red blood cells, while a large part of the haemoglobin was lost into the solution (30–34). The protein-loaded ghost cells were stable in vitro after incubation in appropriate isotonic solutions. Particularly good results were obtained if the field pulse applications were carried out at 4°C in

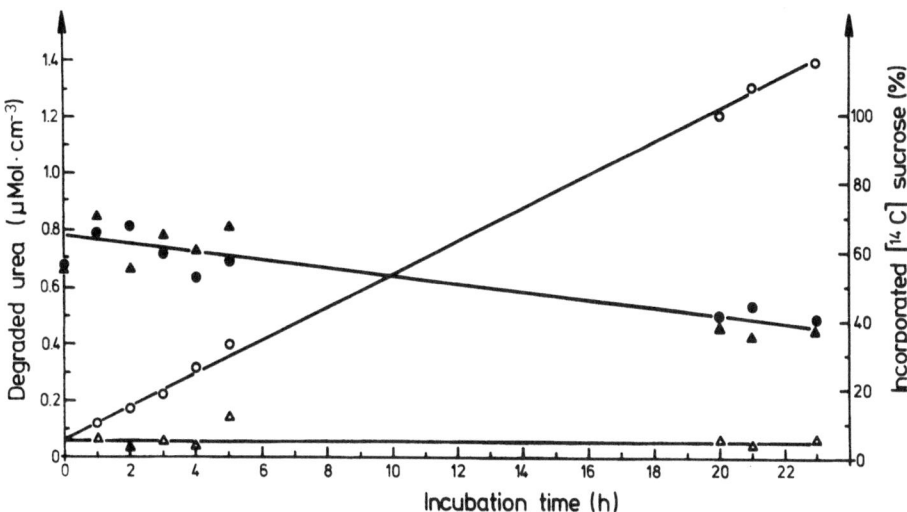

Fig. 10. Catalysed hydrolysis of urea to ammonium and carbon dioxide by urease and (^{14}C)-sucrose loaded human red blood ghost cells using the discharge set-up (Fig. 9). In one set of experiments (○) ghost cells were suspended in a solution containing 145 mmol NaCl, 5 mmol KCl, 4 mmol MgCl$_2$, 7.6 mmol Na$_2$HPO$_4$, 2.4 mmol NaH$_2$PO$_4$, and 10 mmol glucose, pH 7.2 to which 50 mmol urea was added. The analysis of the ammonium produced was based on the so-called Berthelot reaction [see (80)]. The blue color developed was measured photometrically at 550 nm. ● denotes sucrose loss observed during the experiment (measured in the sediment). In another set of experiments urease and (^{14}C)-sucrose loaded cells were incubated in the same solution without urea. ▲ denotes the sucrose loss as function of incubation time measured in the supernatant; △ denotes the results of the urease analysis in the supernatant. It is obvious that urease has not passed the membrane and it is also evident from the figure that sucrose loss is not affected by the urea in the solution since ▲ and ● are on the same line. From (80)

solutions which, corresponding to the composition of the intracellular fluid, contained a high potassium concentration (18). Figure 10 shows experiments with urease-loaded red blood cells which were incubated in isotonic physiological solution containing added urea after the resealing process had been completed (80). Figure 10 shows that urea diffuses through the erythrocyte membrane and that it is broken down into CO$_2$ and ammonia by the urease entrapped in the cells. The degradation of urea is measured experimentally by the release of ammonia.

In these experiments radioactively labelled sucrose was entrapped in the cells together with urease, in order to investigate the cell membrane integrity and its change during subsequent incubation in isotonic physiological solutions. After the resealing process and a 22-h incubation period, the membrane exhibited an increased permeability to sucrose – albeit only a slight one (see Fig. 10). Figure 11 shows the uptake of the dye eosin into mouse thymocytes (18). In the absence of an electrical field the dye is not taken up into the cells. However, when the breakdown

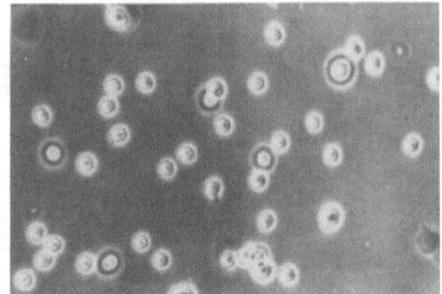
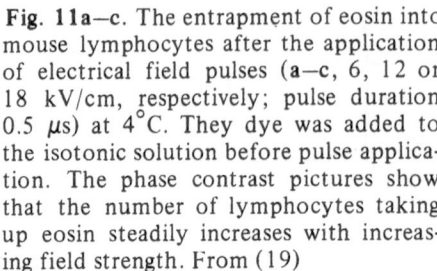

a

Fig. 11a—c. The entrapment of eosin into
mouse lymphocytes after the application
of electrical field pulses (a—c, 6, 12 or
18 kV/cm, respectively; pulse duration
0.5 μs) at 4°C. They dye was added to
the isotonic solution before pulse application. The phase contrast pictures show
that the number of lymphocytes taking
up eosin steadily increases with increasing field strength. From (19)

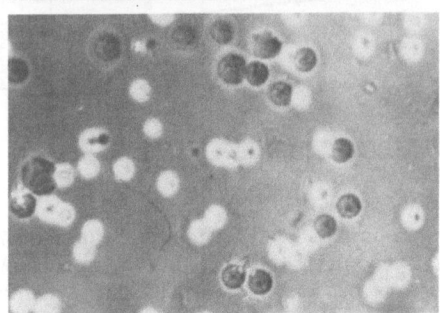

b

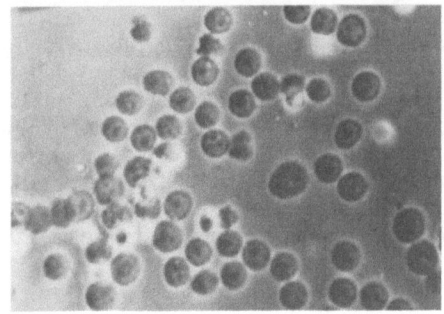

c

voltage for membrane sites oriented in field direction is exceeded and the
field strength is progressively increased, there is an increased uptake of
dye into more and more cells.

The graphic evaluation of such microscopic data in Fig. 12 shows that
because of the radius dependence of the breakdown voltage, about 18 kV/
cm is required in order to stain more than 80% of the cells. In these
experiments, which were carried out in isotonic electrolyte solutions, the
pulse duration had to be about 500 ns, since irreversible damage in the
membrane and cell was observed at longer pulse durations (18).

The resealing process, particularly at higher temperatures, was demonstrated in a modified experiment (Fig. 13). The lymphocytes were subjected to breakdown at 4°C, then the temperature was elevated to 37°C
and the dye was added at different time intervals. It is evident that the
original impermeability of most of the cells towards eosin was restored
after about 2 minutes.

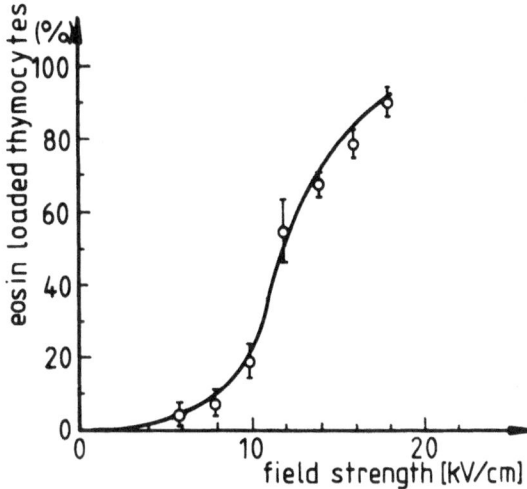

Fig. 12. Graphic evaluation of the eosin uptake in mouse thymocytes (lymphocytes). The cells were suspended at a density of 1:20. The pulse length was 0.5 μs. During field application the temperature was kept constant at 4°C, and after 10 min elevated to 37°C in order to restore the original membrane permeability. Eosin was added in a final concentration of 2.3 mmol to the suspension before field application. The percentage of eosin loaded thymocytes is plotted against increasing field strength of the applied electrical field. Data were taken from phase contrast microscope measurements as shown in Fig. 11. Each data point represents 300 cells of three independent experiments. *Bars* indicate standard deviation. From (18)

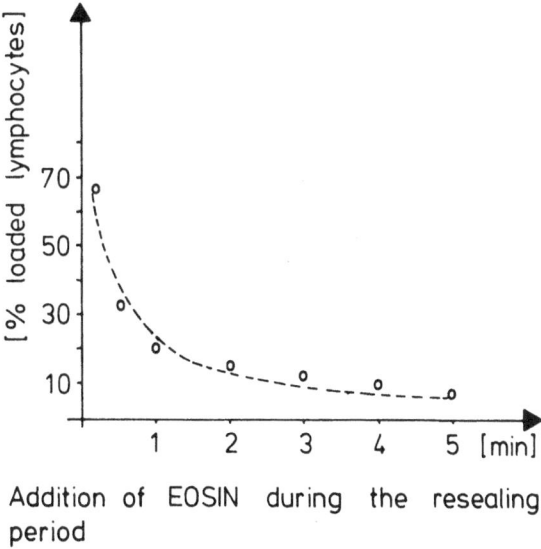

Addition of EOSIN during the resealing period

Fig. 13. Resealing of field-treated mouse thymocytes (lymphocytes). Thymocytes were subjected to a field pulse of 14 kV/cm pulse length 0.5 μs. The temperature of the suspension was raised 10 min after field application at 4°C to 37°C. Eosin (2.3 mmol) was added at different time intervals within a period of 5 min after raising the temperature to 37°C. The number of the cells able to trap eosin decreases from about 70% in the first few seconds to almost 5% after 5 min. This indicates that the original membrane permeability towards eosin has been restored during this time interval. From (18)

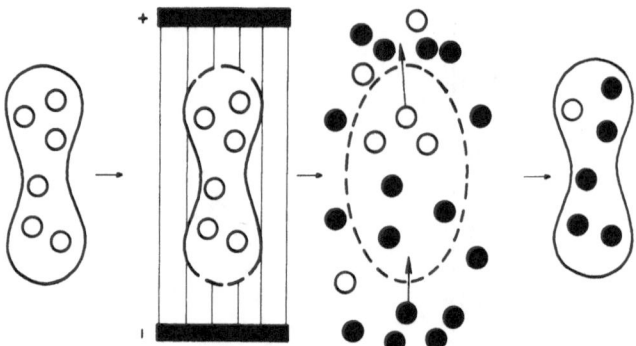

Fig. 14. a Schematic diagram of an erythrocyte prior to the application of an electrical field pulse. Hemoglobin molecules are represented by *open circles*. b An electrical field pulse (6–8 kV/cm field strength and 10–40 µs duration) brings about electrical break-down which leads to the formation of "pores" in the membrane (indicated by the *dashed line;* temperature 4°C). c Hemoglobin molecules diffuse out through the "holes" in the membrane while foreign molecules added to the external medium, e.g., drug molecules (*filled circles*), pass through the erythrocyte membrane into the cell interior. d The membrane "reseals" when the temperature is raised to 37°C

Electropermeabilization was successfully used by Zimmermann and co-workers (18, 34, 96–98), and later by Tsong and co-workers (99, 100), to load red blood cells with drugs (Fig. 14). Drug-loaded cells have great potential for targeting drugs to specific sites in an organism. They can also be used as depot systems in the blood circulation for continuous release of drugs into the blood over long periods of time. The interested reader is referred to the review article by Zimmermann published on this sub-ject matter (34).

However, it is worth mentioning at this point that under the appro-priate experimental conditions, particles of microscopic size can be trans-ported through membranes with the aid of the electropermeabilization technique. Small magnetic particles were incorporated into red blood cells by electrical breakdown for specific targeting of drug-loaded cells (34, 97). Figure 15 shows a typical in vitro experiment with red blood cells which contained both a drug and small magnetic particles (Fig. 15a). The magnetically identifiable erythrocytes were concentrated between the poles of a permanent magnet (Fig. 15b). It was also possible to entrap latex particles in erythrocytes with this method (81, 101).

In the meantime it has also become possible, by the appropriate choice of experimental conditions, to release organelles from plant protoplasts (Fig. 3) into the external solution by electrical means without loss of cell membrane integrity (22, 66, 67). As mentioned above it is even possible to transport smaller cells through the membranes of larger host cells by electrical means (Figs. 4, 5).

Fig. 15. a Human red blood cells which have been loaded with very small magnetic particles (Fe_3O_4) by means of an electrical field pulse are placed between the poles of a permanent magnet. **b** The "magnetically labelled" erythrocytes quickly accumulate in the magnetic field. From (97)

The first proof of DNA transfer with the aid of this electrical method was provided by Auer et al. (102). Using the erythrocyte system these authors demonstated that DNA and viral RNA can be sequestered across the cell membrane and that both DNA and RNA can be entrapped in the cells when the resealing process is completed.

In the meantime there have been other studies of electrically induced DNA transfer. Langridge et al. (103) used electropermeabilization for the transformation of carrot protoplasts with Ti-plasmid DNA from *Agrobacterium tumefaciens*. The uptake of labelled plasmid DNA, 26–200 kbp in size, into carrot protoplasts was stimulated by high field pulses at field strengths from 0.5–4 kV/cm. Protoplast regeneration, somatic embryogenesis and plantlet regeneration were unaffected by the electropermeabilization conditions used for DNA uptake. Plasmid pTiC58 transformed carrot protoplasts to hormone independence. Teratoma-like somatic embryos were detected in electrotransformed carrot cultures 45 days after electropermeabilization, at frequencies greater than 1.5×10^3 transformants per 10^6 somatic embryos per μg pTiC58 DNA. The transformed somatic embryos developed into teratomas which synthesised nopaline. A labelled T-DNA fragment from pTiC58 hybridised to DNA fragments from 4-month-old teratomas regenerated from electrotransformed protoplasts.

Karube et al. (104) recently reported on the successful electrical transformation of yeast cells [see also (20)].

There have also been reports of electrically induced DNA transfer in nucleated mammalian cells from Neumann et al. (75) and Potter et al. (82). However, the yields of DNA transfection are not very high, and it therefore seems worthwhile discussing the experimental conditions used by these authors in more detail, since this discussion will show how important it is to harmonise the individual steps in this method carefully in the light of the preceding considerations. Neumann et al. (75) used trypsinised cells. However, as electrorotation experiments have shown (105, 106), trypsin leads to irreversible changes in the cell membrane and should therefore not be used – at least not for electrical breakdown experiments. Dispase is much more suitable for suspending cells (grown in confluent layers), since neither the viability of the cells nor the electrical properties of the membrane are affected. In addition Neumann et al. (75) performed electrical field applications in solutions containing high concentrations of sodium chloride but no potassium ions. Because potassium ions leak through the membrane permeabilized by the electric field and treated by trypsin, the viability of the cells will be greatly reduced when the resealing process is completed. Furthermore, electropermeabilization was carried out at 30°C. At this temperature the life span of the high permeation state of the membrane is very short. Under these conditions even small molecules like sucrose are only taken up into the cells in small quantities compared with the amounts taken up at lower temperatures (80). The temperature dependence of the resealing process is one of the reasons why Neumann et al. (75) had to apply at least three breakdown pulses in order to achieve gene transfer at all.

The field conditions used by Potter et al. (82) for gene transfer were such that extremely low transfection rates were predictable from the outset. They used field intensities between 4 kV/cm and 8 kV/cm. Considering the steep increase in the breakdown voltage towards lower temperatures, the field strength of the external field pulse must be at least 8 kV/cm in order to achieve sufficient permeabilization of the membrane. A temperature of 0°C, as used in the experiments of Potter et al. (82), should definitely be avoided, not only because the breakdown voltage increases considerably below 4°C, but also because irreversible changes in the membrane are observed during field application at this temperature (40).

Stopper et al. (107) have recently developed a DNA transfection protocol which leads to extremely efficient transformation of animal cells with the aid of the electropermeabilization method. In principle this protocol is based on the originally published experimental conditions (18, 30–34), but it also takes into account the latest findings relating to electrically induced transport of substances through biological membranes.

As mentioned above, the electrical breakdown of the cell membrane is governed by the total potential across the membrane which is made up of the intrinsic membrane potential V'_m and the potential V_g generated in response to the electrical field (87, 88). As shown in Fig. 16, these two potential differences are superimposed in parallel in one cell hemisphere, and antiparallel in the other hemisphere. In theory, we can therefore expect breakdown of the two cell hemispheres to be asymmetric, in addition to the radius and angle dependence of the breakdown voltage. This was indeed observed in plant protoplasts and in animal cells (Stopper and Zimmermann, in preparation). In nonconductive solutions fluorescent dye was found to enter the cell hemisphere facing the anode (see Fig.17). When the ionic strength of the solution was raised (by adding about 45 mmol electrolyte) symmetrical uptake into animal cells was observed, at least for myeloma cells. If the electrolyte concentration was raised still further, very often asymmetric uptake was again observed, however this time by the cell hemisphere facing the cathode. This finding can be explained by the assumption (see above) that changes in the surface potentials with ionic strength can lead to changes in the intrinsic membrane potential (89).

Pretreatment of animal cells with dispase results in a more markedly asymmetric uptake of the fluorescent dye over the entire electrolyte range.

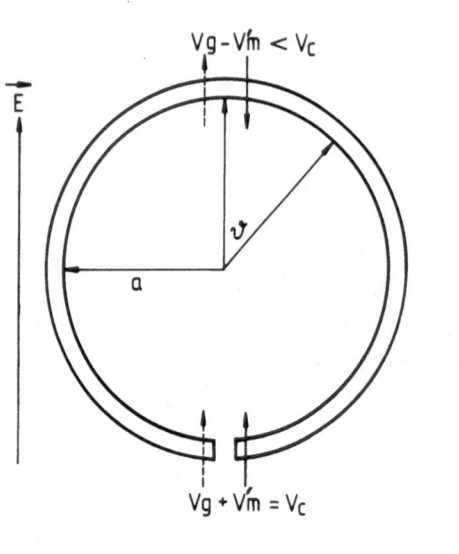

Fig. 16. Schematic diagram of a cell exposed to a uniform electrical field ($\vec{E}$) between two large electrodes. *a*, Radius of the protoplast; V'_m, resting transmembrane potential; V_g, generated (superimposed) membrane potential; ϑ, angle between a given membrane site and the field direction

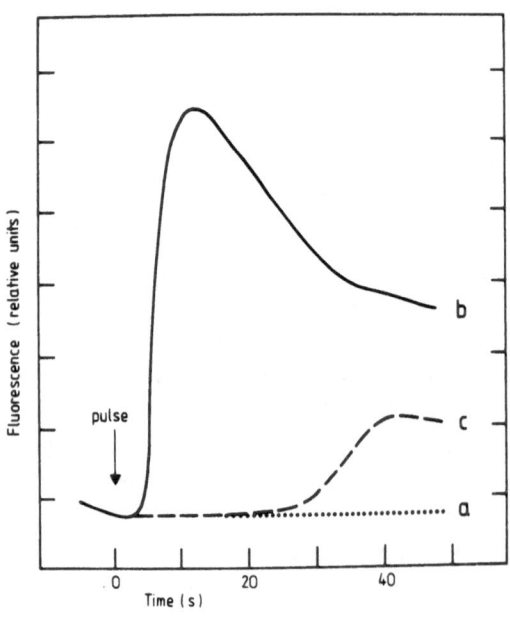

Fig. 17. Microfluorimetric determination of the uptake of ethidium bromide by intact oat mesophyll protoplasts without (*a*) or after (*b*, *c*) treatment with an electrical field pulse (2 kV/cm, 20 μs). After pulse application the time course of the change in fluorescence intensity of the protoplast hemisphere oriented toward the anode (*b*) or toward the cathode (*c*) was determined. From (87)

Table 1. Yield of stably transfected clones of mouse L-cells under various field conditions, obtained with circular plasmid DNA

Exp. no.	Field intensity (kV/cm)	Number of 5 μs pulses	Number of cell clones	Clones per 10^6 cells	Clones per 10^6 cells and 1 μg DNA	DNA concentration (μg/ml)
8-1	8	1	320	71.7	23.9	
8-5	10	1	243	71.9	24.0	
13-1	10	1	72	122	40.7	
10-1	10	1	105	128	42.7	
8-2	15	1	358	65.6	21.7	
8-3	20	1	306	63.6	21.2	
8-6	10	2	397	58.8	19.6	
13-2	10	3	93	88	29.3	
8-7	10	3	504	86.8	28.9	
Average clone number:			266	84	28	
9-1	10	1	65	8.9 (130)	29.7	0.1
9-2	10	1	240	41.4 (9.6)	2.8	5
9-3	10	1	240	44.4 (48)	14.7	1

Unless otherwise stated the DNA concentrations was 1 μg/ml. Numbers in parentheses refer to the number of clones per 1 μg DNA. In control experiments, in which cells were subjected to the same experimental procedure without pulse application, either no or only a single clone was found. From (107)

We can therefore presume that electrically induced gene transfer is influenced by the ionic strength of the solution. Stopper et al. (107) found that high yields of transformed mouse L-cells could be obtained in an isotonic inositol solution containing 30 mmol KCl, provided that the cells had been pretreated with dispase. One interpretation of this would be that the chances of survival of the cells are increased if breakdown occurs in only one cell hemisphere while the other remains relatively intact.

In the experiments of Stopper et al. (107) electrical transfection of mouse L-cells was carried out with the aid of the plasmid pSV-neo which confers resistance to the antibiotic G-418. As shown in Tables 1 and 2 up to 500 clones were obtained, whereby the yield was much higher when linearised DNA was used rather than a circular one.

As in the development of electrofusion, we can expect further optimisation of the electrical parameters and media conditions to lead to even higher yields of transformed cells. However, one thing is already clear, namely that the electropermeabilization method is well able to compete with the conventional chemical transfection methods.

For the sake of completeness, we should mention that there are reports of DNA transfection by other physical means. Zimmermann (108) showed that macromolecules like proteins and drugs can be incorporated by

Table 2. Comparison between the yield of stable transformants of mouse L-cells obtained with circular and linearised plasmid DNA. The linearised DNA was obtained from the circular form by digestion with *Eco*R 1 (107). The results were obtained in parallel experiments using 1 μg/ml of DNA and 1.3 and 2.4 × 10^6 cells/ml for the circular and linear form of DNA respectively

Form of DNA	Field intensity (kV/cm)	Number of 5 μs pulses	Number of cell clones	Clones per 10^6 cells	Clones per 10^6 cells and 1 μg DNA
Circular	10	1	112	22	5.5
Linear	10	1	4153	439	109.8

osmotic processes, at least as far as red blood cells are concerned. Auer and Brandner (109) also reported on the successful loading of human red blood cells with DNA and RNA. The osmotic technique seems to be restricted to the red blood cell system because of the peculiar properties of these cells (73).

A new physical method of DNA transfection was recently published by Tsuhakoshi et al. (110). These authors used laser microbeams to permeabilise cell membranes and incorporate the splicing genes from *Escherichia coli* (*Eco-gpt*) into rat kidney cells. They claimed a high yield of transfection. However, in contrast to electropermeabilization, the requirements in terms of equipment seem to be very high. Only time will tell which technique has the greater potential for cell biology and technological applications.

Finally, I would like to point out that the incorporation of whole cells or organelles into other host cells may have great potential in future. It is conceivable that this technique can be used to introduce artificial "organelles" into a living system. For example, Ringsdorf and co-workers (111, 113–115) and, independently, Chapman and co-workers (112) described vesicles made up of artificial lipids which can be polymerised by UV or other agents. Incorporation of substances into such vesicles prior to polymerisation and subsequent entrapment have also been demonstrated. It is thus conceivable that vesicles entrapped in a host cell with the aid of electrical breakdown could act as a pool of substances which could be used to control a number of processes, provided that there is a continuous and controllable release of these substances from those vesicles. The author of this review believes that there is great potential for such a system, e.g., in the field of drug delivery systems as discussed above. Although it has been possible to load red blood cells with drugs and these were able to circulate in the blood of a mouse for more than two weeks (34), such systems exhibit severe limitations as to the type of drug they can carry. Most of the drugs interact with the cell membrane and cause it to become leaky. However, the limitations of this and other

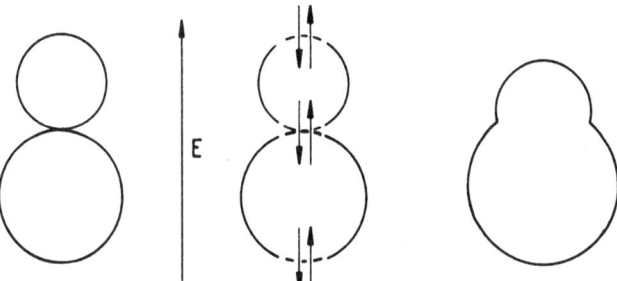

Fig. 18. Diagrammatic representation of electrofusion of two cells adhering to each other in the direction of the external electrical field by dielectrophoresis or other physical forces (*left*). Electrical breakdown occurs in the membrane area in the zones of contact which are oriented vertically with respect to the field lines (see *arrows*). Under the given conditions electrical breakdown does not only induce exchange of substances between the cells and their medium but also intercellular exchange. As a result fusion occurs (*right*)

carrier systems could be overcome by incorporating the drug in polymerised vesicles and subsequently entrapping these drug-loaded vesicles in red blood cells. The actual concentration of the drug in the cytoplasm of the red blood cell would be very low on average so that severe side effects on the membrane would be minimised.

Such shuttle systems would also have an enormous range of applications, e.g., in the study of membrane structure or in triggering processes within the cell.

3 Fusion with Dielectrophoresis

Electrical breakdown will trigger fusion of cells, provided that the membranes are in close contact with each other. In the standard electrofusion technique (Fig. 18), close membrane contact is achieved by cell movement, orientation and alignment in an inhomogeneous alternating electrical field of relatively low intensity (Fig. 19) compared to the field strengths involved in electropermeabilization (range of 1–100 V/cm depending on the cell radius). This effect (116–154) was termed dielectrophoresis, although the underlying mechanism is completely different from that involved in electrophoresis (Fig. 19a). Since this term now seems to be well established, we will continue to use it here.

The effect of dielectrophoresis has been discussed from various aspects in a number of reviews (19–28). I will therefore restrict myself to a brief description of this phenomenon and devote more attention to recent findings and new approaches in this field. Briefly, dielectrophoresis is the

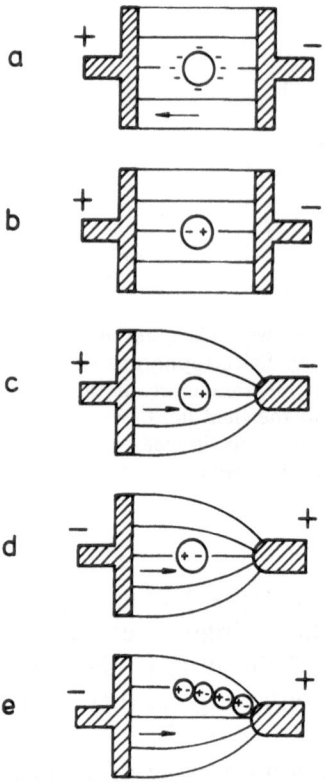

Fig. 19a—e. Diagrammatic representation of electro- and dielectrophoresis. Because of the net charge on the outer surface of the membrane, cells or charged particles migrate in the direction of an electrode (electrophoresis) in a homogeneous electrical field (plate capacitor) (**a**). The direction of migration depends on the sign of the charge and on the direction of the external voltage. In general, cells have a negative surface charge. In addition, a dipole is induced in a dielectric particle or in a cell (regardless of whether these particles carry a net charge or not) (**b**). Since the electrical field intensity is equal on both sides of the particle or cell (**b**), this induced charge does not contribute to movement. Uncharged particles are thus unable to migrate in a homogeneous field. In an inhomogeneous electrical field, on the other hand, uncharged particles are also able to migrate because the electrical field exerts a net force on the electrically induced dipole (**c**); the field intensity is not equal on both sides of the particle, resulting in a net force acting on the particle. This effect is known as dielectrophoresis. In contrast to the direction of migration during electrophoresis, the direction of the dielectrophoretic migration of the particles and cells is not reversed when the external electrical voltage between the two electrodes is reversed (**d**). The cells adhere to each other and form chains at the electrodes along the electrical field lines, because the dipoles induced in the cells attract each other (**e**)

motion of neutral (or charged) bodies in a nonuniform field. As shown in Fig. 19 the presence of the electrical field leads to the generation of a dipole due to the charge separation process mentioned above (see also Figs. 1 and 2). For a spherical body the relaxation time of the charging process is given by Eq. 3. In a nonuniform field, the field strengths acting

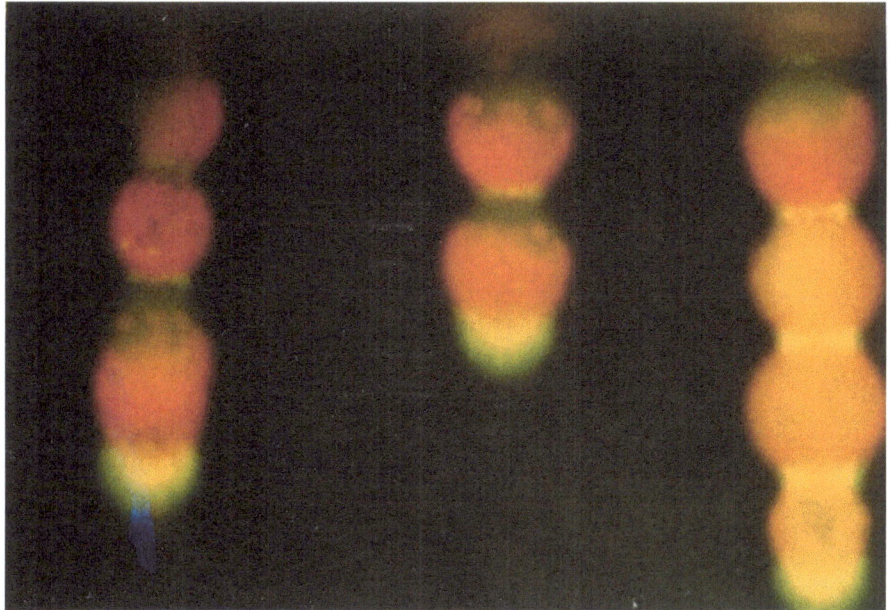

Fig. 20. Pearl chain formation of mesophyll protoplasts of *Avena sativa* in an inhomogeneous alternating electrical field between two cylindrical parallel electrodes (frequency 1 MHz, field strength 200 V/cm, isotonic mannitol solution). The distortion of the field lines in the neighborhood of the protoplast chains was made visible by addition of small bacteria cells. It is obvious that the electrical field across the gap between adjacent cells is intensified greatly by current constriction; this makes nonlinear charge transport on the outer membrane surface likely [see also (21)]. (W. Mehrle, R. Hampp and U. Zimmermann, unpublished data)

on the separated charges on the two sides of the cell will be different. This gives rise to a net force which pulls the particle in the direction of the highest field intensity, provided that the complex dielectric constant of the particle is higher than that of the surrounding medium. Otherwise the particle will move towards the region of the weakest field intensity. If particles approach each other during their migration, they will attract each other because the attractive forces arising from the dipoles are much higher than the repulsion forces arising from the net surface charge, from Brownian motion and from the hydration forces (Figs. 19 and 20). These will only become dominant at very close distances between the cells, because of the a^{-4} dependent dipole-dipole interaction. The overall result is the formation of "pearl chains" in which the cells are in very close contact with each other. Breakdown will occur predominantly in the contact zone if the field strength is just high enough to induce electropermeabilization. The latter immediately leads to cell fusion.

In contrast to electrophoresis, cell alignment (dielectrophoresis) will also operate in an alternating field, as is evident from Fig. 19. The effect

Fig. 21a,b. Up to a frequency of 7 MHz turkey red blood cells suspended in an isotonic mannitol solution form chains, in which the longest axis of the cells is parallel to
the field lines (**a**). Above 25 MHz the shortest axis of the cells is parallel to the field
lines (**b**). This results in cell stacks which can be fused together. From (22)

Fig. 22. Formation of cell chains at high suspension densities of a blood sample containing normal red blood cells and sickle cells. In contrast to the orientation of normal cells, sickle cells (*arrowed*) are oriented perpendicular to the field if an alternating field is applied (5 MHz, 250 V/cm). The electrodes appear as *dark lines* (× 300). From (153)

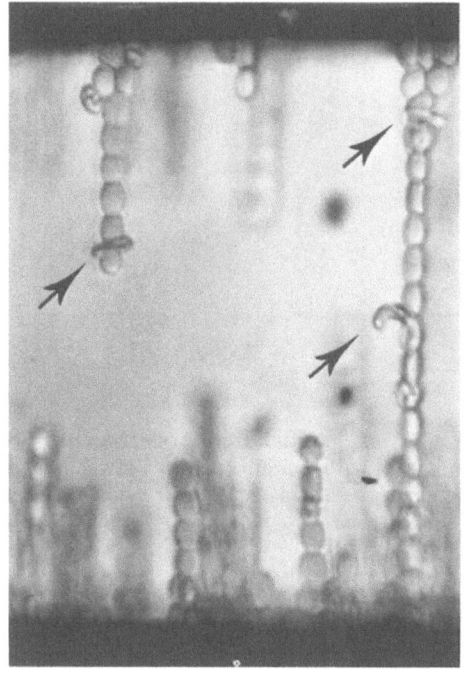

of electrophoresis is thus masked if an alternating field is applied. Normally alternating fields of frequencies between 10 kHz and 3 MHz are used. The lower limit of the frequency range is determined by the occurrence of electrolysis.

The orientation of nonspherical particles in an alternating electrical field is dependent on the frequency (20–28, 129–143), as shown in Fig. 21 for turkey red blood cells (20) and in Fig. 22 for a mixture of normal human red blood cells and sickle cells (153). In the kilohertz range orientation of the longest axis in parallel with the field lines is generally favoured, whereas above a certain turnover frequency in the megahertz range (depending among other things on the external specific conductivity) the shortest axis is orientated in field direction so that stacks of particles are formed. With turkey erythrocytes the situation is somewhat more complicated because these cells have three different semi-axes and a flattened ovoid shape [see (22)]. Semi-quantitatively the frequency-dependent orientation arises from the radius- (axis-) dependent membrane potential and from the radius-dependent charging time which is different in the three directions of a nonspherical cell (Eqs. 3 and 4).

The different types of orientation result in varying degrees of membrane contact. For example, turkey erythrocytes can only be fused if they are oriented at 30–60 MHz frequency (Fig. 21).

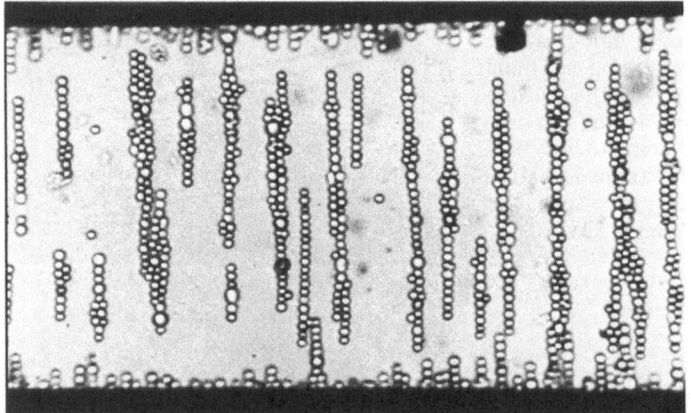

Fig. 23. Dielectrophoretic collection of Friend cells. The cells were suspended in 0.3 M mannitol solution and placed between the electrodes. After application of an inhomogeneous alternating field (100 V/cm, 2 MHz) the cells arrange like pearl chains between the electrodes

Fusion is usually performed with cell chains which are in contact with the electrode surface or which have formed on the glass surface at the bottom of the electrode chamber (Fig. 23). Electrostatic attraction between the charged glass surface and the surface charges of the cells prevents the cell chains from migrating towards the electrodes. Contact between the glass surface and the cells can be varied by pretreating the cells (e.g., with polylysine or a polylysin-pronase mixture) in such a way that the mobility of the cells in the chains remains sufficiently high to enable the fused cells to become rounded (68). Performing cell fusion without electrode contact (93) can sometimes be advantageous when the cells tend to adhere too strongly to the electrode surface. Fusion without electrode contact also prevents the cells within a pearl chain from being exposed to markedly different field strengths. It should be borne in mind that the field strength increases dramatically near the electrodes, sometimes by 2 orders of magnitude. Although studies of the hybridisation of yeast cells (90) and the production of hybridoma cells (91) have shown that such problems rarely occur, it may nevertheless be sometimes necessary to carry out large-scale fusion without electrode contact.

This could be achieved with the aid of appropriate chambers in combination with dielectric levitation (144–147). Although experiments of this kind have still to be performed, we would like to consider in more detail this interesting modification of dielectrophoresis for cell fusion, since dielectric levitation has already been extensively studied for dielectric particles (Fig. 24). In general the term "levitation" means that a stable cell suspension with no physical contact is established in a suitable field of forces. This can be achieved by holding the particles in a static equilibrium

Fig. 24. Typical electrical field plot for ring disk electrode showing cusped electrical field. From (146)

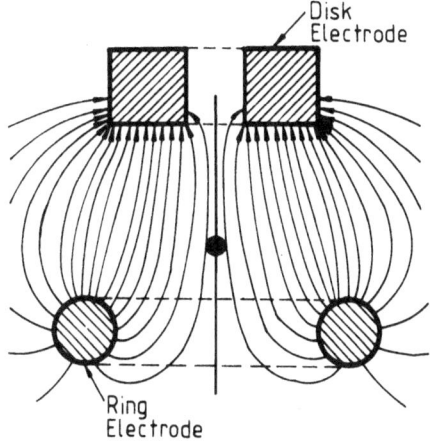

against gravity, using static (or quasi-static) electrical or even magnetic fields. In a static equilibrium the physical forces compensate the gravitational forces so that the particles are suspended freely in space. Several authors (144–146) have identified the general requirements for stable electrostatic and magnetohydrostatic levitation for dielectric material (like droplets and bubbles contained in liquids and gaseous media). Lin and Jones (146) demonstrated that electrostatic levitation can be achieved particularly easily with the aid of cusped electrical fields (Fig. 24). They also set up the appropriate equations so that the necessary conditions for cusped electrical fields can be calculated. The interested reader is referred to the relevant literature. Here we will only consider the conditions that give rise to stable levitation for dielectric particles. In principle at least, these considerations also apply qualitatively to cells, although, as we have already said, cells are particles exhibiting dielectric losses.

The dielectrophoretic force, $\vec{F}$, exerted on an uncharged dielectric body in a nonuniform field is given by Eq. 7 for a spherical particle:

$$\vec{F} = 2\,\pi a^3 \cdot \epsilon_0 \cdot \epsilon_1 \left(\frac{\epsilon_2 - \epsilon_1}{\epsilon_2 + 2\,\epsilon_1} \right) \quad \nabla \mid E \mid^2 \tag{7}$$

where ϵ_1 and ϵ_2 are the relative dielectric constants of the medium and the particle respectively, and E is the imposed electrostatic field. Equation 7 indicates that if $\epsilon_1 < \epsilon_2$, the particle will seek out the maxima of electrical field intensity, whereas if $\epsilon_1 > \epsilon_2$, the particle will seek out the minima. The dielectrophoretic force is independent of the electrical field polarity (Fig. 19). Its direction depends only on the gradient of E and the relative magnitude of ϵ_1 and ϵ_2. If E is a sinusoidal time-dependent electrical field, then $\vec{F}$ reduces to a constant (average) term plus a time-dependent term (at twice the frequency of the electrical field). The average force is determined by using the root-mean-square electrical field magnitude in Eq. 7. If dielectrophoretic levitation of a spherical particle (or bubble, etc.) in the gravitational field is considered, the net force is:

$$\vec{F}_{net} = F + \frac{4}{3}\pi a^3 \ (\rho_2' - \rho_1')g \tag{8}$$

where $\rho_{1,2}'$ are the mass densities of the medium and the particle, respectively, and g is the local gravitational acceleration.

Levitation occurs if the dielectrophoretic force is balanced by the opposing gravitational force.

Virtually all dielectric materials exhibit some conductive behaviour which can influence the effective dipole moment generated in the field. Benguigui and Lin (147) arrived at a compact form for the time average force that must now replace Eq. 7:

$$\vec{F} = 2 \ \pi a^3 \epsilon_1 \left(\frac{\epsilon_2 - \epsilon_1}{\epsilon_2 + 2 \ \epsilon_1} + \frac{3 \ (\epsilon_1 \sigma_2 - \epsilon_2 \sigma_1)}{\tau \ (\sigma_2 + 2 \ \sigma_1)^2 \ (1 + \omega^2 \tau^2)} \right) \nabla \ E_{rms}^2 \tag{9}$$

where ω is the angular frequency of the electrical field and

$$\tau = \frac{\epsilon_2 + 2 \ \epsilon_1}{\sigma_2 + 2 \ \sigma_1}$$

$\sigma_{1,2}$ are the specific conductivities of the medium and the particles, respectively.

This equation is easily interpreted with the aid of Table 3 which shows the generalised dielectric levitation criteria.

It is evident for dielectric particles that stable levitation over the entire frequency range can be only achieved when $\epsilon_2 < \epsilon_1$. For the case of $\epsilon_2 > \epsilon_1$, stable levitation is expected only for smaller frequency ranges, provided that the conductivity of the particle is less, than that of the external medium. Indeed, Benguigui and Lin [e.g., (144, 147)] have reported success in levitating particles with $\epsilon_2 > \epsilon_1$ in a quasi-two dimensional electrode configuration presumably using an alternating voltage (see Fig. 24).

In any case, since there are frequency ranges in which negative dielectrophoresis is observed (e.g., 10 kHz for yeast cells), it is possible, in principle, for dielectrophoretic levitation to occur with living cells. However, it is difficult to predict which frequency range will give rise to levitation of cells because the equations are not valid for particles and media exhibiting dielectric losses. The equations can therefore only be used for a qualitative discussion of the effects to be expected.

Table 3. Criteria for dielectrophoretic levitation

	$\epsilon_2 < \epsilon_1$	$\epsilon_2 > \epsilon_1$
$\sigma_2 > \sigma_1$	stable levitation for $\omega > \omega_0$ only	no stable levitation at any frequency
$\sigma_2 < \sigma_1$	stable levitation for all frequencies	stable levitation for $\omega < \omega_0$ only

$$\omega_0^2 = \frac{\sigma_1 - \sigma_2}{\epsilon_2 - \epsilon_1} \cdot \frac{1}{r}$$

This generally applies to the derivation of Eq. 7 and to the quantitative analysis of the effects involved in pearl chain formation given by Krasny-Ergen (120) and, particularly, by Saito and Schwan (118). Generally speaking, these analyses are based on the relations for the body force which were derived from the Helmholtz energy principle of electrostatics (see above). As Sauer (150) recently pointed out, this principle fails and cannot be applied to media and particles with losses (dielectric losses, conducting media). In this case, it is necessary to calculate the forces with the help of the electromagnetic momentum balance via the Maxwell stress tensor. This leads to other complications because of the intimate connection of the Maxwell stress tensor with the pressure tensor. Sauer (150) has given an approximate solution of the problem and was able to explain for the first time the trajectories of the particles (cells) in a non-uniform sinusoidal field. The mathematics of his theory are very demanding, and the interested reader is referred to the relevant paper by Sauer (150).

Within the framework of this review we can conclude that the future application of Sauer's theory to dielectric levitation of cells will undoubtedly improve the present techniques of establishing membrane contact by dielectrophoresis.

The electrode arrangements used at present for dielectrophoresis in the standard electrofusion technique are described in detail elsewhere (19–28). Some chambers used for electrofusion so far are shown in Fig. 25. I would also like to point out that the socalled helical chamber (see Fig. 26) and the rotational chamber (Figs. 27 and 28) have proved to be extremely efficient for the large-scale production of hybrids by electrofusion. Axial symmetric fields such as are established between two concentric metal electrodes are also very suitable for cell fusion by electrical fields (21, 151) (Fig. 25c). The advantage of such fields is that they enable the precise calculation of the field strength. For this reason they were used by Dimitrov and co-workers (151, 152) for the exact experimental measurement of the dielectrophoretic mobilities of cells.

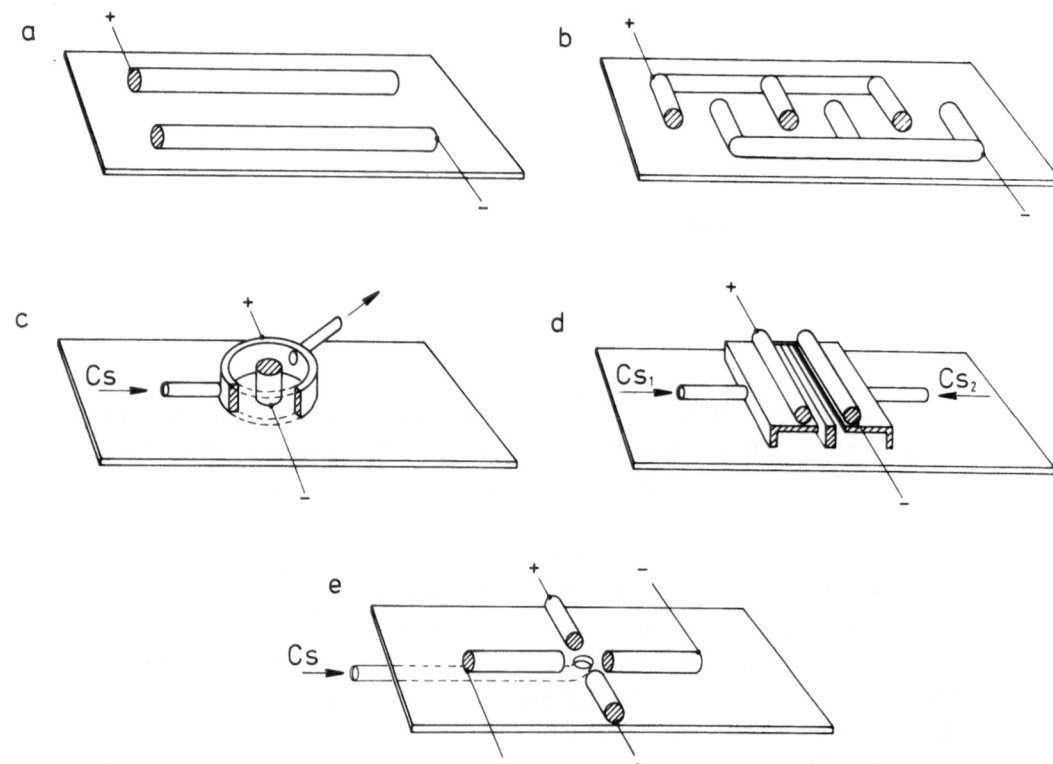

Fig. 25a–e. Set-up used for electrically induced fusion. **a** Two cylindrical platinum wires are mounted in parallel on a slide. The gap between the two electrodes has to be matched to the size of the cells to be fused. **b** Electrode arrangement for the production of large amounts of fused cells; so-called meander chamber. **c** Stronger divergence of the field can be achieved with a set-up in which a central wire (or cylindrical electrode) is surrounded by an outer cylindrical electrode of larger diameter. **d** Flow chamber system for the production of cell hybrids of different species in high yield. Two highly diluted cell suspensions are successively sucked through the slit between the electrodes. **e** Flow chamber system in which the cells enter centrically the gap between 4 electrodes arranged crosswise. Cells move to the region of highest field intensity between the individual electrodes due to dielectrophoresis

Dielectrophoresis should, if possible, be performed in weakly conductive solutions. Normally, sugars are used to maintain isotonicity. Certain conditions have to be fulfilled so that the cells can survive in this environment. The problems involved have been discussed in several review articles (20, 22, 28). However, we can confirm that, if the individual stages of electrofusion are carried out correctly, there are no adverse side effects, not even when the cells are incubated in weakly conducting solutions [see also (154)]. This conclusion is supported by similar findings in free flow electrophoresis where weakly conductive solutions are used for cell separation (155–158).

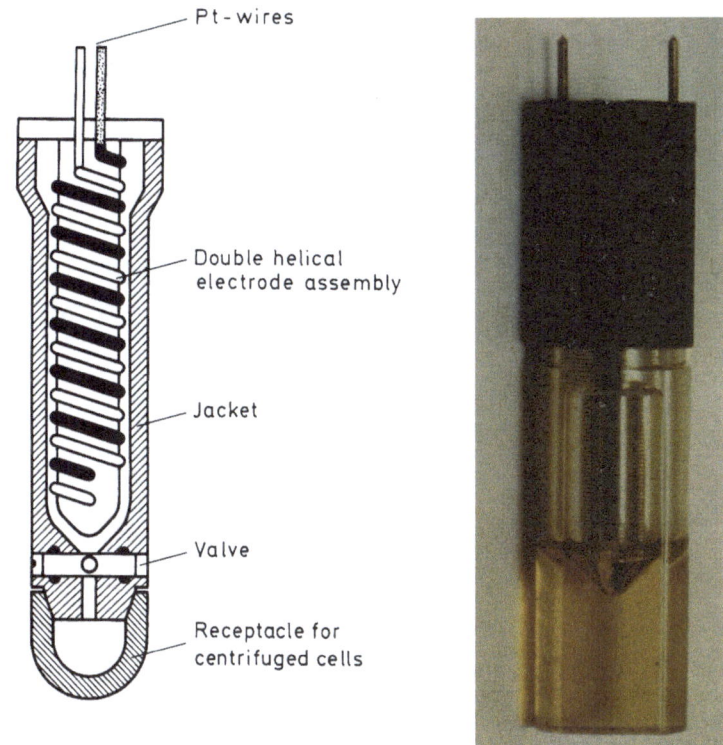

Fig. 26. Large-scale production of fused cells. In principle the helical chamber consists of a hollow Perspex tube around which are wound two cylindrical Pt-electrodes at a distance of, say, 200 μm (total length about 1 m). The Perspex tube can be filled with cooling medium in order to avoid temperature increases during the field application. The Perspex tube with the Pt wires is introduced into a cylindrical jacket which already contains the cell suspension. The cell suspension rises up through the intervening space. Analogous to the arrangment in Figs. 19, 20 and 25, cell chains are formed between the parallel electrode wires in response to an alternating field. After application of a breakdown pulse and subsequent fusion, the cells can be collected by centrifugaton into a receptacle by way of a valve. The receptacle is disposable and can very easily be removed for cloning of the fused hybrids. The whole procedure can be performed under sterile conditions. From (67); the photograph was given by courtesy of GCA Corporation, Chicago, Illinous, USA

Finally, I would like to discuss the degree of inhomogeneity of the field required for optimum fusion. An inhomogeneous field can be established by using either appropriately shaped electrodes (Figs. 25–28) or a homogeneous field with the appropriate suspension density. In the latter case, which also represents the first successful method for fusing cells by electrical means (16), a homogeneous field is established between the plates of a capacitor. The inhomogeneity of the field required for the establishment of close membrane contact is produced by the cells which are introduced into the field. Each cell causes a distortion of the field lines in its

Fig. 27. In principle the rotational fusion chamber consists of a cylindrically shaped casing which may be made to act as a centrifuge by spinning it around its axis of symmetry. In practice the floor of the chamber consists of a Plexiglass carrier plate onto which electrodes have been previously vacuum-evaporated. Of the various electrode configurations, Fig. 27 (*left, below*) shows a configuation which has proved suitable for the mass production of electrically fused yeast hybrids and hybridoma cells. The electrodes cover the circular area of the carrier disk, with electrode pairs (two electrodes running in parallel at a distance of 125 μm) radiating out from the center (see Fig. 28). The thickness of the vacuum-evaporated electrodes is about 1 μm. The circular electrode-plate (or disk) is surrounded by a lower-lying, annular channel into which the fused cells can be centrifuged after the fusion process has been completed. The channel is filled with postfusion medium, when hybridoma are produced [see (91)].

The figure on the right side shows a drive unit consisting of a rotor driven by a motor. After the electrofusion event the entire chamber can be inserted into the rotor for centrifugation of the fused cells. After centrifugation an outlet system using a valve allows the cells and the medium to be collected directly from the annular channel into culture vessels. During the fusion process the kinetics of fusion can be monitored through the carrier plate by means of an inverted microscope.

This new chamber (available commercially from GCA Corporation, Chicago, Illinois, USA) not only permits large-scale production of hybrids, but also allows rapid screening of optimal media components and other system parameters (e.g., the ratio of cell numbers when fusing cells of different species). For this purpose droplets of cell suspension are applied to different points on the disk area. Because of the cohesive forces of the water, the droplets do not spread out, so that it is possible, for example, to test quickly the influence of different ions in the fusion medium on the yield of fusion

immediate surroundings. If another cell is present in the immediate vicinity of the first cell, i.e., if high suspension densities are used, the cells attract each other. High suspension densities thus cause the field between the capacitor plates to become strongly inhomogeneous. The distinction

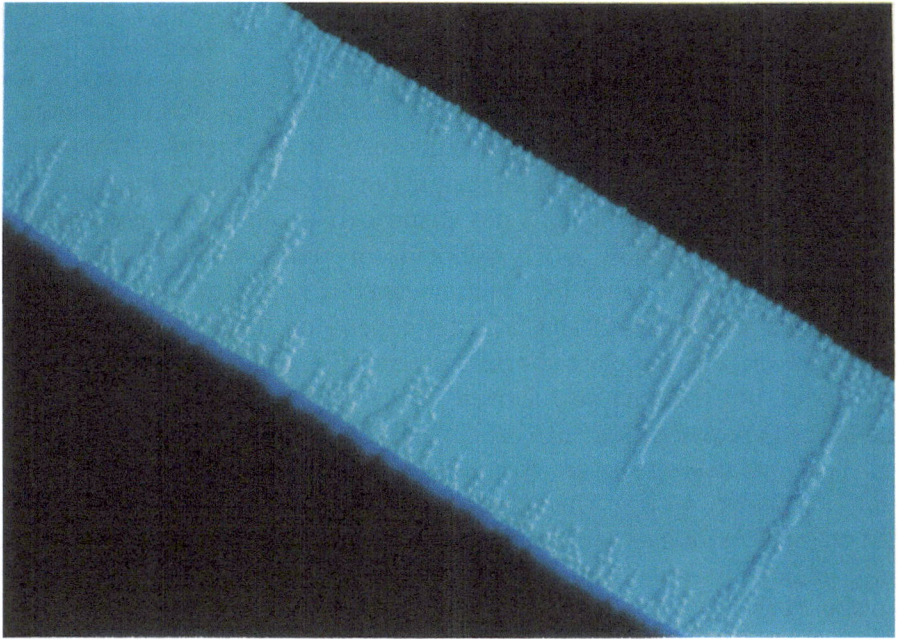

Fig. 28. Fusion of yeast cells in the rotational fusion chamber is shown after application of the breakdown pulse [see (90, 91) for experimental conditions]. The chamber can be sealed with a lid, allowing the entire fusion process to be completed under sterile conditions

between homogeneous and inhomogeneous fields (157) is therefore misleading from a physical point of view. This aspect has already been discussed in detail elsewhere (20).

Which method is chosen depends on the nature of the problem to be solved and on the cells under investigations as well as on the cell suspension density. If a large number of cells is available, it is possible to use a field in a plate capacitor for the fusion of the cells in combination with a discharge capacitor (see Fig. 9). Under these conditions the electrical breakdown pulse may already be sufficient to achieve migration over a short distance during its application. Though relatively short, this pulse is of high intensity.

Alternatively, the alternating field need only be applied for a very short period of time (a few seconds) in order to establish intimate membrane contact. The critical suspension density required for this experimental procedure can be calculated with Eq. 6 in (20). However, in this particular experimental set-up, there is no provision for controlled fusion between cells of different species. Fusion between cells occurs at random (statistically). Where few cells are available, or in experiments where controlled membrane contact between one parental cell of one species and one parental cell of a second species is required, it is necessary to use an appropriate inhomogeneous alternating field with appropriately shaped electrodes (Fig. 25). The aim here is to reduce the distance between the cells by making them migrate further, so that they come within the sphere of attraction of the appropriate partner cell. In producing mouse hybridoma cells, Vienken and Zimmermann (94) were able to show that the inhomogeneity of the field in the vicinity of a cell can be used for the preferential production of lymphocyte/myeloma combinations (94). When geometrically dependent inhomogeneous fields are used, cell chains are formed whose length and position depend on the suspension density and duration of the applied alternating field. Cell chains emanate from the electrodes, but they are also formed in the centre of the electrode gap as mentioned above (see Fig. 23).

Analogous cell chains are formed when the suspension density in the plate capacitor is high. The statement by Watts and King (159), that in the presence of geometrically dependent inhomogeneous fields fusion were only possible between cells adhering to the electrode surface is incorrect. Fusion between adhering cells occurs throughout the entire electrode space, for example in the case of two parallel cylindrical Pt electrodes glued onto a microslide. With this electrode arrangement, the field is only weakly inhomogeneous so that the equations for the field calculation, for the induced membrane potential and for the relaxation time of the establishment of the membrane potential, which were derived for a homogeneous field, are virtually wholly applicable. Focussing through the depth of the electrode space with a microscope shows clearly that most of the cells are arranged in chains in the electrode gap and that they fuse in this location. In cases where the density of the cells is greater than that of the medium, fusion of the cell chains occurs preferentially on the floor of the fusion chamber, i.e., on the surface of the microslide (Fig. 23).

We can therefore establish the following practical guidelines for the electrofusion of cells of one or more species:

1. The dielectrophoretic force or the attractive forces between two cells depend on the dielectric properties of the medium and the cell. These in turn are dependent on the frequency. The dielectrophoretic force is proportional to the volume of the cells and to the square of the field strength and increases with increasing inhomogeneity of the field (divergence of the field) (Eq. 7). For cells with a large diameter (e.g., plant protoplasts), alternating fields with only very weak field strengths are required in order to achieve dielectrophoresis and membrane contact (157). With high suspension densities, membrane contact may already be achieved during the short duration of a breakdown pulse. Fusion is possible under these conditions. The field may be "geometrically" homogeneous.

2. In the case of smaller cells (e.g., most animal cells), the field should be weakly inhomogeneous, particularly if only low suspension densities are available. It is possible to use fields that would still be effective with large cells (protoplasts), but either higher field strengths would be required or the alternating field would have to be applied for longer periods of time at a weaker intensity in order to achieve membrane contact. Both procedures could have adverse side effects on the viability of the fused cells, because the presence of an alternating field or a square field pulse causes interactions with the membrane components by way of the tangential components of the field (20). The result is a redistribution of proteins (and probably also of charged lipid molecules). If there is extensive redistribution over large areas of the membrane, the effect may be irreversible. Irreversible clustering of proteins and other molecules influences the viability of the fused cells in a negative way. It is therefore better to increase the divergence of the field rather than the field strength itself. The divergence should be chosen so that cell contact is achieved within 20–60 s.

3. In the case of very small cells (bacteria and organelles), more strongly divergent fields [e.g., a central electrode wire surrounded by a cylindrical electrode (Fig. 25c)] must be used in order to establish adequate contact between the membranes. If weaker inhomogeneous fields and correspondingly higher field strengths are used, there is a danger that the dielectric breakdown voltage of the cells is reached or even exceeded during the application of the alternating field. As discussed elsewhere (20), this has a lethal effect on the cells. The commercial fusion devices supplied by GCA Corporation, Chicago, Illinois, USA, and Bachofer, Reutlingen, Federal Republic of Germany, therefore include an alarm which is triggered as soon as this critical field range in the alternating field is reached.

4 Fusion without Dielectrophoresis

Membrane contact with subsequent application of field pulses can also be achieved by other chemical and physical means. In general, the use of chemicals (usually PEG) does not allow controlled membrane contact to be established between suspended cells, whereas the standard electrofusion technique does (nonuniform alternating field and breakdown pulse). One exception is the establishment of membrane contact by specific chemical reactions (e.g., avidin-biotin binding). This interesting technique is dealt with later. In this section we will only discuss methods that use physical (i.e. vectorial) forces to establish membrane contact. The simplest procedure is to use high suspension densities (16, 20), discussed in the previous section. In the extreme case, fusion and production of hybrids are also possible in a cell pellet obtained by centrifugation (20). The high suspension density technique has also been employed by Watts and King (159) for plant protoplast fusion. However, the theoretical background of this paper is rather poor. The authors suggest that some kinds of cooperative effects are responsible for membrane contact at high suspension densities because they recorded no fusion below a critical suspension density.

The following represents another very efficient way of fusing cells with electrical field pulses (28, 160): free suspended cells at low suspension densities or at very high dilutions are subjected to field pulses followed by centrifugation. Under these conditions fusion occurs in the pellet. In this way it was possible to produce yeast hybrids and also hybridoma cells (by fusing myeloma cells with lymphocytes). Yields tended to be high if the whole process was carried out at a low temperature because the life span of the field-induced perturbations in the membrane was longer. However, higher field strengths were required because the reduced temperature increased the breakdown voltage (see above). These experiments also show that the simple pore model for electrical breakdown is not appropirate because the "pores" would not be in the correct orientation to each other in the pellet.

Both techniques share the disadvantage that the fusion process cannot be monitored by optical means.

Cells can be pushed together mechanically by micropipettes and then subjected to a fusion pulse, as shown by Senda et al. (84) and by Schnettler and Zimmermann for yeasts (unpublished data).

This method is restricted to the fusion of only a few cells. More cells can be fused if the filter technique is used (28). Cells of each species are sucked into the pores of filters or artificial membranes or are bound electrostatically to the surface of artificial membranes. The cell-loaded filters (or membranes) are placed between two parallel flat electrodes. The distance between the two electrodes or filters can be reduced with the aid of a suitable device (e.g., a micrometer screw) until contact is made between the cells in the two sheets (combination with dielectrophoresis is also possible). Preliminary experiments have shown that high yields of hybridoma cells can be obtained with this technique (160).

Using Fe_3O_4 particles, Kramer et al. (161) were able to produce magnetically coated erythrocytes (161) and yeast cells (unpublished data). These coated cells could be attracted into a small volume by means of crossed magnetic fields (Fig. 29). The force directed towards the centre of this volume was sufficient (the fields had strong gradients) to establish

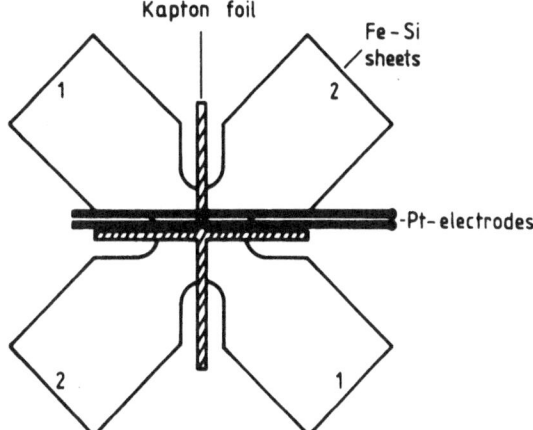

Fig. 29. Scheme of the chamber used for magneto-electro-fusion. Thin Fe-Si sheets are glued on top of a support. The constant distance of 12.5 μm between the four sheets is achieved by means of Kapton films. On top of the sheets two Pt-electrodes (diameter 150 μm) are arranged about 200 μm apart from each other and separated from the metal sheets by means of an insulating foil (not shown). A magnetic-field gradient is induced in the area of the cross by applying 5 V (for fusion of red blood cells) to two independent coils with iron cores (number of windings, 1700; R = 32 Ω) which "sit" at points denoted by 1,1 and 2,2. Any magnetic particle (e.g., a "magnetic" blood cell) experiences forces in these magnetic gradients so that it moves towards the "gaps." The geometry not only causes an alignment of magnetic particles but also gives rise to close contact between the particles because there is a moderate force directed to the midpoint. The fusion of the magnetically collected cells was triggered by two electrical breakdown pulses. From (161)

contact between the cells so that the application of two field pulses produced fusion.

If the fusion products contain magnetic particles (due to uptake during the intermingling process), then this may be used to advantage in a subsequent magnetic isolation of fused cells, allowing for example the isolation of hybrids formed from commercially important yeast cells (which possess no usable genetic markers). The breakdown voltage of coated cells was found to be 40% higher than that of normal cells, and the fusion pulse had to be considerably stronger and longer than expected. Possibly the magnetic coating acts to attenuate shorter pulses or to prevent optimal contact between the cell membranes.

A further advantage of this technique is that fusion can be performed in conductive solutions because of the absence of heat production. The amount of heat developed during the breakdown pulse is so small as to be negligible.

In a sonic field forces are exerted on the particles in a way that is analogous to dielectrophoresis (162, 163). The force exerted in the sonic field depends on the difference in density between particles and medium,

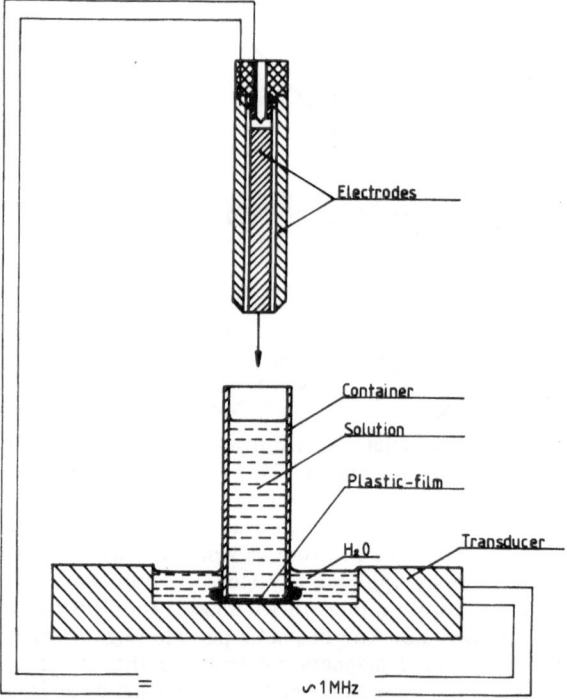

Fig. 30. Set-up for electro-acoustic fusion. For further details, see (164, 165)

whereas in dielectrophoresis the difference in dielectric constants is the important factor.

Sound wavelengths have been used which are much smaller than the fusion chamber. This allows not only the production of pearl chains (in a purely propagating wave) but also the formation of very high concentrations of cells at standing-wave pressure maxima. Vienken et al. (164, 165) were able to use 1 MHz ultrasound (1-mm wavelength) to concentrate erythrocytes and myeloma cells. The standing-wave maxima of 0.5 mm periodicity were developed between the coaxial electrodes of the apparatus shown in Figs. 30 and 31. Fusion followed the application of a pulse consistent with normal electrofusion practice, so presumably good membrane contact was achieved. By varying the cell density it was possible to achieve preferential formation of two-cell or multi-cell fusion products (Fig. 32). Fusion can also be performed in a conductive solution.

The author believes that this technique may be of great value in the future because of the simplicity of the procedure and because there is no need to lable the membranes of the cells. It should be noted that only a piezoelectric crystal is needed in connection with the fusion device (GCA Corporation, Chicago, Illinois, USA, or Bachofer, Reutlingen, Federal Republic of Germany.

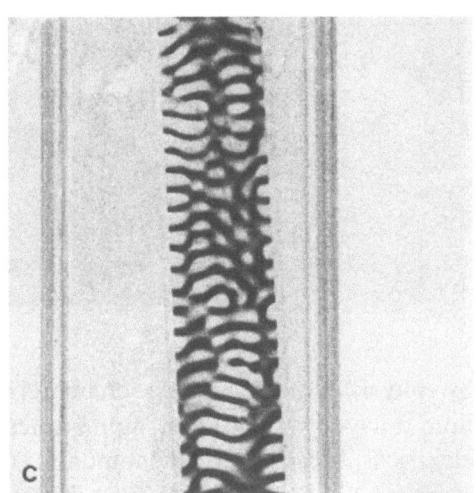

Fig. 31a–c. To illustrate the banding of red blood cells following sound irradiation a Camlab-Microslide is filled with a suspension of erythrocytes and inserted into the container of the electro-acoustic fusion set-up (**a**) (see also Fig. 30). Two seconds after the beginning of irradiation, cells are no longer suspended homogeneously (**a**), but show a banding over the whole length of the microslide (**b**) due to the sound aggregation forces. At higher magnification (**c**) the concentration of red blood cells at planes separated by half a wavelength and at right angles to the direction of sound propagation is clearly to be seen

Finally, it may well become possible to bring cells together by means of laser beams. This has already been done successfully by Ashkin and Dziedzic (166) with oil drops 7 μm in diameter.

5 Electrofusion

From the very beginning of electrofusion, Zimmermann and co-workers (e.g. 19–21) have suggested that this technique should be a general, predictable fusion method applicable to all living cells, liposomes and planar lipid bilayer membranes. The proposed universality of the method is based on the fact that both electrical breakdown and membrane contact

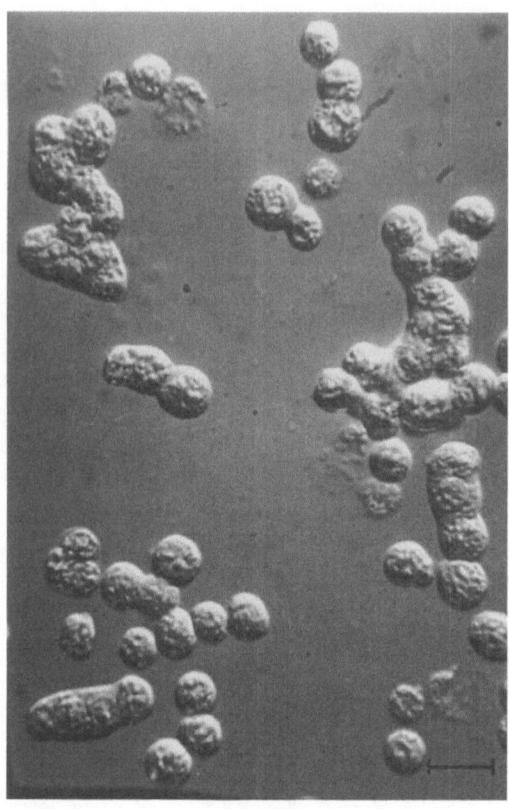

Fig. 32. Electro-acoustic fusion of myeloma cells. After a 10-s period of sound irradiation three electrical field pulses of 10 μs duration and 60 V amplitude are applied. Fusing cells are now maintained in contact by a further sound irradiation of 50 s and then pipetted on a microslide. The elongated cell aggregates will become spherical after the addition of PBS approximately 10 min later (interference phase micrograph, *bar* 20 μm). From (164, 165)

by physical means reflect characteristic features of living and artificial lipid bilayer membranes. Membranes of dead cells do not exhibit the electrical breakdown phenomenon (167). In the meantime, so many studies have been carried out that we can safely assume that any given cell or liposome/lipid membrane system can be fused with the aid of electrical fields.

Results obtained up to the beginning of 1984 have been extensively reviewed (19—28). In this review we shall therefore restrict ourselves to a brief summary of these data and discuss in more detail the most recent findings on this subject matter.

5.1 Planar Lipid Bilayer Membranes

Electrofusion of black lipid membranes has so far only been reported by Melikyan et al. (168). Using a special Teflon cell, two planar lipid bilayer membranes were brought into juxtaposition at a distance of about 1 mm. The application of a small hydrostatic pressure difference forced the two

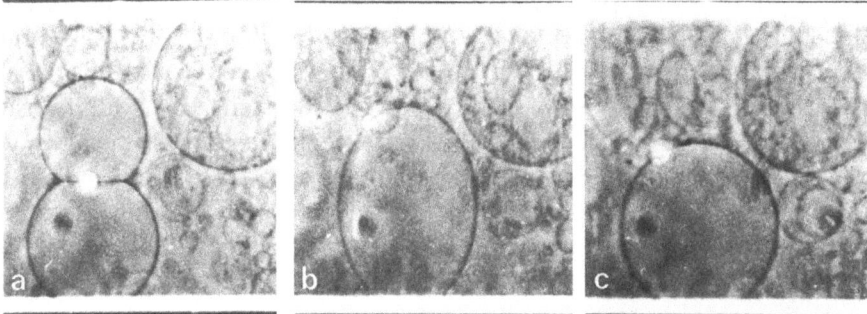

Fig. 33a–c. Phase-contrast photographs of the fusion process of lipid vesicles, prepared from a 1:1 mixture of polymerizable lipid butadien derivate/cholesterol (171); diameter of the vesicles, 37 μm and 45 μm, respectively; **a** alignment in an alternating electrical field of about 200 V/cm; **b** elongated fused liposome, 1 s after application of a pulse of 3 kV/cm strength and 30 μs duration; **c** spherical new vesicle after turn off of the alternating field; diameter of the new liposome 51 μm

membranes to bulge towards each other until they made contact. Subsequent injection of an electrical breakdown pulse gave rise to the formation of a cylindrical membrane, i.e., the two bilayers fused. The membranes were made up of dioleoyl-phosphatidylcholine (DOPC), DOPC/cholesterol (9:1 w/w), azolectin or phosphatidylethanolamine (PE). If symmetrical charged or uncharged membranes were used, e.g. two DOPC/cholesterol or two azolectin membranes, fusion could only be obtained by external field pulses. On the other hand, it a charged membrane made up of azolectin (surface potential 87 mV) was brought into contact by the hydrostatic pressure difference with a neutral bilayer made up of DOPC/cholesterol, spontaneous fusion was observed. Melikyan et al. (168) interpreted these findings in terms of the hypothesis of Gingell and Ginsberg (169) and Zimmermann and Scheurich (170), namely that in the latter case the intrinsic field is strong enough to fuse the membranes (see also above). Measurements of the surface charge revealed that the cylindrical membrane formed by either spontaneous or electrostimulated fusion of charged or uncharged bilayers has an intermediate surface charge which correlates with the area ratio of the noncontacting parts of the two individual membranes in the "stuck" state prior to fusion.

5.2 Liposomes

Büschl et al. (171) described fusion of large liposomes made up of natural and polymerizable lipids, using the standard electrofusion technique (Fig. 33). Both preparation of the large liposomes (average diameter about 5 μm) and electrofusion were performed in distilled water (172). The data has been reviewed [e.g. (22)], and we will only mention the fact that liposomes without surface charge (e.g., DOPC liposomes) exhibit negative dielectrophoresis in the kHz frequency range, whereas either positively or negatively charged liposomes (e.g., phosphatidylserin, PS) exhibit positive dielectrophoresis in the kilohertz range. Electrofusion of liposomes with cells is also possible (173). Because of the presence of a cell partner, such experiments have to be carried out in isotonic sugar solutions. It is interesting to note that liposomes made up of charged lipids also exhibit negative dielectrophoresis in such solutions. Positive dielectrophoresis of such liposomes, which is required for membrane contact between liposomes and cells, can be re-established if electrolyte is entrapped inside the liposomes during their preparation. However, liposomes cannot be obtained in the presence of high electrolyte concentrations. On the other hand, preparation in 6-carboxyfluorescein gives rise to liposomes which show positive alignment (173). At the same time it is easily possible to monitor the fusion of such stained liposomes with cells under a fluorescence microscope. Büschl (173) was able to show that liposomes made up of polymerisable lipids can be electrofused with cells under these conditions. These studies also pave the way to future experiments in which DNA encapsulated in liposomes can be entrapped in cells with the aid of electrofusion.

5.3 Bacteria

Shivarova et al. (174) reported fusion of *Bacillus thuringiensis* by combined chemical-field treatment. These authors agglutinated protoplasts of two gram-positive mutant strains of *B. thuringiensis*. One strain, containing the plasmid pUB 110, was resistant to kanamycin; the second was sensitive to kanamycin and produced a brown pigment. The agglutinated protoplasts were subjected to three short consecutive electrical field pulses with an intensity of 20 kV/cm. The authors reported that colonies were obtained in selective media which were kanamycin-resistant and able to produce the brown pigment. With PEG alone as the fusogenic chemical, no colony formation was observed.

The first electrofusion of bacteria was recently reported by Ruthe and Adler (175). These authors fused spheroplasts of *Escherichia coli* (*E. coli* AW405 and *E. coli* DH1) and *Salmonella typhimurium* (*S. typhimurium* TH32) using the standard electrofusion technique. Alignment of the normal gram-negative cells of *E. coli* or *S. typhimurium* was observed in an inhomogeneous alternating field, but electrofusion was not possible, not even at a field strength of 25 kV/cm, presumably because of the rigidity of the peptidoglycan layer of the bacterial cell envelope.

Spheroplasts (size between $2-10$ μm) obtained by growth of the cells in the presence of penicillin (an inhibitor of peptidoglycan synthesis) were also not fusable with electrical field pulses. Neither was electrofusion possible with spheroplasts ($0.5-1.0$ μm in diameter) obtained by the treatment of the bacteria with Tris/EDTA/lysozyme to hydrolyse the peptidoglycan layer. However, electrofusion was easy when giant spheroplasts were used. For this purpose the bacteria were first grown into long filamentous cells, $50-150$ μm long, in the presence of cephalexin, a β-lactam antibiotic. These filaments were then treated with Tris/EDTA/lysozyme to form giant, nonseptate spheroplasts (average diamater 7 μm). More than 90% of the spheroplasts of $E.$ $coli$ and $S.$ $typhimurium$ can revert to the bacterial form.

Alignment of the spheroplasts was found to be optimal at a field strength of 1 kV/cm and a frequency of 1 MHz. High fusion yield (more than 60%) of the aligned spheroplasts was observed when the cells were subjected to a sequence of 5 pulses, 5 kV/cm in strength and 15 μs in duration, provided that fusion was carried out in a medium containing 0.8 M sucrose and 1 mmol MgCl. The addition of Ca^{2+} was not as useful as Mg^{2+} and tended to inhibit fusion. Pronase was also found to be of little use in these biological systems. Viability and reversion of the fusion products were demonstrated by electrofusion of a mixture of cells with two different genetic markers.

In one set of experiments Ruthe and Adler (175) used $E.$ $coli$ DH1 because its properties F^-, λ^- should reduce the spontaneous exchange of genetic material. $E.$ $coli$ DH1 was transformed by non-self-transmissible plasmids carrying eithr tetracycline or kanamycin resistance. Giant spheroplasts of the two transformed strains were electrofused and transferred to selection medium. Five to 200 colonies of $E.$ $coli$ which were resistant to both tetracycline and kanamycin were counted on plates containing both antibiotics. Genetic markers for $E.$ $coli$ DH1 were found for all doubly resistant colonies tested. In a second set of experiments these authors electrofused spheroplasts of $S.$ $typhimurium$ TH32 with those of $E.$ $coli$ DH1. Being sensitive to kanamycin, the first strain is highly motile and can grow on citrate as the sole carbon source, while the second one (nonmotile) cannot utilise citrate, but when transformed with the kanamycin resistance plasmid, it can grow in the presence of kanamycin. Electrofusion of a mixture of these two strains yielded about 100 colonies in each of the seven experimental runs which were able to grow on citrate as the sole carbon source and in the presence of kanamycin. The colonies exhibited motility characteristics of $S.$ $typhimurium$ TH32.

5.4 Fungi

Electrofusion of fungal protoplasts has recently been demonstrated by
Schnettler and Zimmermann (unpublished results). The experimental
conditions, particularly the fusion medium, are almost identical with
those developed for the fusion of yeast by Schnettler and Zimmermann
(90).

5.5 Yeast

Chemical (PEG)-induced fusion of two different genetically marked
strains of the yeast *Saccharomyces cerevisiae* has been reported by Weber
et al. (176, 177). In their experiments the fusion rate of yeast protoplasts
was enhanced by a factor of 200 by the external electrical field pulse,
compared with the yield obtained with PEG alone. At an optimal field
strength of 2.5 kV/cm, these authors obtained 233 colonies per plate containing the minimal medium, compared with the initial 108 protoplasts
subjected to fusion. The low field strength of the pulse is very surprising,
because with a mean cell radius of 3 μm for yeast cells this field strength
would only just be sufficient to reach the breakdown voltage. With the
standard electrofusion technique (see below) higher field strengths are
required in order to achieve high yields of fusion products (see below).
Weber et al. (176, 177) also reported higher field strengths for inter-
genetic PEG/field fusion (*Saccharomyces lipolytica/Loderomyces elongi-
sporus*), but not for the intra-specific PEG/field fusion of *S. lipolytica.*
The very low field strengths required for the fusion of *S. cerevisiae*
mutants may be due to the natural properties of these cells. Weber et al.
(177) used strains with the phenotype a and α which tend to exhibit
spontaneous fusion anyway if the cell wall is not removed.

  Halfmann et al. (178) were the first to report that yeast protoplasts
can be fused with high yields by means of the standard electrofusion
technique (Fig. 34). These authors fused the two heterozygous diploid
strains of *S. cerevisiae* 3441 and 2114 with each other. Strain 3441 carries several auxotrophic markers, whereas strain 2114 is a respiratory
deficient mutant. Before electrofusion the cells were field-stabilised by
pronase pretreatment. Alignment was achieved at a field strength of
1 kV/cm and at a frequency of 2 MHz. Fusion was initiated by the application of two consecutive breakdown pulses, 7–8 kV/cm in strength and
40 μs in duration. An anylsis of the colonies of the fusion products in
selection medium showed that a high rate of plasmogamy had occurred
($1-2 \times 10^{-2}$ compared with about 10^{-5} when conventional fusion techniques were used), but that no karyogamy had taken place. The absence

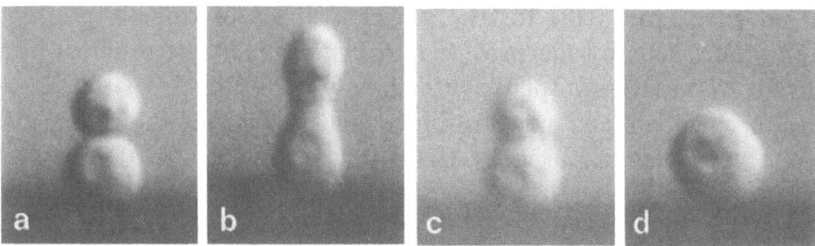

Fig. 34a–d. Electrofusion of yeast protoplasts of two different strains (*Saccharomyces cerivisiae 2114* and *3441*). **a** Dielectrophoresis was performed in an alternating electrical field (1.8 kV/cm, 2 MHz). **b** Fusion was initiated by two subsequent breakdown pulses (7 kV/cm, 40 μs duration) applied at an interval of about 10 s. **d** The final stage is reached after about 3 min. From (178)

of karyogamy seems to be an inherent feature of the heterozygous diploid strains. It is well known that during the life cycle of yeasts, copulation with subsequent karyogamy takes place only between haploid cells or between diploid cells that are homozygous for mating type. In a follow-up paper, Halfmann et al. (179) were in fact able to show that electrofusion of protoplasts of the haploid partners *S. cerevisiae* 21a and 111a (identical mating type) resulted in karyogamy. By appropriate labelling of the strains it was possible to distinguish, after electrofusion, between plasmogamy and plasmogamy followed by karyogamy. In these experiments the protoplasts were not pretreated with pronase. Alignment conditions were the same as before, except that the field strength and duration of the two consecutive breakdown pulses were 11 kV/cm and 7 μs respectively. The fusion frequency of viable hybrids produced by plasmogamy was estimated to be 5×10^{-2}, that of hybrids produced by karyogamy 5×10^{-3}. Halfmann et al. (180) were also the first to demonstrate that it is possible to fuse yeast partners of different sizes. They electrofused electrical-field-aligned cells of the haploid straind *S. cerevisiae* 111a with the polyploid (triploid) strain 93 (baker's yeast) by the application of two breakdown pulses, 8 kV/cm in strength and 40 μs in duration.

The hybrids obtained after transfer to selection medium were heterokaryons which gave rise to spontaneous parental-type segregants on nutritionally complete medium. Stable hybrids could be obtained after several passages on selection medium. Genetic analysis of these hybrids showed that recombination had taken place with partial or even complete elimination of the chromosome set. The authors suggested that there was a chromosome imbalance in one hybrid.

Recently Schnettler and Zimmermann [(90), see also (181)] considerably improved the electrofusion technique of yeast cells, so that now large numbers of yeast hybrids can be produced with the aid of this technique. The work of these authors simultaneously led to the formulation of advanced protocols for the efficient and reproducible production of hybridoma cells by means of electrofusion (91). The progress in electrofusion of yeast protoplasts was mainly attributable to the use of new fusion chambers and new fusion media. Instead of the simple two electrode wire chamber used by Halfmann et al. (178–180). Schnettler and Zimmermann (90, 181) used the helical chamber (Fig. 26) for electrofusion and showed that about 60 hybrids could be obtained per experimental run when electrofusion was performed in isotonic sorbitol solution (181). However, certain additives to the fusion solution considerably increased the number of hybrids. In particular, the addition of 0.1 mmol calcium acetate, 0.5 mmol magnesium acetate, and 1 mg/ml bovine serum albumin resulted in about 4000 hybrids per experimental run in the helical chamber (90). In the meantime, further improvements are producing yields of about 7000 hybrids (Tsoneva, Schnettler and Zimmermann, unpublished data).

The field conditions were similar to those described by both Halfmann et al. (178–180) and Schnettler et al. (90). This demonstrates that the application of an alternating field and high-intensity field pulses in the appropriate manner does not lead to any severe side effects within the cells and the membrane, but rather that more consideration needs to be given to the composition of the fusion media and to the system parameters (chambers, temperature, etc.).

The studies of Schnettler and Zimmermann (90, 181) are worth mentioning from another point of view, because these authors were able to show for the first time that vector-bound DNA in yeast cells can be transferred with the aid of electrofusion. The authors used two yeast strains of the same mating type: *S. cerevisiae AH22* and *AH215* (both of mating type a). Strain *AH22* is characterised by the double mutation leu 2-3, 2-112, and the single mutation his 4-519. The transformation of strain *AH22* with plasmid pADH 040-2 results in a compensation of the leu 2 mutation. Plasmid pADH 040-2 encodes the leu 2 gene of yeast and the β-lactamase gene from *E. coli* which allow a quick detection of plasmid positive cells. Strain *AH215* carries the same double mutation leu 2-3, 2-112 and in addition the double mutation his 3-11, 3-15. Hybrids of *AH215* and *AH22* (pADH 040-2) were able to grow on minimal media containing neither leucine nor histidine because of the non-allelic his mutations and the presence of the plasmid.

5.6 Plant Protoplasts

A large number of papers have been published on the electrofusion of plant protoplasts (17–28, 87, 160, 170, 182–200) (Fig. 35). This is hardly

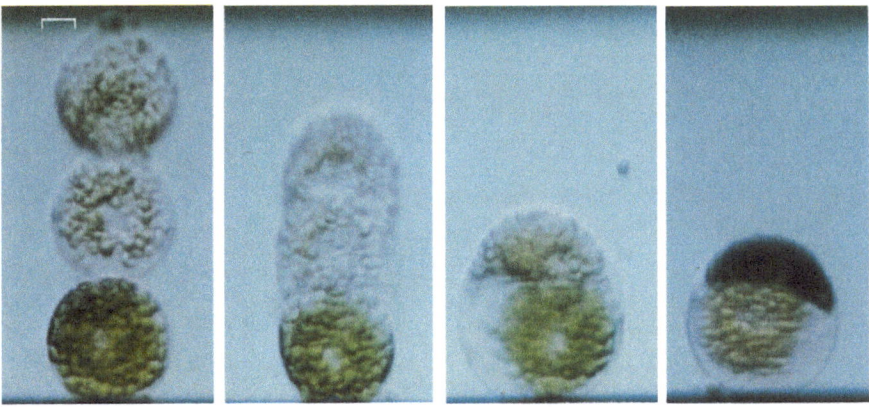

a,b   c, d

Fig. 35a–d. Electrofusion between a green mesophyll protoplast and two etiolated protoplasts of *Avena sativa*. Both kinds of protoplasts were suspended in 0.5 *M* sorbitol solution and aligned by dielectrophoresis (frequency 1 MHz, field strength 75 V/cm). The fusion process began immediately after pulse application (750 V/cm field strength and 20 μs duration). Time course of fusion: **a** before field pulse application; **b** 30 s; **c** 8 min and **d** 10 min after pulse application. Apparently the cytoplasms of the original cells do not mix, as chloroplasts and etioplasts remain in distinct areas for some time (about 30 min). *Bar* in **a** represents 10 μm. From (184)

surprising, given the enormous potential of this fusion technique for plant breeding and the fact that plant protoplasts can easily be fused by electrical means. Because of their membrane properties, plant protoplasts can be fused under virtually any field conditions and in a great variety of different media without any particular know-how. However, one factor that is frequently overlooked is that the system parameters determine the viability and regenerative capacity of the obtained fusion products. Senda et al. (84, 85), for example, fused two plant protoplasts brought together mechanically by two micropipettes by applying a field pulse of millisecond duration. As described above, this inevitably leads to irreversible changes in the membrane.

In most of the work published to date on the electrofusion of protoplasts, the standard electrical field technique (dielectrophoresis and breakdown pulse) was used. However, for the sake of completeness, we should mention that Senda et al. (84) published a follow-up paper (85) in which they describe the use of the discharge technique of Zimmermann and Pilwat (see Fig. 9) for the fusion of protoplasts isolated from cultured cells of *Rauwolfia serpentina* var. *bentham* and from mesophyll cells of *Hordeum vulgare* L., *Gose shikoku* and *Nicotiana tabacum* var. *samsum*. In these experiments the authors also used exposure times in the millisecond range for cells in the electrical field.

Alignment of plant protoplasts is normally achieved at frequencies between 20 kHz and 2 MHz. Generally a breakdown pulse of 0.5–5 kV/cm

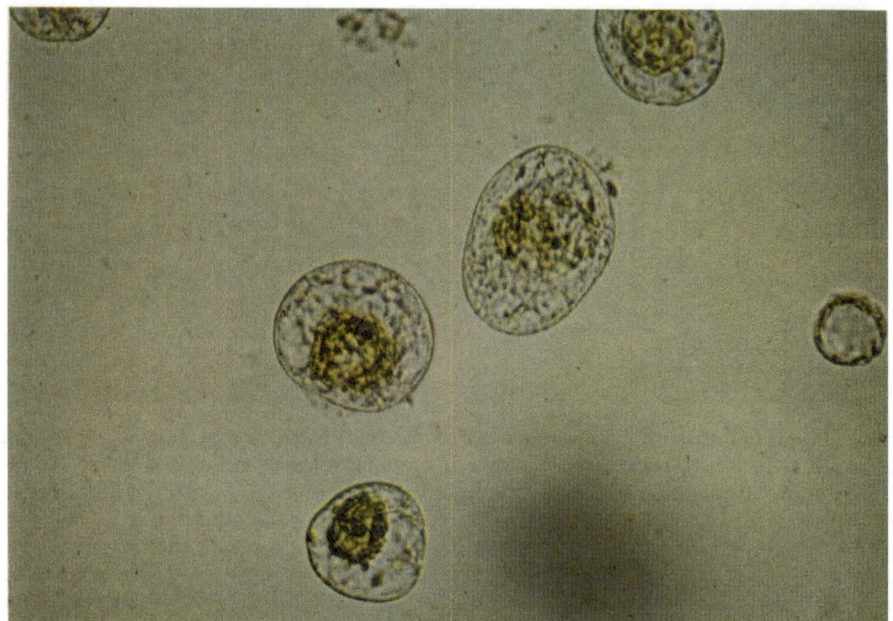

Fig. 36. Parental protoplasts (*S. tuberosum* cv. *Maris Piper*, mesophyll and *Nicotiana sylvestris* Aec R suspension) were electrically fused (pulse duration 50 μs, 1.2 kV/cm strength). The figure shows several heterokaryons after isolation by micropipette and culture within a nutrition medium

strength (depending on the radius of the protoplast) and 20–50 μs duration is sufficient.

Electrofusion is reported for protoplasts isolated from the mesophyll cells of leaves, from stomata and root tips, and from suspension cultures of *Avena sativa, Petunia inflata, Vicia faba, Kalanchoë daigremontiana, Vigna, Brassica napus, Solanum brevidens, Triticum aestivum, Hordeum vulgare, Nicotiana sylvestris, N. rustica, N. tabacum, Datura innoxia, S. brevidens, Lycopersicon esculentum*, etc. (17–28, 87, 160, 170, 182–200).

Gaff et al. (199) recently demonstrated electrofusion of protoplasts prepared from desiccation-tolerant "resurrection grasses" of the genera *Eragrostis* and *Sporobolus* (*E. nindensis, S. festivus, S. lampranthus* and *S. stapfianus*) and from desiccation-sensitive grasses of the same genera (*E. tef* and *S. pyramidalis*).

Saga et al. (192) electrically fused protoplasts isolated enzymatically from the marine alga *Enteromorpha intestinalis*. Their work could be important because of the potential of somatic hybridisation for the development of new marine crop plants.

Cell wall regeneration of electrofused plant protoplasts was first demonstrated for *Petunia inflata* by Zimmermann et al. (19).

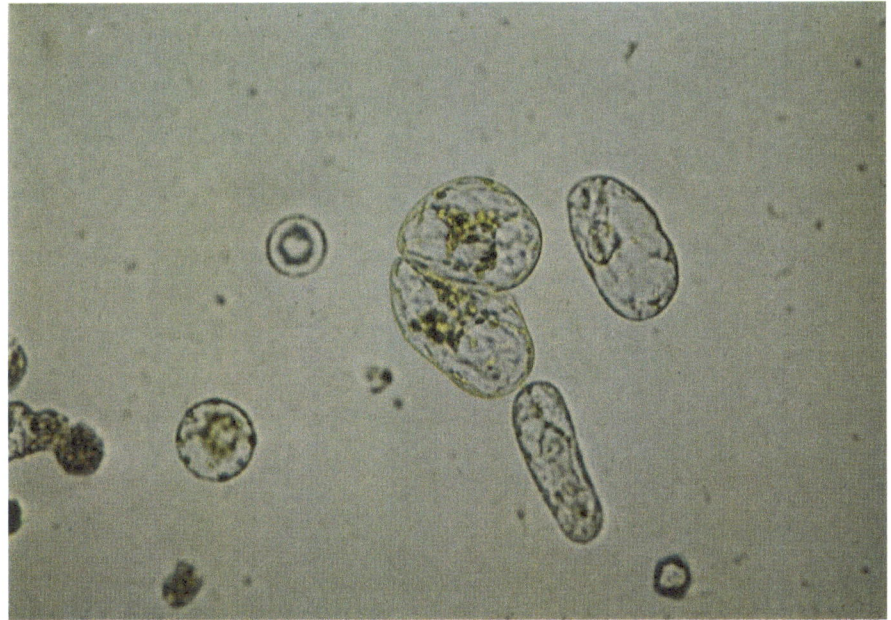

Fig. 37. Dividing heterokaryon of the electrofusion experiment described in Fig. 36

Electrical-field mediated interkingdom fusion was recently described by Salhani et al. (200). These authors fused a plant protoplast of *Petunia hybrida* cv. *comanche* with a Friend cell. Properties of both parents were expressed in the fusion product. Cell wall regeneration was observed, and haemoglobin synthesis could be induced with the addition of DMSO (dimethylsulfoxide).

Proof that electrically fused plant protoplasts are able to regenerate to the callus stage was first provided by Koop et al. (195, 196) who fused two mesophyll protoplasts of *N. tabacum* cv. *xanthi* by electrical means. The fusion products were transferred in microdroplets of nutrient medium and cultured individually. From 100 individually cultured fusion products, Koop et al. (195, 196) succeeded in regenerating 62 microcalli. The ability to regenerate to the callus stage was not significantly different from that of unfused protoplasts from the same course. Tempelaar and Jones (190) demonstrated that regeneration of shoots from calli obtained by electrofusion is also possible. These authors produced heterokaryons by electrofusion of mesophyll protoplasts and suspension cells of different species (Figs. 36–38). Kohn et al. (197) reported somatic hybridisation of electrofused protoplasts isolated from suspension cultures of two nitrate reductase deficient mutant cell lines of *N. tabacum* cv. *gatersleben*. They aligned protoplasts in an alternating field of 400 V/cm and a frequency of 0.9 MHz. Optimal results were obtained by injecting a

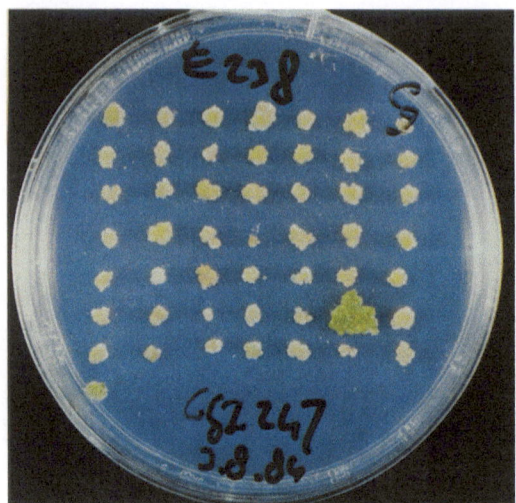

Fig. 38. Colonies from the experiments described in Figs. 36 and 37 about 6 weeks after electrofusion. Photographs (Figs. 36–38) by courtesy of Dr. M.G.K. Jones and Dr. M.J. Tempelaar, Dept. of Biochemistry, Rothamsted Experimental Station, Harpenden, Herts, U.K.

pulse of 50 μs duration and 1.5 kV/cm strengths. Somatic hybrids were selected by their vigorous growth on a nitrate-containing culture medium and were found to have an active nitrate reductase system. The authors concluded that somatic hybrids derived from electrofused plant protoplasts were viable and capable of regenerating into calli and exhibiting organogenetic responses by shoot differentiation.

Although high yields of fused plant cells (capable of whole-plant regeneration) have been obtained by electrofusion, it is thought that their viability may be compromised by the vacuoles. These may release salts, acids and lytic enzymes during the fusion process, and their large size may act to inhibit cell division. The electrofusion process itself is retarded by the presence of vacuoles. Zimmermann and Vienken (26) have reported that vacuoles only fuse about a minute after the intermingling process of the plasmalemma membranes. The fused cells can only become rounded when the vacuoles have fused. Isolated vacuoles, on the other hand, fuse in a fraction of a second after the application of the breakdown pulse.

Bearing these problems in mind, Griesbach and Sink (201) developed a centrifugal method of vacuole removal. Hampp et al. (198) subsequently demonstrated the electrofusion of such evacuolated plant protoplasts. It is interesting to note that the field strength required for electrofusion of evacuolated protoplasts is half that required for fusion of normal protoplasts (if the different radii are taken into account). This result is to be expected if the two membranes, tonoplast and plasmalemma, are switched in series – from an electrical point of view – in normal cells.

In any case, the work of Hampp et al. (198) may provide a new way of achieving better regeneration of fused hybrids of different origin for plant breeding purposes.

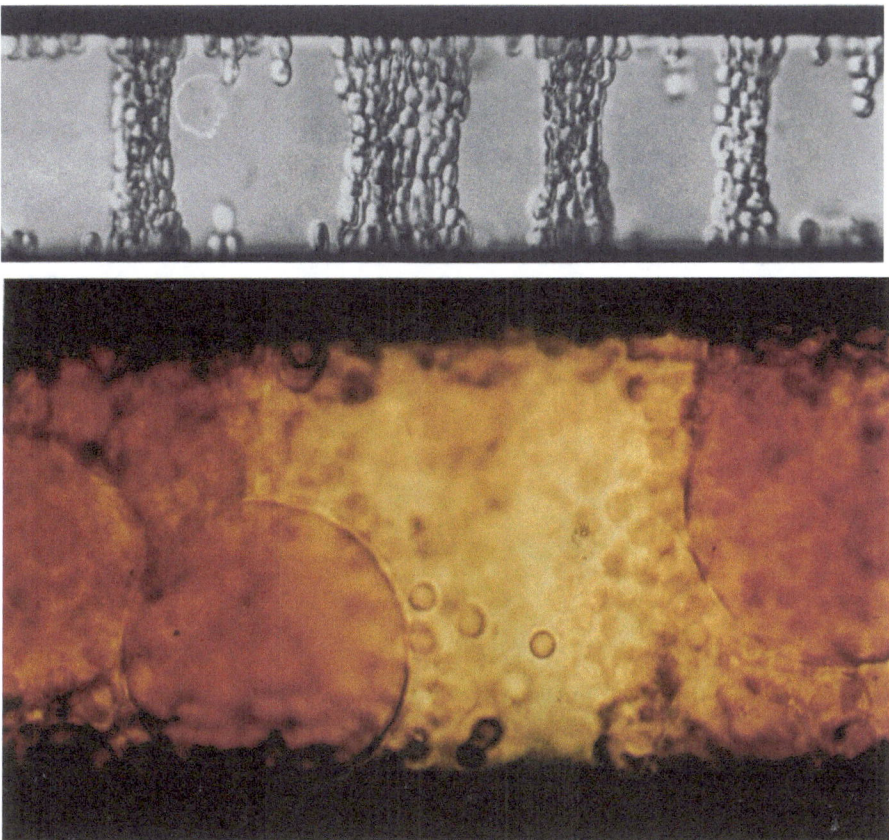

Fig. 39. Fusion of human red blood cells. *Top:* Cell chain formation of human erythrocytes at high suspension densities (400 V/cm, 1 MHz, 0.3 mol mannitol solution). *Bottom:* Fusion was induced by three subsequent electrical field pulses of 6 kV/cm strength and 5 μs duration. The giant cells are surrounded by erythrocytes which were not exposed to the electrical field. From (80)

5.7 Animal Cells

Zimmermann and Pilwat (16) were able to demonstrate as early as 1978 that discharging a high voltage capacitor through a human red blood cell suspension leads to fusion of cells. However, electrofusion using alternating fields for alignment is more efficient. The controlled fusion of thousands of red blood cells with the standard electrofusion technique (Fig. 39), giving rise to giant cells with a diameter of 30 μm to 1 mm (19−28, 202), is a spectacular example. The production of giant nucleated cells by electrical means is also possible (19−28, 203) (Fig. 40). Ultrasound field fusion and magneto-electrofusion of human red blood cells at high suspension densities favour fusion of aggregates of 2 or 3 red blood cells by field pulses, especially if ultrasound is used to establish cell contact (164, 165).

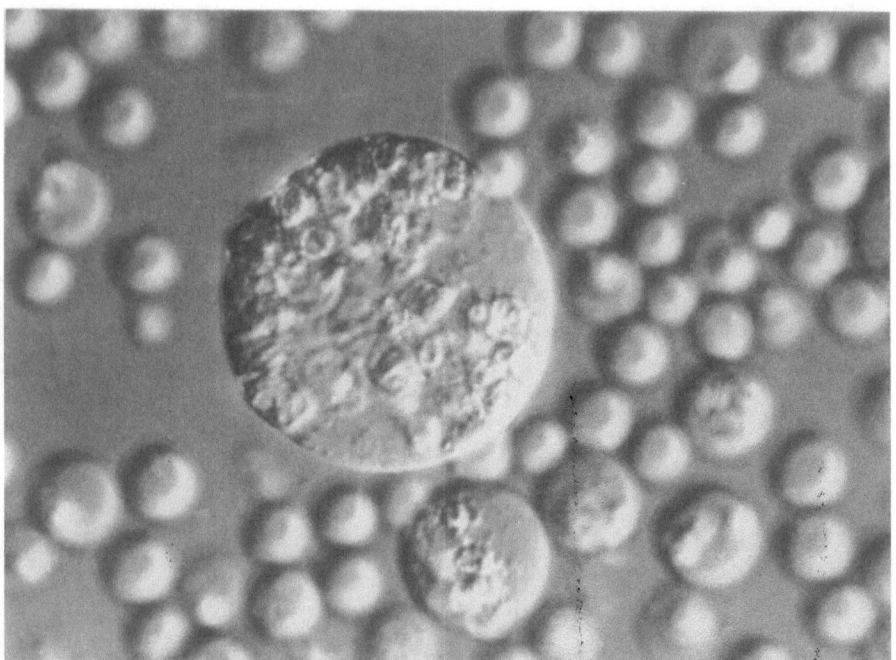

Fig. 40. Formation of giant Friend cells of different sizes obtained by electrofusion of many pearl chains attached to each other (not shown, but see also Fig. 23). Fusion was induced by the application of a field pulse of 2.7 kV/cm strength and 20 μs duration. The cells were suspended in isotonic mannitol solution. About 1 min after field pulse application electrolyte solution was added to induce the multinucleated fused cells to become spherical

Electrofusion of human red blood cells requires pretreatment of the cells with neuraminidase, pronase or dispase, whereas nucleated cells like Friend cells, myeloma cells and lymphocytes can be electrically fused in the absence of these digestive enzymes (provided that the appropriate solutions are used).

In general, however, the fusion rate of animal cells and eggs is increased by a brief enzyme pretreatment. As mentioned elsewhere (203), enzyme pretreatment increases the stability of the cells in high field strengths so that the critical value at which reversible breakdown turns into irreversible mechanical destruction of the membrane becomes significantly higher. The various possible mechanisms responsible for this effect of the digestive enzymes are discussed elsewhere (20). Recently the facilitation of electrofusion by proteolytic enzymes has been reinvestigated by Ohno-Shosaku and Okada (204, 205) who found that electrofusion of mouse lymphoma (L 5178 Y) was also enhanced by dispase, pronase and trypsin treatment. Protease inhibitors (aprotinin and p-tosyl-L-lysine chloromethylketone) suppressed the effect of trypsin. On the other hand, the protease inhibitor PMSF (phenylmethylsulfonyl fluoride) failed to abolish the pronase effect on electrofusion of mouse lymphoma cells. This finding agreed with the results reported by Vienken et al. (68) who found that fusion between SP 20 and X 63 mouse myeloma cells on the surface of microslides treated with poly(L)-lysine and pronase is not affected by the presence of PMSF. Ohno-Shosaku and Okada (204, 205) argued that the discrepancy between the different effects of inhibitors on trypsin and pronase can easily be explained by the heterogeneous composition of the pronase enzyme. They pointed out that PMSF may not inhibit all components of pronase. Their conclusion was supported by the experimental finding that heat treatment of pronase E (90°C, 3 min) significantly reduces its effect on electrofusion of mouse lymphoma cells. Ohno-Shosaku and Okada (204, 205) therefore concluded that the facilitation of electrofusion by proteases is attributable to their proteolytic activities. However, they pointed out that very low concentrations of free Ca^{2+} ions are a necessary prerequisite, having found that Ca^{2+} (but not Mg^{2+}) ions at more than millimolar concentrations are indispensable for cell fusion.

The author of this review believes that the positive effect of pronase on electrofusion may be due to changes in the surface charge and hence the surface potential, caused by the proeolytic reaction. As discussed in previous sections, a change in the surface potential leads to changes in the intrinsic membrane potential (field). If we assume that the intrinsic field is reduced, then higher field strengths are required for breakdown.

In a follow-up paper, Ohno-Shosaku and Okada (206) showed that extracellular application of phospholipase A or C suppressed the fusion yield. This finding could indicate that phospholipid domains play a crucial role in electrofusion and that the proteolytic reaction gives rise to protein-free lipid domains. This has also been suggested by Zimmermann et al. (19–28) with the comment that this possible interpretation need not necessarily contradict the "surface potential" hypothesis. The two effects may be connected or occur simultaneously. Further studies are undoubtedly necessary in order to explain the effect of these enzymes.

Normal human lymphoblasts (HSC93) and mouse leukaemic lymphoblasts (MCN151) which have a similar phenotype to Fanconi's anaemia cells have been successfully fused with the standard electrofusion technique [field conditions (204), alternating field 0.8 kV/cm in strength, duration 10 μs]. Differential staining of the two cell types with Janus Green and Neutral Red revealed that about 40% of viable fused cells represented heterokaryonic fusion.

Podesta et al. (207) recently reported steroid hormone and cyclic AMP production in adrenal-Leydig cell hybrids generated by electrofusion. Leydig cells were prepared by collagenase digestion of adult rat testis and fused with rat adrenocortical cells obtained by trypsin digestion from the zona fasciculata-reticularis. It is interesting to note that alignment of combinations of the two cell types in an alternating field of 650 kHz frequency and 2.5 kV/cm field strength took 9 min and that fusion could only be achieved by injecting 9 pulses of 15 μs duration and 3.75 kV/cm field strength. The resulting hybrids simulatenously synthesised testosterone and corticosterone when stimulated with lutropin or adrenocorticotropin.

The groups of Tsong and Teissie (208–210) reported electrical pulse-mediated fusion of cells in monolayer cultures. In these experiments natural cell-to-cell contact in confluent was used. Teissie et al. (208) fused mouse 3T3-C2 fibroblasts by electrical means. The application of 5 successive pulses of 1 kV/cm field strength and 50 μs duration caused approximately 20% of the cells to fuse in the presence of 1 mmol magnesium. Multinucleated cells were also obtained. On the other hand, electrofusion of free suspended 3T3 cells, using the standard technique, resulted in a high yield of fusion products, but high field strengths (7 kV/cm) were required to obtain viable products as verified by Trypan Blue staining [see review (20)]. Blangero and Teissie (210) fused Chinese hamster ovary cells (CHO) in monolayers. They claimed yields of up to 80%; the viability was also checked with Trypan Blue.

Using the same pulse technique, Finaz et al. (209) fused a co-culture of clone ID and CH rodent cell lines (using 5 successive pulses of 1.5 kV/cm strength and 50 μs duration). Hybrids were obtained in selection medium with a frequency of 1×10^{-3}, which is 100 times higher than with PEG. Seventeen of the 325 hybrid clones obtained in this manner were propagated for 2 months in selective medium and karyotyped and analysed for their enzymatic markers to confirm their hybrid nature.

Electrical pulse-induced fusion of a special biological system is described by Neumann et al. (211) who electrically fused clustered cells of the mould *Dictyostelium*.

5.7.1 Hybridoma

Hybridoma cells are capable of permanent production of an antibody specifically directed against a given antigen (so-called monoclonal antibody) and therefore have enormous potential in medicine (serology, diagnostics and therepeutics) and technology (212, 213). The production of hybridoma cells which secrete monoclonal antibodies is based on the fusion of myeloma cells and spleen cells from appropriately immunised animals (usually mice or rats).

Electrofusion of mouse myeloma cells with lymphocytes was first reported by Vienken and Zimmermann (94), who used a special electro-hydraulic procedure in order to obtain high yields of pronase-pretreated myeloma/lymphocyte pairs. Production of human hybridoma cells by electrofusion was reported at the same time by Bischoff et al. (214). These studies have already been reviewed in detail, so there is no need to discuss them at length here. However, we should mention that Hübner et al. (215) have also recently reported on the successful production of human hybridoma cells with the aid of electrofusion.

Recent progress in electrical field-induced production of hybridoma cells has been achieved by improving the fusion medium and the resealing conditions. Vienken and Zimmermann (91) found that, as in the case of electrofusion of yeast cells (90), divalent cations played an important role and that the sugar included in the nonelectrolyte solution in which electro-fusion is performed influenced the viability of the fusion products. Optimal fusion in the helical chamber with very high yields of hybridoma cells was obtained when fusion was performed in isotonic isositol solu-tions containing 0.1 mmol Ca^{2+}-acetate, 0.5 mmol Mg^{2+}-acetate and 1 mmol phosphate buffer. Thirty seconds after electrofusion the tempera-ture was raised to 37°C, and so-called postfusion medium was added to improve the resealing process of the membranes of the resulting hybridoma cells. The composition of the postfusion medium was: 132 mmol NaCl, 8 mmol KCl, 0.1 mmol Ca^{2+}-acetate, 0.5 mmol Mg^{2+}-acetate, and phos-phate buffer. A 30-min incubation of the fusion products in this solution and subsequent transfer to selection (HAT) medium led to yields of hybridoma cells which were 2–3 orders of magnitude higher than those obtained by PEG fusion. The hybridoma cells obtained with lymphocytes stimulated in vitro by lipopolysaccharides or by immunisation of mice with sheep red blood cells were found to secrete monoclonal antibodies (Schmitt and Zimmermann, unpublished data). It should be noted that pretreatment of the parental cells with digestive enzymes was not required in these experiments.

Karsten et al. (216) and Brown et al. (217) were also able to produce hybridoma cells capable of secreting monoclonal antibodies, using the standard electrofusion technique and conditions described by Vienken and Zimmermann in their first two papers (94, 214). Even though these authors pretreated the parental cells with pronase, the yield of secreting hybridoma cells was satisfactory. It was a little higher than in the control experiments where PEG had been used for fusion.

Further improvement of hybridoma electrofusion is expected if it becomes possible to preselect actively secreting lymphocytes prior to fusion. Arnold et al. (105, 106, 218) showed that it is possible to distin-guish actively secreting lymphocytes from nonsecreting cells by means of the technique of cell rotation in a rotating electrical field. When a rotating field of variable frequency is applied to a cell suspension, the maximum rotation rate occurs at a specific frequency which depends (among other things) on the cell membrane properties (105, 106, 218, 219). The differences in the specific frequencies detected between secreting and non-secreting lymphocytes may therefore reflect changes in the membrane capacitance or resistance or in the internal conductivity of the cells. This technique is sensitive to all these changes (219). Preselection of other physical stages of cells may well be possible with this technique, thus allowing for more successful and precise cell fusion.

Another approach of combined preselection and electrical pulse-mediated fusion of myeloma cells with lymphocytes was reported by Lo et al. (220). These authors were able to ensure proximity between B lymphocytes and myeloma cells by synthesising a selective cross-bridge between the two. The B cell is bound by the antigen against which it is specialised (the surface of the B cell has receptors for this antigen). The myeloma cell surface is modified with biotin and therefore binds to avidin molecules which are covalently linked to the antigen. It is then possible to achieve fusion between myeloma cells and those B lymphocytes secreting a particular antibody. Electrically fused hybridoma cells have shown to produce monoclonal antibodies.

Although the above technique is interesting, it does require the time-consuming multistep synthesis of a new link molecule for each new antibody to be produced. In addition, the antigen must be strongly interactive.

5.7.2 Blastomeres and Eggs

Berg (221) described electrical field-mediated fusion of 2- to 8-cell mouse blastomeres using attached microelectrodes. However, no mention is made of the viability of the fusion products.

Electrofusion of blastomeres of 2-cell mouse embryos with an intact zona pellucida was also reported by Kubiak and Tarkowski (222). Fusion was most frequent with the field strength of 1 kV/cm and field pulses of 100–250 μs duration. An electrolyte solution (PBS) was used instead of a nonelectrolyte solution (0.3 M mannitol). The viability of blastomeres fused in these two types of solution is similar. Fused 2-cell blastomeres develop into tetraploid blastocysts but die after implantation. Embryos in which blastomeres failed to fuse despite the treatment (diploid controls) developed till term. The technique was also be applied to 3- and 4-cell embryos and to zona-free oocytes and blastomeres.

Electrofusion has also been used to fuse eggs. Richter et al. (223) fused denuded eggs of the sea urchin species *Paracentrotus lividus* (Fig. 41). Because of the size of the eggs, special electrofusion chambers with a large space between the electrodes were requited. Egg chains were obtained in an alternating field of 2 MHz. In well-pigmented eggs, pigment capping was observed in the areas of contact. Fusion was initiated by an initial short electrical field pulse (400 V/cm, 50–400 μs duration) followed by a

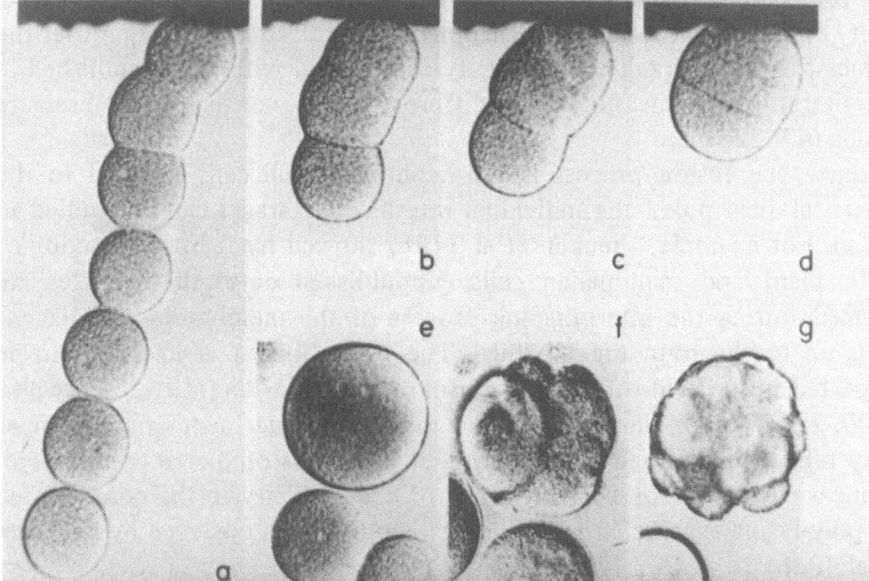

Fig. 41a—g. Time course of electrofusion between three sea urchin eggs attached to the electrode by dielectrophoresis in nonconductive, isotonic solution (frequency 2 MHz, field strength about 200 V/cm). Fusion was triggered by the application of two field pulses (400 V/cm, 50 μs). Photographs were taken after pulse application: **a** 1 min, **b** 2 min, **c** 3 min, **d** 7 min and **e** 15 min. In **e—g** the eggs were transferred into artificial sea water. **f, g** irregular cleavage patterns of fused two-egg stages after fertilisation by sperm. From (223)

second pulse 3 min later. The fused eggs had intact nuclei and could be fertilised, but underwent abortive cleavage.

Savage et al. (83) also recently reported the successful electrofusion of Xenopus eggs. Pulse lengths in the millisecond range were required because of the radius-dependent relaxation time for the build-up of the membrane potential (see above).

6 Electrofusion as a Tool in Membrane Research

Electrofusion has great potential not only in somatic hybridisation but also in membrane research.

The controlled production of giant cells by fusing thousands of individual cells enables microelectrodes to be introduced into these giant cells for measurement of the electrical properties of cell membranes (Figs. 39 and 40). This opens the way to direct studies of the interactions of membrane-active substances (including pharmaceuticals).

For example, Salhani et al. (224) investigated the hydraulic conductivity (water permeability) of fused plant protoplasts. They found that the
values for the hydraulic conductivity agreed well with those published in
the literature and measured with different techniques such as the pressure
probe (47, 225).

Since the fusion process is synchronous for all cells exposed to the
electrical field pulse, the individual intermediate stages can be studied in
detail. For example, Vienken et al. (185) showed by light microscopy of
both plant and mammalian cells (unpublished data) that vesicles are
formed during the intermingling process of the membranes of adjacent
cells up to the rounding-off stage. The formation of vesicles was to be
expected in the light of the mechanism of electrofusion [(19−28), see also
(220, 227)]. It should also be noted that the arrangement of the vesicles
may reflect a periodicity in the contact points, as predicted by the mem-
brane-wave theory of Dimitrov (64, 65). Periodicity in the contact area
of poly-lysine aggregated erythrocytes has also been observed by Hewison
et al. (228).

Studies of the rate of rounding-off of the cells after application of the
breakdown pulses can also yield information on the factors involved.
Verhoek-Köhler et al. (186, 191) were thus able to demonstrate that the
rate of rounding-off of mesophyll protoplasts of *A. sativa* was affected by
the energy charge of the cells which could be influenced by various
metabolic inhibitors. The authors were also able to show on various fused
multiples of plant protoplasts that the ATP/ADP ratio did not change.
This indicated that no leakages had occurred during the fusion process.

Hampp et al. (198) showed that various physical factors such as viscosity and surface tension of the cell membrane may also be involved in
the kinetics of electrofusion. That the effective viscosity of the cell contents is sometimes limiting is evident from the observation that, compared
to protoplasts, fused cells without vacuoles round off very slowly, whereas
vacuoles (184) fuse very rapidly [as do liposomes (171), see also above].
Hampp et al. (198) also found that the threshold voltage for fusion of
protoplasts is almost double that of evacuolated cells. We can therefore
deduce that the vacuolar membrane is electrically in series with the plasmalemma under fusion pulse conditions.

The combination of freeze-fracture electron microscopy and electrofusion is likely to be of particular value for membrane studies. Sowers
(229, 230) demonstrated this for the first time for vesicle-shaded mito-
chondrial inner membranes and for red blood cells. He was able to show
that it is possible to precisely co-ordinate quick-freezing of the suspension
with electrofusion at a known time after fusion had been initiated electrically. In two recent papers (231, 232). Sowers studied the movement of
a fluorescent lipid label (DiI), following electrofusion, from a labelled

erythrocyte membrane to an unlabelled one. The great advantage of the electrofusion technique, namely the synchronicity of the fusion process, is highlighted by this author. The average lateral diffusion coefficients measured by Sowers (231, 232) compare very favourably with the results of the photobleaching studies of the lateral diffusion of DiI in erythrocyte membranes.

Donath and Arndt (233) also measured diffusion constants by using the electrofusion technique. Their approach is quite ingenious. They fused neuraminidase- or trypsin-pretreated human erythrocytes with untreated cells. The resulting doublets had a permanent disappearance of the dipole moment which disappeared as a function of time. Since the disappearance of the dipole moment is due to protein interdiffusion, it is possible, in principle, to determine the lateral diffusion coefficient. The authors achieved this by fixing fused cells at different time intervals after the electrical pulse and by measuring the reorientation of fixed doublets in a static field. They found a surprisingly high diffusion constant for proteins (3×10^{-9} cm^2/s).

Even though we are only just beginning to apply this technique to membrane research, it has become evident that electrofusion is an invaluable tool for the elucidation of membrane processes and structure.

7 The Possible Role of Electrofusion in Evolution

There is now strong experimental evidence (234–237) to support the hypothesis of Zimmermann (67) that electrofusion was an important step in evolution after the first simple cells or analogous entities had come into being. High electrical field of millisecond duration, such as are required for the electrical breakdown of the cell membrane, continue to occur on Earth even now. The critical breakdown voltage could easily be reached if the electrical field was suddenly increased by a lightning discharge or even as a direct effect of the currents radiating from a nearby stroke to ground. The typical duration of lightning is $1-100$ μs which would be sufficient to produce the occasional viable hybrid. Large electrical fields can also be built up during the freezing process of aqueous solutions, the so-called Workmann-Reynold effect [see review (23)]. For example, a voltage potential of 20 V can be generated in the interphase between unfrozen solution and ice during the freezing of 1 mmol NaCl solution. Experimental evidence supports the view that these high voltages arise from charge separation. This seems to be the result of selective rejection from the ice rather than selective inclusion of ions in the ice crystals.

High electrical fields (sufficient to cause coronal discharge) within the Earth's surface are also thought to be generated by severe earthquakes, perhaps by a mechanism involving rapid vaporisation of water within the fracture zone (238). However, other mechanisms which possibly operate during earthquakes, e.g. piezoelectricity (239), could also be candidates. The short time constant of the piezoelectric effect in rocks of realistic conductivities is no disadvantage, because membrane breakdown occurs in 10 ns or less (240).

Zimmermann and Küppers (234–236) performed simulated lightning experiments in the laboratory and were able to show that mesophyll protoplasts of *A. sativa* could be fused, provided that the suspension density was sufficiently high. Under these conditions, dielectrophoresis with subsequent membrane contact is established for the duration of the field exposure. Dielectrophoresis could also be induced by electromagnetic waves such as are emitted by the sun. Subsequent simulated lightning discharge also resulted in the fusion of plant protoplasts.

Broda and Zimmermann (237) have also been able to demonstrate that simulated lightning discharges lead to fusion between yeast cells (*S. cerevisiae*) with different genetic markers, analogous to the experiments of Schnettler and Zimmermann (90). Viable hybrids resulting from such experiments could be obtained with the aid of selection media.

Zimmermann and Küppers (234–236) were able to demonstrate most convincingly that cells could be fused between pieces of ore-containing rocks if the cells were exposed to electrical discharges in gaps between the rocks (Fig. 42). Theoretical estimates of Küppers et al. (236) show that under certain conditions, electrical migrating waves in the earth can lead to fusion in pond water at distances of up to half a kilometre from the lightning strike. Moreover, Broda and Zimmermann (237) have demonstrated that lightning strikes in water can lead to fusion by way of electric fields caused by lightning discharge into the water.

In the light of these experimental findings, the question is to what extent present-day atmospheric conditions can be extrapolated to the beginnings of evolution, i.e., how frequent were lightning storms during evolution? There are hardly any archaeological traces of such storms (sand and stone fulgurites), and in the case of possible fossil fulgurite finds, the possibility of a more recent origin cannot be ruled out.

Nevertheless, we can safely assume that lightning storms frequently occurred in the primitive atmosphere, given that creation of a lightning storm only requires a dense atmosphere in which charge separation can take place. The composition of the atmosphere plays a completely subordinate role in this process. Planets in our solar system such as Venus, Jupiter, and Saturn, which fulfil the criteria of a dense atmosphere, also exhibit atmospheric electrical discharges. No discharges are observed on Mars because the atmosphere is too thin.

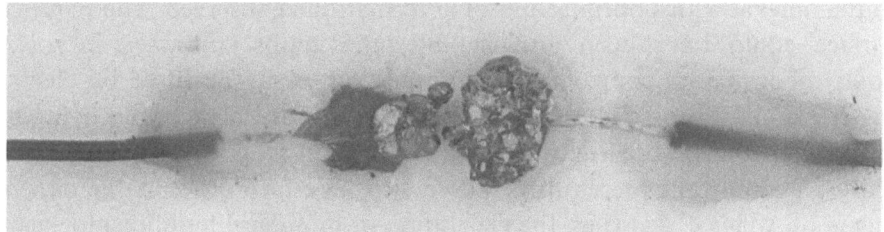

Fig. 42. Set-up of the "evolution" experiments. Plant protoplasts were fused between two pieces of ore-containing rocks. Fusion was iniated by a spark discharge (not shown)

If we further consider that lightning storms are particularly common in the warm humid regions of the Earth, then the hypothesis of lightning-induced electrofusion has to be accorded a certain amount of credibility. At the very least, this hypothesis has the advantage over other hypotheses currently under discussion that it can be supported by experiments in the laboratory.

It is also interesting to note that Hui et al. (241) described a fusion method for liposomes based on freezing and thawing. It is conceivable that the basic step in the fusion of liposomes with this procedure is an electrical field effect arising from the high field strength at the membrane/water interface. Freezing and thawing may therefore represent an alternative method of generating short-lived electrical fields of high intensity. Steponkus et al. (242) also recently reported that they have strong evidence for the involvement of electrical breakdown in the cryoinjury of isolated plant protoplasts.

If this is true, we can speculate that transient high electrical fields occur during freeze-fracture which may alter the membrane structure, with the consequence that freeze-fracture electron microscopy does not reflect the natural state of the membrane or the cell interior, but rather the state after breakdown has occurred.

In this context it is also worth mentioning again that organelles and even whole cells can be sequestered through membranes with the aid of electrical breakdown, without influencing cell integrity (Figs. 3–5). The hypothesis that chloroplasts and mitochondria in eukaryotic cells originated from bacteria taken up into the cells is strongly supported by these laboratory experiments. Furthermore, Hub et al. (172) found that uni-lamellar liposomes of cell size are able to form when lipid solutions are exposed to an elevated temperature of 70–90°C for about 4–6 h. In particular, high yields of cell-sized liposomes are produced when the surface area is increased, for example by adding small glass beads to the lipid solution (243). It is likely that such temperatures occurred in the primaeval soup. In addition, the tidal times of 6 h (low tide and high tide) correspond to the times at which optimal liposome formation is observed. The required surface could have been provided by sand grains contained in rocky pools of primaeval soup. Since liposomes can easily be fused by electrical fields, it is possible that electrical fields were also involved in the creation of the first cell entity for which a lipid membrane was undoubtedly an important prerequisite. If we also take into account the experiments of Miller (244) (i.e., the generation of biological building blocks by means of lightning discharges), we may postulate that life can only be created if the atmosphere of a planet has all the prerequisites for lightning discharges.

8 Conclusions

The considerable amount of activity in the field of electrofusion and electropermeabilization is very promising from the point of view of new insights into biomembranes and new technologies in the future for the production of new compounds and modification of cell systems for nutrition, energy production and the removal of waste products. It is particularly gratifying to see how basic science has provided the foundation for a useful technology, although in some cases the time needed to develop an application is very long. In other cases, it is necessary to overcome the difficulties posed by existing schools of thought which have been shown to be wrong. It is fascinating to observe the many developments and discoveries in the areas of physics, material science, space technology and electronics which are just waiting to be applied to biological systems. An increased interdisciplinary collaboration between physicists and biologists could provide considerable impetus to biology and its application in technology. However, this can only be achieved if basic research into biological membranes is accelerated. The techniques for electrical breakdown, electropermeabilization and electrofusion could be an important tool in this process, since we cannot rule out the possibility that the high electrical fields occurring naturally in the membrane play an important role in the selective transport of substances across the membrane as well as in natural regulatory processes.

Acknowledgement. I am very grateful to Mrs. I. Stürmer for typing the manuscript.

Appendix: Mechanism of Electrical Breakdown

The molecular processes and structural reorganisations leading to electrical breakdown and occurring during the subsequent resealing process still have to be fully elucidated. However, in recent years a number of theoretical concepts about the mechanism of electrical breakdown have been developed which represent the first stages towards an understanding of the field-induced molecular processes. Instabilities of the membrane which arise as a result of electrical or osmotic (mechanical) stress and ultimately lead to reversible breakdown, can be described by various thin film models using a linear stability analysis (35, 37, 40–50, 64, 65).

Crowley (44) developed a thin film model for irreversible electrical breakdown of bilayer membranes. In this model, the membrane is regarded as an isotropic elastic layer sandwiched between two semi-infinite, electrically conducting liquids. Crowley (44) assumed that the instability in response to a membrane potential developed as a monotonic growth of some small initial disturbances, as opposed to an oscillation of increasing amplitude. The calculations of the resulting changes in the electrical and elastic stresses and the analysis of the equations obtained showed the existence of a critical membrane potential at which rupture occurs. The calculated breakdown voltages of about 200–300 mV were too low to explain reversible electrical breakdown at about 1 V (but see below), but they agreed very well with the voltage values at which irreversible mechanical rupture of cell membranes and artificial bilayer membranes had been observed [see also (37)]. The belief was therefore at the time that a membrane cannot be polarised beyond, say, 200–400 mV.

On the basis of Crowley's model, Zimmermann et al. (35) and Zimmermann and Coster (40, 41, 45–47) developed an electromechanical model for reversible electrical breakdown, in which they included mechanical external forces in the analysis, in addition to the electrical forces.

Briefly, the membrane can be regarded as a capacitor filled with a dielectric material of a low dielectric constant (of the order of 2) compared with that of water (Fig. 1). As a result, free charges can be accumulated at both membrane surfaces, but cannot exist in high concentrations in the membranes, at least not within the hydrocarbon layer. The application of the field pulse leads to an accumulation of more surface charges or to an increase in the transmembrane potential (see above). The generated charges are of opposite signs and attract each other. This attraction gives rise to a compression pressure which acts on the membrane surface and causes the membrane thickness to decrease (Fig. 1b). The distance between the opposite charges on the two membrane surfaces is thus reduced, and the compression pressure is correspondingly raised because the electric forces increase with decreasing separation. An elastic or viscoelastic (for rapid thickness changes in response to high electric field pulses of very short duration) restoring force opposes the electrocompression of the membrane thickness. Since these electric compressive forces increase more rapidly with decreasing membrane thickness than the elastic restoring compression forces, a local breakdown of the membrane occurs at a certain external electrical field strength (compression pressure). This is because any local perturbations of the membrane surface shape (due to thermal fluctuations or other causes) will grow spontaneously in an electrical field which has sufficient intensity to overcome the opposing rheological forces.

A theoretical analysis shows (40) that the breakdown voltage is represented by the following equation:

$$V^2_{c\ (P=0)} = \frac{0.369\ \gamma_m}{\epsilon_m \cdot \epsilon_o} \cdot d^2_o \tag{10}$$

where $V_c\ (P=0)$ = breakdown voltage at pressure zero, γ_m = elastic compressive modulus. ϵ_m = dielectric constant of the membrane, ϵ_o = dielectric constant of the vacuum and d_o = thickness of the membrane in the unstressed state.

The incorporation of mechanical compression forces arising from hydrostatic pressure differences or osmotic differences across the membranes (which in plant cells are stabilised by a cell wall) leads to the following dependence of the breakdown voltage on the hydrostatic pressure difference:

$$V_c = V_{c\ (P=0)}\ \exp(-P/\gamma) \tag{11}$$

where γ = effective elastic modulus [see (225)].

Equation 11 shows that the breakdown voltage should decrease as the pressure gradient increases. This prediction has been confirmed experimentally for a pressure range of a few bar by combined direct pressure and breakdown measurements in giant algal cells of *Valonia utricularis* (41). The predicted decrease of the breakdown voltage with increasing pressure gradient can be understood qualitatively, since precompression of the membrane by mechanical forces must lead to an equivalent reduction of the electrical compression forces. If Eq. 11 is extrapolated to high pressure values, we would expect the mechanical precompression of the membrane to become so high in the extreme case that the breakdown voltage is achieved by the natural resting potential. Indeed, Zimmermann et al. (49, 50) were able to demonstrate that a uniform hydrostatic pressure of about 600 bar leads to a reversible breakdown of the membrane, as measured by the increase of the potassium efflux. This pressure range was predicted by the electromechanical model (by making some assumptions about the effect of uniformly acting pressure and of the pressure dependence of the elastic modulus).

The electromechanical model of Zimmermann et al. (35, 41) has the advantage that all parameters are experimentally accessible and that a number of experimental findings can be explained. Moreover, by analysing the breakdown data in terms of electromechanical compression, it was postulated for the first time that the membrane can be compressed perpendicular to the normal, i.e., that the membrane thickness in walled cells is a function both of the turgor pressure and the transmembrane potential and so should be dependent on those ions in the solution that determine the potential (225, 245).

Nowadays, it is generally accepted that the biological membrane is compressible.

However, this simple thin film model could not explain – at least not quantitatively – the pulse length dependence of the breakdown voltage in the μs range which was discovered by Benz and Zimmermann for the membranes of *V. utricularis* and for artificial bilayer membranes [see review (19, 246–248)]. The electromechanical model had also failed to take into account the contribution of the surface tension and the viscous properties of the membrane.

The kinetics of reversible breakdown and in particular the pulse length dependence of the breakdown voltage were recently explained by Dimitrov (64) and Dimitrov and Jain (65) who generalised the thin viscoelastic film model by using the disjoining pressure approach. This means that the effects of the long-range intermolecular forces are incorporated into the boundary conditions instead of being introduced into the equation ·of motion (body force approach) as was originally done by Jain and Maldarelli (249) and Sanfeld and co-workers (250).

The linear stability analysis consisted of an analysis of the behaviour of small perturbations (due to thermal fluctuations or external constraints like the electrical field) of the shape of the two membrane surfaces. An arbitrary shape perturbation can be represented by a superposition of waves characterised by their wave numbers k, where the wave number is inversely proportional to the wavelength. The wave number k of a perturbation that corresponds to marginal stability is referred to as the critical wave number k_c. In most cases, waves with smaller wave numbers than the critical one lead to membrane instability. Among these waves there are one ore more for which the so-called growth factor has a maximum value, which means that the perturbation will grow. These waves are called dominant because they will break and/or bend the membranes faster than the other waves. In this way, the time of rupture or bending can be estimated as being proportional to the inverse value of the growth coefficient of the dominant wave. An analysis of the equation relating the frequency to the wave number (i.e., the disperson equation) shows that the perturbation motion can be split into two fundamental modes of vibration: a squeezing mode in which the two surfaces are out of phase (phase shift 180°) and the film exhibits local thinning (as in the simple electromechanical model), and a stretching and bending mode in which the surfaces are in phase and the film thickness remains constant. One or both of the modes may be present during membrane instability, leading to membrane rupture and/or bending.

The disjoining pressure approach of Dimitrov (64) has the advantage of being simpler than the body force approach of Jain and Maldarelli (249). Because of the lack of experimental data for more complicated viscoelastic models, Dimitrov (64) analysed the simple case of a Kelvin body. The theory predicts a critical potential which is given by Eq. 12 for the boundary condition that the membrane tension is zero:

$$V_c = \left(\frac{24 \cdot \sigma_t \cdot \gamma_m \cdot d_o^3}{\epsilon_o^2 \cdot \epsilon_m^2} \right)^{1/4} \tag{12}$$

where σ_t = surface tension.

Equation 12 holds for zero membrane tension.

This equation differs from the predictions of the electromechanical model of Zimmermann and Coster (40, 41, 45–47) by the factor

$$\left(\frac{8 \cdot \sigma_t}{d_o \cdot \gamma_m} \right)^{1/4} \tag{13}$$

(see also Eq. 10).

It is evident that the breakdown voltage is proportional to $\sigma_t^{1/4}$ in Dimitrov's model. A reduction of the surface tension, e.g., by the addition of detergents and local anaesthetics, thus leads to a decrease in the breakdown voltage even though the elastic compressible modulus γ_m may be very high. As demonstrated by Pilwat et al. (78) this is indeed the case, at least for the local anaesthetic benzyl alcohol at high concentrations. The theory also yields the following relationship between the breaking time, which is inversely proportional to the growth factor of the dominant wave, and the apparent critical breakdown voltage:

$$\tau_b^{-1} = \overline{V}^{-4} - 1 \tag{14}$$

where τ_b is the normalised time constant of membrane breakdown in this particular case, the time after which an applied electrical field will induce a high conductance state; $\overline{V} = V_m/V_c$.

In other words, Eq. 14 states that higher voltages must be established across the membrane at shorter pulse lengths, i.e., the breakdown voltage increases compared to the values at longer exposure times. In contrast, the electromechanical model explains the increase of the breakdown voltage at shorter pulse lengths qualitatively by the assumption that the elastic compressive modulus γ_m increases with shorter pulse lengths because of the viscoelastic properties of the membrane, with the consequence that the breakdown voltage actually increases according to Eq. 10. A combination of Eqs. 12 and 14 gives rise to Eq. 15:

$$\overline{V} = (1 + 10^{-\lg\tau_b})^{1/4} \tag{15}$$

Equation 15 was used to fit the experimentally established dependence of the breakdown voltage on pulse length (246–248). Using the appropriate values, Dimitrov (64) was able to show that the data of Benz and Zimmermann (246–248) could be fitted very well over a wide range of pulse lengths with the exception of pulse lengths below 100 ns. The data of Abidor et al. (51) on the other hand could only be fitted over a narrow range of longer pulse lengths by the theoretical equation. However, this disagreement is understandable in the light of the experimental approach used by these latter authors.

On the basis of the good agreement between the theory and the experimental data of Benz and Zimmermann (19, 246–248), Dimitrov (64) suggested the following sequence of events involved in electrical breakdown: (1) growth of the membrane shape fluctuations, (2) molecular rearrangements leading to a discontinuity of the membrane (which is not as yet clarified), and (3) expansion of the pores resulting in the mechanical breakdown of the membrane.

The first process lasts microseconds or less and is described by the thin film model. The second stage is very short (nanoseconds). The last stage takes milliseconds or more and may be described by the phenomenological statistical theory of Chizmadshev and co-workers (51–59) or by the stochastic model of Sugar and Neumann (63).

The statistical model of Chizmadshev and co-workers (51–59) is based on the assumption that the existing defects in the membrane structure can increase in size under the influence of the electric field. This process leads to irreversible destruction of the membrane. The disadvantage of this theory is that the final equations contain some parameters which are not accessible experimentally.

Petrow et al. (60) investigated the formation of hydrophobic pores and their transition to hydrophilic ones in the course of their growth. By analogy to Chizmadshev et al. (51--59). Petrow et al. (60) examined the influence of a radius- and field-dependent edge energy upon pore stability. As he pointed out, these relationships are very important.

The field dependence of edge energy due to the favorable orientation of head group dipoles was also studied by Petrow et al. (60) and Sugar (62).

In this context, it should be noted that Petrow et al. (60) also demonstrated that curvature-induced polarisation may make a considerable contribution to the membrane potential [see also (251)]. Such effects may also be involved in electrofusion (20).

Finally, we would like to discuss the question of whether pores actually do occur in the membrane as a result of electrical breakdown. The term electrical breakdown means — when analogous phenomena in solid state physics are taken into account — that transmembrane defects occur which can be considered as pores, channels, irregular leaks or orientation defects (251).

Evidence for the formation of pores comes from electrical breakdown experiments on planar lipid bilayer membranes. Benz et al. (37) showed that reversible electrical breakdown in lipid membranes made up of oxidised cholesterol leads to a change in the resistance of the lipid bilayer by more than 8 orders of magnitude in the presence of 1 mol KCl solution on both sides. Such dramatic changes in the membrane resistance can apparently only be explained by assuming the presence of pores filled with conductive solution. Similar changes in the membrane resistance, once reversible breakdown has occurred, have also been observed in lipid bilayer membranes of different composition (which were partly modified by the addition of uranyl ions) (252, 253). Such marked membrane resistance changes have not as yet been observed in cell membranes. Direct measurements on giant algal cells of *V. utricularis* by means of intracellular and extracellular electrodes have revealed changes in the membrane resistance of only about 1 order of magnitude (40, 79). In a lipid bilayer membrane, the resealing times of the generated perturbations are very short (about 2–20 μs) compared with the long-lived defects in biological membranes which can persist for 10 min or more at low temperatures. We therefore have to conclude that conformational changes of proteins or nonuniform aggregation of proteins as a result of the tangential components of the electrical fields in the membrane give rise to these long-lived permeabilization stages of the lipid-protein membrane. In addition, rearrangements of the membrane components should occur very easily because of the high diffusion coefficients of the components in a membrane exhibiting a large number of defects [see (233)].

Schwister and Deuticke (254) showed in erythrocytes that the electrically induced leaks do not discriminate very well between the alkali ions, but that in contrast the sodium halides permeate at increasing rates in the order of NaF < NaCl < NaBr < NaJ, the sequence of decreasing hydrated radii. Organic anions are discriminated according to size and charge. The common properties of these electrically induced defects and of chemically induced leaks (diamide, periodate, t-butylhydroperoxide) in the erythrocyte membrane led these authors to conclude that there are common similarities in the molecular organisation of the electrically and chemically induced defects. They pointed out that the electrically induced perturbations may be comparable to the "leaky patches" thought to be formed in membranes round the complement complex after its insertion into the target membrane.

Schwister and Deuticke (254) also suggested that a redistribution of proteins in the electrical field may be responsible for the occurrence of such leaky patches [see also (20)].

Dimitrov (64) also argued that it is still not clear whether these defects can be regarded as hydrophilic or hydrophobic pores in the lipid bilayer membrane of the red blood cells or whether they represent some sort of induced mismatch between lipids and intrinsic proteins. This is the key point which leads us back to the question formulated at the beginning, namely whether we can regard the long-lived defects induced in cell membranes as pores. At the present time it is difficult to make a decision, although the author of this review believes, in the light of the results and theoretical considerations of Sackmann (251) and the arguments presented above, that this concept is too simple and that the structural changes are considerably more complex. It is worth noting that Deuticke et al. (74, 77) have pointed out the parallels in the molecular properties or pores and flip sites. These authors have also shown that the dielectric breakdown of the erythrocyte membrane enhances transbilayer mobility of phospholipids. Their results clearly suggest the formation of reversible structural changes in the lipid domain of cell membranes after breakdown. The concept that such structural defects may represent pores as well as flip sites has been proposed by Van Deenen and co-workers (255) for artificial lipid and lipid-protein systems. An enhanced flip-flop could be connected with the formation of pores, as suggested by Sackmann (251) and McLaughlin and Haray (256).

As was first described by McLaughlin and Haray (256), the voltage drop across the bilayer provides an important driving force for the inside/outside distribution of charged lipids. These lipids determine the charge density of the inner and outer membrane surface. According to Boltzmann's law, we would expect:

$$\sigma_o'/\sigma_i' = \exp\left(-eV_m/kT\right) \tag{16}$$

where the indices i and o stand for the charge density σ' in the inner and outer lipid-water interface respectively. For normal flip-flop times of some 5 h, the lipid distribution could not follow changes in the membrane potential. However, if hydrophilic pores were able to form, the inside/outside lipid distribution could change within a time scale of 1 μs. This is the time a lipid would need to cross the pore by lateral diffusion. Thus, such effects can operate in the membrane once breakdown has occurred.

References

1. Ringerts NR, Savage RE (1976) Cell hybrids. Academic, New York
2. Poste G, Nicholson GL (1978) Cell surface reviews membrane fusion, vol 5. Elsevier, Amsterdam
3. Beers RF Jr, Bassett EG (1984) Cell fusion: gene transfer and transformation. Raven, New York
4. Evered R, Whelan J (1984) Ciba Foundation symposium 103. Cell fusion. Pitman, London
5. Wilschut J, Hoekstra D (1984) Trends Biochem Sci 9:479
6. Lucy JA (1982) In: Chapman D (ed) Biological membranes. Academic, London, p 367
7. Fodor K, Alföldi L (1976) Proc Natl Acad Sci USA 73:2147
8. Gabor MH, Hotchkiss RD (1979) J Bact 137:1346
9. Spencer JFT, Spencer DM (1981) Curr Genet 4:177
10. Ferenczy L (1984) In: Beers RF Jr, Bassett EG (eds) Cell fusion: gene transfer and transformation. Raven, New York, p 145
11. Melchers G, Labib G (1974) Mol Gen Genet 135:277
12. Cocking EC (1984) In: Beers RF Jr, Bassett EG (eds) Cell fusion: gene transfer and transformation. Raven, New York, p 139
13. Köhler G, Milstein C (1975) Nature 256:495
14. Croce CM, Linnenbach A, Hall W, Steplewski Z, Koprowski H (1980) Nature 288:488
15. Küster E (1909) Dt Bot Ges 27:589
16. Zimmermann U, Pilwat G (1978) Sixth Int Biophys Cong Kyoto Abstr IV-19(H):140
17. Zimmermann U, Vienken J, Scheurich P (1980) In: Gersonde K (ed) Biophysics of structure and mechanism, vol 6. Springer, Berlin Heidelberg New York, p 86
18. Zimmermann U, Vienken J, Pilwat G (1980) Bioelectrochem Bioenerg 7:553
19. Zimmermann U, Scheurich P, Pilwat G, Benz (1981) Angew Chem 93:332
19a. Zimmermann U, Scheurich P, Pilwat G, Benz R (1981) Int Ed 20:325
20. Zimmermann (1982) Biochim Biophys Acta 694:227
21. Zimmermann U, Vienken J (1982) J Membrane Biol 67:165
22. Zimmermann U, Vienken J, Pilwat G (1984) In: Chayen J, Bitensky L (eds) Investigative microtechniques in medicine and biology, vol 1. Dekker, New York, p 89
23. Zimmermann U, Büchner K-H, Arnold WM (1984) In: Allen MJ, Usherwood PNR (eds) Charge and field effects in biosystems. Abacus, Tunbridge Wells, p 293
24. Zimmermann U, Vienken J, Pilwat G, Arnold WM (1984) In: Evered R, Whelan J (eds) Cell fusion: Ciba foundation symposium 103. Pitman, London, p 60

25. Arnold WM, Zimmermann U (1984) In: Chapman D (ed) Biological membranes, vol 5. Academic, London, p 389
26. Zimmermann U, Vienken J (1984) In: Beers RF Jr, Bassett EG (eds) Cell fusion: gene transfer and transformation. Raven, New York, p 171
27. Zimmermann U, Vienken J (1984) In: Stern NJ, Gamble HR (eds) Hybridoma technology in agricultural and veterinary research. Rowman and Allanheld, Totowa, p 173
28. Zimmermann U, Vienken J, Halfmann J, Emeis CC (1985) In: Misrahi A, van Wezel AL (eds) Advances in biotechnological processes 4. Liss, New York, p 79
29. Zimmermann U, Schultz J, Pilwat G (1973) Biophys J 13:1005
30. Zimmermann U, Pilwat G, Riemann F (1974) Z Naturforsch 29c:304
31. German patent application no. P 24 05 119, KFA Jülich, filed February 2, 1974, Inventors: Zimmermann U, Pilwat G, Riemann F; German patent no. 24 05 119; British patent no. 1 481 480; US patent no. 4 081 340; French patent no. 75 02743
32. Zimmermann U, Pilwat G, Beckers F, Riemann F (1976) Bioelectrochem Bioenerg 3:58
33. Zimmermann U, Pilwat G, Riemann F (1975) Biochim Biophys Acta 375:209
34. Zimmermann U (1983) In: Goldberg E (ed) Targeted drugs. Wiley, New York, p 453
35. Zimmermann U, Pilwat G, Riemann F (1974) Biophys J 14:881
36. Zimmermann U, Pilwat G, Riemann F (1974) In: Zimmermann U, Dainty J (eds) Membrane transport in plants. Springer, Berlin Heidelberg New York, p 146
37. Benz R, Beckers F, Zimmermann U (1979) J Membrane Biol 48:181
38. Schwan HP (1983) In: Advances in biological effects of dosimetry of low energy electromagnetic fields. Plenum, New York, p 213
39. Jeltsch E, Zimmermann U (1979) Bioelectrochem Bioenerg 6:349
40. Coster HGL, Zimmermann U (1975) J Membrane Biol 22:73
41. Zimmermann U, Beckers F, Coster HGL (1977) Biochim Biophys Acta 464:399
42. Benz R, Conti F (1981) Biochim Biophys Acta 645:115
43. Gauger B, Bentrup FW (1975) J Membrane Biol 48:249
44. Crowley JM (1973) Biophys J 13:711
45. Coster HGL, Zimmermann U (1976) Z Naturforsch 31c:461
46. Coster HGL, Steudle E, Zimmermann U (1977) Plant Physiol 58:636
47. Zimmermann U (1978) Annu Rev Plant Physiol 29:121
48. Zimmermann U (1980) In: Gilles R (ed) Animals and environmental fitness. Pergamon, Oxford, p 441
49. Zimmermann U, Pilwat G, Pequeux A, Gilles R (1980) J Membrane Biol 54:103
50. Pequeux A, Gilles R, Pilwat G, Zimmermann U (1980) Experientia 36:565
51. Abidor IG, Arakelyan VB, Chernomordik LV, Chizmadzhev YuA, Pastushenko VF, Tarasevich MR (1979) Bioelectrochem Bioenerg 6:37
52. Pastushenko VF, Chizmadzhev YuA, Arakelyan VB (1970) Bioelectrochem Bioenerg 6:53
53. Chizmadzhev YuA, Arakelyan VB, Pastushenko VF (1979) Bioelectrochem Bioenerg 6:63
54. Pastushenko VF, Chizmadzhev YuA, Arakelyan VB (1979) Bioelectrochem Bioenerg 6:71
55. Arakelyan VB, Chizmadzhev YuA, Pastushenko VF (1979) Bioelectrochem Bioenerg 6:81
56. Pastushenko VF, Arakelyan VB, Chizmadzhev YuA (1979) Bioelectrochem Bioenerg 6:89
57. Pastushenko VF, Arakelyan VB, Chizmadzhev YuA (1979) Bioelectrochem Bioenerg 6:97
58. Chizmadzhev YuA, Abidor IG (1980) Bioelectrochem Bioenerg 7:83
59. Chernomordik LV, Abidor (1980) Bioelectrochem Bioenerg 7:617
60. Petrov AG, Mitov MD, Derzhanski AI (1980) In: Lajos Bata (ed) Edge energy and pore stability in bilayer lipid membranes. Pergamon, Oxford, p 695

61. Pastushenko VF, Petrov AG (1984) School Proc:69
62. Sugar IP (1983) In: Spach G (ed) Physical chemistry of transmembrane ion motions. Elsevier, Amsterdam, p 21
63. Sugar IP, Neumann E (1984) Biophys Chem 19:211
64. Dimitrov DS (1984) J Membrane Biol 78:53
65. Dimitrov DS, Jain RK (1984) Biochim Biophys Acta 779:437
66. Zimmermann U, Küppers G, Salhani N (1982) Naturwissenschaften 69:451
67. Zimmermann U (1983) Trends Biotech 1:149
68. Vienken J, Zimmermann U, Fouchard M, Zagury D (1983) FEBS Lett 163:54
69. Sowers AE (1985) Biophys J 47:171a
70. Lucy J (1986) Biochem Soc Trans (to be published)
71. Ohki S (1984) J Membrane Biol 77:265
72. Stoicheva N, Tsoneva I, Dimitrov DS, Panaiotov I (1985) Z Naturforsch 40c:92
73. Zimmermann U, Pilwat G, Holzapfel Chr, Rosenheck K (1976) J Membrane Biol 30:135
74. Kinosita K Jr, Tsong TY (1977) Proc Natl Acad Sci USA 74:1923
75. Neumann E, Schaefer-Ridder M, Wang Y, Hofschneider PH (1982) EMBO J 1:841
76. Deuticke B, Poser B, Putkemeier P, Haest CWM (1983) Biochim Biophys Acta 731:196
77. Dressler V, Schwister K, Haest CWM, Deuticke B (1983) Biochim Biophys Acta 732:304
78. Pilwat G, Zimmermann U, Riemann F (1975) Biochim Biophys Acta 406:424
79. Zimmermann U, Beckers F, Steudle E (1977) In: Thellier M, Monnier A, Demarty M, Dainty J (eds) Transmembrane ionic exchanges in plants. CNRS, Paris, p 155
80. Zimmermann U, Riemann F, Pilwat G (1976) Biochim Biophys Acta 436:460
81. Vienken J, Jeltsch E, Zimmermann U (1978) Cytobiology 17:182
82. Potter H, Weir L, Leder P (1984) Proc Natl Acad Sci USA 81:7161
83. Pilwat G, Zimmermann U (1985) Biochim Biophys Acta 820:305
84. Savage JS, Grey RD (1983) Poster, Am Soc Cell Biol, San Antonio, Texas
85. Senda M, Takeda J, Abe S, Nakamura T (1979) Plant Cell Physiol 20:1441
86. Senda M, Morikawa H, Takeda J (1982) Plant Tissue Culture 615
87. Mehrle W, Zimmermann U, Hampp R (1985) FEBS Lett 185:89
88. Farkas DL, Korenstein R, Malkin S (1980) FEBS Lett 120:236
89. McLaughlin S (1977) In: Bronner, Kleinzeller (eds) Current topics in membranes and transport. Academic, New York, p 71
90. Schnettler R, Zimmermann U (1985) FEMS Microbiol Lett 27:195
91. Vienken J, Zimmermann U (1985) FEBS Lett 182:278
92. Zimmermann U, Pilwat G, Richter H-P (1981) Naturwissenschaften 68:577
93. Pilwat G, Richter HP, Zimmermann U (1981) FEBS Lett 133:169
94. Vienken J, Zimmermann U (1982) FEBS Lett 137:11
95. Zimmermann U, Vienken J, Greyson J (1984) Biotech 84 Europe 1:231
96. Zimmermann U, Pilwat G, Esser B (1978) J Clin Chem Clin Biochem 16:135
97. Zimmermann U (1983) Labo-Pharma Probl Tech 31:69
98. Zimmermann U, Pilwat G (1976) Z. Naturforsch 31c:732
99. Kinosita K Jr, Tsong TY (1977) Nature 268:438
100. Kinosita K Jr, Tsong TY (1978) Nature 272:258
101. Schüssler W, Ruhenstroth-Bauer G (1984) Blut 49:213
102. Auer D, Brandner G, Bodemer W (1976) Naturwissenschaften 63:391
103. Langridge WHR, Li BJ, Szalay AA (1985) In: Biotechnology in plant science: relevanve to agriculture in the eighties. Program and abstracts for an international symposium. June 23–27, 1985, p 25
104. Karube I, Tamiya E, Matsuoka H (1985) FEBS Lett 182:90
105. Arnold WM, Zimmermann U (1982) Z Naturforsch 37c:908

106. Arnold WM, Wendt B, Zimmermann U, Korenstein R (1985) Biochim Biophys Acta 813:117
107. Stopper H, Zimmermann U, Wecker E (1985) Z Naturforsch 402c:133
108. Zimmermann U (1973) British patent 22965/74. Biochemical synthesis or degradation using entrapped enzymes, filed May 23
109. Auer D, Brandner G (1976) Z Naturforsch 31c:149
110. Tsukakoshi M, Kurata S, Nomiya Y, Kasuya T (1984) Appl Phys 35:2884
111. Day D, Hub H-H, Ringsdorf H (1979) Isr J Chem 18:325
112. Johnston DS, Sanghera S, Pons M, Chapman D (1980) Biochim Biophys Acta 602:57
113. Hub H-H, Hupfer B, Koch H, Ringsdorf H (1980) Angew Chem Int Ed Engl 19: 938
114. Bader H, Dorn K, Hupfer B, Ringsdorf H (1985) In: Gordon M (ed) Polymer membranes. Springer, Berlin Heidelberg New York Tokyo, p 1
115. Elbert R, Laschewsky A, Ringsdorf H (1985) J Am Chem Soc 107:4134
116. Muth E (1927) Kolloid Z 41:97
117. Schwarz G, Saito M, Schwan HP (1965) J Chem Phys 43:3562
118. Saito M, Schwan HP, Schwarz G (1966) Biophys J 6:313
119. Liesbesny P (1939) Arch Phys Ther 19:736
120. Krasny-Ergen W (1986) Hochfrequenztechn Elektroakustik 40:126
121. Sher LD (1968) Nature 220:695
122. Schwan HP, Sher LD (1969) J Electrochem Soc 116:170
123. Saito M, Schwan HP (1961) In: Peyton MF (ed) Biological effects of microwave radiation. Plenum, New York, p 85
124. Schwan HP, Sher LD (1966) J Electrochem Soc 116:22c
125. Crane JE, Pohl HA (1972) J theor Biol 37:15
126. Mason BD, Townsley PM (1971) Can J Microbiol 17:879
127. Lin IJ, Kaplan BZ, Zimmels Y (1983) Sep Sci Tech 18:683
128. Benguigui L, Lin IJ (1984) J Phys D: Appl Phys 17:9
129. Füredi AA, Valentine RC (1962) Biochim Biophys Acta 56:33
130. Heller JH, Teixeira-Pinto AA (1959) Nature 183:905
131. Wildervanck A, Wakin KG, Herrick JF, Krusen FH (1959) Arch Phys Med 40: 45
132. Ludloff K (1985) Pflügers Arch 59:525
133. Verworn M (1986) Pflügers Arch 62:415
134. Mast SO (1931) Z Wissenschaftl Biol, Abt C: Z Vergl Physiologie 15:309
135. Teixeira-Pinto AA, Nejelski LL, Cutler JL, Heller JH (1960) Ex Cell Res 20:548
136. Griffin JL, Stowell RE (1966) Exp Cell Res 44:684
137. Novak B, Bentrup FW (1973) Biophysik 9:253
138. Sher LD (1963) Ph D Thesis, University of Pennsylvania, Philadelphia
139. Jennings BR, Morris VJ (1974) J Colloid Interface Sci 49:89
140. Morris VJ, Rudd PJ, Jennings BR (1975) J Colloid Interface Sci 50:379
141. Shinar R, Druckmann S, Ottolenghi M, Korenstein R (1977) Biophys J 19:1
142. Sher LD, Kresch E, Schwan HP (1970) Biophys J 10:970
143. Füredi AA, Ohad J (1964) Biochim Biophys Acta 79:1
144. Jones TB, Bliss GW (1977) J Appl Phys 48:1412
145. Kallid GA, Jones TB (1980) IEEE Trans Ind Appl IA-16:69
146. Lin IJ, Jones TB (1984) J Electrostat 15:53
147. Benguigui L, Lin IJ (1981) J Appl Phys 53:1141
148. Stoicheva N, Tsoneva I, Dimitrov DS (1986) Z Naturforsch (to be published)
149. Tsoneva ICh, Zhelev DV, Dimitrov DS (1984) Cell Biophys 6
150. Sauer FA (1985) In: Chiabrera A, Dinolini C, Schwan HP (eds) Interactions between electromagnetic fields and cells. Plenum, New York, p 181
151. Dimitrov DS, Tsoneva I, Stoicheva N, Thelev D (1984) J Biol Phys 12:26
152. Dimitrov DS, Stoicheva N, Tsoneva I, Zhelev DV (1985) In: Biotech 85 (Europa). Online Publ, Pinner, UK, 677

153. Vienken J, Zimmermann U, Alonso A, Chapman D (1984) Naturwissenschaften 71:158
154. Förster E, Emeis CC (1985) FEMS Microbiol Lett 26:65
155. Hannig K (1969) In: Bier M (ed) Electrophoresis, theory, methods and applications. Academic, New York, p 423
156. Zeiller K, Löser R, Pascher G, Hannig K (1975) Hoppe-Seyler's Z Physiol Chem 356:1225
157. Hannig K, Wirth H, Schindler RK, Spiegel K (1977) Hoppe-Seyler's Z Physiol Chem 358:753
158. Hannig K (1978) J Chromatograph 159:183
159. Watts JW, King JM (1984) Biosci Rep 4:335
160. Büchner K-H, Zimmermann U (1986) (to be published)
161. Kramer I, Vienken K, Vienken J, Zimmermann U (1984) Biochim Biophys Acta 772:407
162. Nyborg WL (1978) In: Fry FJ (ed) Ultrasound: its applications in medicine and biology, part 1. Elsevier, Amsterdam, p 1
163. Nyborg WL, Miller DW, Gershoy A (1974) In: Michaelson SM et al. (eds) Fundamental and applied aspects of nonionizing radiation. Plenum, New York, p 277
164. Vienken J, Zimmermann U, Zenner HP, Coakley WT, Gould RK (1985) Naturwissenschaften 72:441
165. Vienken J, Zimmermann U, Zenner HP, Coakley WT, Gould RK (1985) Biochim Biophys Acta 820:259
166. Ashkin A, Dziedzic JM (1985) Phys Rev Lett 54:1245
167. Coster HGL, Zimmermann U (1975) Z Naturforsch 30c:77
168. Melikyan GB, Abidor IG, Chernomordik LV, Chailakhyan LM (1983) Biochim Biophys Acta 730:395
169. Gingell D, Ginsberg L (1978) In: Poste G, Nicholson GL (eds) Cell surface reviews, membrane fusion, vol 5. Elsevier, Amsterdam, p 791
170. Zimmermann U, Scheurich P (1981) Biochim Biophys Acta 641:160
171. Büschl R, Ringsdorf H, Zimmermann U (1982) FEBS Lett 150:38
172. Hub H-H, Zimmermann U, Ringsdorf H (1982) FEBS Lett 140:254
173. Büschl R (1984) Thesis, Johannes Gutenberg Universität, Mainz
174. Shivarova N, Grigorova R, Förster W, Jacob H-E, Berg H (1983) Bioelectrochem Bioenerg 11:181
175. Ruthe H-J, Adler J (1985) Biochim Biophys Acta 819:105
176. Weber H, Förster W, Jacob H-E, Berg H (1981) In: Stewart GG, Russell I (eds) Current developments in yeast research. Pergamon, Toronto, p 219
177. Weber H, Förster W, Berg H, Jacob H-E (1981) Curr Genet 4:165
178. Halfmann HJ, Röcken W, Emeis CC, Zimmermann U (1982) Curr Genet 6:25
179. Halfmann HJ, Emeis CC, Zimmermann U (1983) Arch Microbiol 134:1
180. Halfmann HJ, Emeis CC, Zimmermann U (1983) FEMS Microbiol Lett 20:13
181. Schnettler R, Zimmermann U, Emeis CC (1984) FEMS Microbiol Lett 24:81
182. Scheurich P, Zimmermann U, Schnabl H (1981) Plant Physiol 67:849
183. Zimmermann U, Vienken J, Pilwat G (1982) Stud Biophys 90:177
184. Vienken J, Ganser R, Hampp R, Zimmermann U (1981) Physiol Plant 53:64
185. Vienken J, Zimmermann U, Ganser R, Hampp R (1983) Planta 157:331
186. Verhoek-Köhler B, Hampp R, Ziegler H, Zimmermann U (1983) Planta 158:199
187. Herman EB (1984) Agric Rep 3:25
188. Jacob H-E, Siegemund F, Bauer E (1984) Biol Zentralbl 103:77
189. Tempelaar MJ, Jones MGK (1985) Plant Cell Rep 4:92
190. Tempelaar MJ, Jones MGK (1985) Planta 165:205
191. Verhoek-Köhler B (1984) Erarbeitung biochemischer Parameter zur Bestimmung der physiologischen Integrität von Hybridzellen nach elektrisch induzierter Fusion. Thesis, Technical University, Munich

192. Saga N, Polne-Fuller M, Gibor A (1987) Proceedings of a workshop on the present status and future directions for biotechnologies based on algal biomass production. University Colorado Press, Boulder
193. Bates G, Gaynor J, Shetzhawat N (1983) Plant Physiol 72:1110
194. Bates GW, Hasenkampf CA (1985) Theor Appl Genet 70:227
195. Koop HU, Dirk J, Wolff D, Schweiger HG (1983) Cell Biol Int Rep Z:12
196. Koop HU (1984) Mitteilung Botanikertagung. Wien, p 72
197. Kohn H, Schieder R, Schieder O (1986) Plant Sci Lett 38:121
198. Hampp R, Steingraber M, Mehrle W, Zimmermann U (1985) Naturwissenschaften 72:91
199. Gaff DF, Ziegler H, Zimmermann U (1985) J Plant Physiol 120:375
200. Salhani N, Vienken J, Zimmermann U, Ward M, Davey MR, Clothier RH, Balls M, Cocking EC, Lucy JA (1985) Protoplasma 126:30
201. Griesbach RJ, Sink KC (1983) Plant Sci Lett 30:297
202. Scheurich P, Zimmermann U (1981) Naturwissenschaften 68:45
203. Pilwat G, Richter HP, Zimmermann U (1981) FEBS Lett 133:169
204. Ohno-Shosaku T, Okada Y (1984) Biochem Biophys Res Commun 120:138
205. Ohno-Shosaku T, Hama-Inaba H, Okada Y (1984) Cell Struct Function 9:193
206. Ohno-Shosaku T, Okada Y (1985) J Membr Biol 85:269
207. Podesta EJ, Solano AR, y Vediat LM, Paladini A Jr, Sanchez ML, Torres HN (1984) Eur J Biochem 145:329
208. Teissie J, Knutson VP, Tsong TY, Lane MD (1982) Science 216:537
209. Finaz C, Lefevre A, Teissie J (1984) Exp Cell Res 150:477
210. Blangero C, Teissie J (1983) Biochem Biophys Res Commun 114:663
211. Neumann E, Gerisch G, Opatz K (1980) Naturwissenschaften 67:414
212. Köhler G, Milstein C (1975) Nature 256:495
213. Stern NJ, Gamble HR (1984) Hybridoma technology in agricultural and veterinary research. Rowman and Allanheld, Totowa
214. Bischoff R, Eisert RM, Schedel I, Vienken J, Zimmermann U (1982) FEBS Lett 147:64
215. Hübner GE, Trawinski J, Zembrod A, Opitz U, Bödeker BGD, Hewlett G, Schlumberger HD (1985) Biochem Soc Bull 7:102
216. Karsten U, Papsdorf G, Roloff G, Stolley P, Abel H, Walther I, Weiss H (1985) Cancer Clin Oncol 21:733
217. Brown SM, Ahkong QF, Sage AD, Lucy JA (1986) Biochem Soc Trans (to be published)
218. Arnold WM, Zimmermann U (1983) German Patent DE 3323 425 C2, June 29
219. Arnold WM, Wendt B, Zimmermann U, Korenstein R (1985) Biochim Biophys Acta 813:117
220. Lo MMS, Tsong TY, Conrad MK, Strittmatter SM, Hester LD, Snyder SH (1984) Nature 210:792
221. Berg H (1982) Bioelectrochem Bioenerg 9:223
222. Kubiak JZ, Tarkowski AK (1985) Exp Cell Res 157:561
223. Richter H-P, Scheurich P, Zimmermann U (1981) Dev Growth Diff 23:479
224. Salhani N, Schnabl H, Küppers G, Zimmermann U (1982) Planta 155:140
225. Zimmermann U, Steudle E (1978) Adv Bot Res 6:45
226. Pinto da Silva P, Nogueira ML (1977) J Cell Biol 73:161
227. Pinto da Silva P, Shimizu K, Parkison C (1980) J Cell Sci 43:419
228. Hewison LA, Coakley WT, Tilley D (1986) Biochem Soc Trans (to be published)
229. Sowers AE (1983) Biochim Biophys Acta 735:426
230. Sowers AE (1983) Biophys J 41:361a
231. Sowers AE (1984) J Cell Biol 99:1989
232. Sowers AE (1985) Biophys J 47:519
233. Donath E, Arndt R (1984) Gen Physiol Biophys 3:239
234. Zimmermann U, Küppers G (1983) Naturwissenschaften 70:568

235. Küppers G, Zimmermann U (1983) FEBS Lett 164:323
236. Küppers G, Diederich KJ, Zimmermann U (1984) Z Naturforsch 39c:973
237. Broda HG, Zimmermann U (1986) Production of viable yeast hybrids by electrical discharge: the role of electrofusion during evolution (to be published)
238. Lockner DA, Johnston MJS, Byerlee JD (1983) Nature 302:28
239. Finkelstein D, Powell J (1970) Nature 228:759
240. Benz R, Zimmermann U (1980) Biochim Biophys Acta 597:637
241. Hui SW, Stewart TP, Boni LT (1981) Science 212:921
242. Steponkus PL, Stout DG, Wolfe J, Lovelace RVE (1985) J Membr Biol 85:191
243. Dimitrov DS, Li J, Angelova M, Jain RK (1984) FEBS 1908:398
244. Miller SL (1953) Science 117:528
245. Puchkova TV, Putvinskii AV, Vladimirov YuA, Parnev OM (1981) Biophys 26: 268
246. Zimmermann U, Benz R (1980) J Membr Biol 53:33
247. Benz R, Zimmermann U (1980) Biochim Biophys Acta 597:637
248. Benz R, Zimmermann U (1980) Bioelectrochem Bioenerg 7:723
249. Jain RK, Maldarelli C (1982) Ann N Y Acad Sci 404:89
250. Gallez D, Prevost M, Steinchen A, Sanfeld A (1983) Ann N Y Acad Sci 404:108
251. Sackmann E (1984) In: Chapman D (ed) Biological membranes, vol 5. Academic, London, p 105
252. Abidor IG, Chernomordik LV, Sukharev SI, Chizmadzhev YuA (1982) Bioelectrochem Bioenerg 9:141
253. Chernomordik LV, Sukharev SI, Abidor IG, Chizmadzhev YuA (1982) Bioelectrochem Bioenerg 9:149
254. Schwister K, Deuticke B (1985) Biochim Biophys Acta 816:332
255. Blok MC, van der Neut-Kok ECM, van Deenen LLM, de Gier J (1975) Biochim Biophys Acta 406:187
256. McLaughlin S, Harary H (1974) Biophys J 14:200

Subject Index

acetylcholine 111
adenosine 10
adrenoceptors, on blood cells 6
–, dynamic regulation 6
– and sympathetic nerve activity 6
– see also under alpha and beta
adrenergic neuron, defective, in hypertension 69
– stimulation, pulmonary circulation 146
adrenocortical hybrid cells 233, 234
afferents, sympathetic, autonomic reflexes 35
–, vagal 35, 37
age, baroreceptor sensitivity 29
–, chemoreflexes 34
aldosterone, plasma level 40
aldosteronism, primary 65
algal cells, giant 182
alpha adrenergic drugs 110
– adrenoceptors 10, 150
– –, responsiveness 64
amyl nitrite 18
angina pectoris 122
angiotensin II 70
– receptors 10
–, vascular capacitance 149
antibody, monoclonal, production of 234–236
antidiuretic hormone, reflex control 39–42
anxiety 52
aortic stenosis and cardiovascular reflex control 60, 61
arrest, circulatory 148, 151
arrhythmia 6
–, cardiac 64
–, respiratory 19, 20
arousal 27
ATP 111
atrial receptors 40, 114, 153
atriopeptin 41, 42
–, immunoreactive neurons 42
atrioventricular conduction and baroreflex 21
atrophy, multiple-system 56, 57
autonomic failure, chronic, definition 56
– –, progressive 10, 11
– –, reflex control in 56, 57
– hyperreflexia 72, 73
avidin-biotin binding 236
axon, local sympathetic 37

BaCl$_2$ 111
bacteria, electrofusion 186, 222, 223
baroreceptor(s), activation 30
–, aortic, discharge from 112, 120
– control, resetting 66
– deactivation 21, 59
–, distension of 123
–, dynamic action 24, 26
–, heart rate responses, hypertension 66
– in man, venous response 122, 123
–, methods of study 13–16
– and plasma K$^+$ 30
– reflex see baroreflex
–, sodium sensitivity 42, 43
–, threshold, increase in 68
baroreflex 11
–, abdominal circulation 130, 133, 141
– and ADH 40
– in aged subjects 29, 30
–, aortic 29
– arc 12
–, asymmetry of 65
–, buffer of blood pressure changes 23–25
–, cardiac vs. blood pressure components 66–68
–, carotid, transmural pressure 22
–, –, onto various effector systems 17ff.
–, central control, loss of 56, 58
–, – processing time 23
– on central blood volume 152
–, chronic absence 32
– control, selective 12
–, depressor response 21
–, emotional stress 53
– and exercise 30
– feedback, interruption of 54
– in hypertension 65ff.
–, impairment 65
–, intestinal vascular 135
– mechanisms, interference 62
– and muscle work 43ff., 49ff.
–, plasma catecholamines 8
– responsiveness, oscillation of 19
– sensitivity 66, 67, 71
–, – circadian rhythm 29
– – during exercise 16
–, splanchnic 139
–, splenic 139
– threshold 18

baroreflex, vagal component 26, 27, 31, 53
–, venomotor responses 112, 113, 120
beta adrenergic agonists 110, 111, 137
beta adrenoceptors 10, 37, 64, 70, 149, 150
– –, density 6
– – on lymphocytes 57
Bezold-Jarisch reflex 114
biotin 236
blastomeres, electrofusion of 236
blood flow, cerebral 55
– –, cholinergic mechanism 49
– – in contracting muscles 46
– –, coronary 3
– –, –, reduction in 61
– –, cutaneous, and blood pressure 4
– –, –, reflex control 27
– –, –, temperature regulation 4, 5, 47–49
– – in human limbs 3
– –, muscular 4
– –, –, upon exercise 4
– –, –, reflex changes 4, 25
– –, myocardial 3
– –, renal 35
– –, –, orthostasis 28
– pressure, arterial 29
– –, – and baroreceptors 31
– –, – and baroreflexes 21–23
– –, –, circadian variation 56
– –, – and emotions 51ff.
– –, – during exercise 31, 32
– – regulation, muscle work 43ff.
– – –, orthostatic 41
– – and splanchnic vascular resistance 28
– – variations 12, 13
– vessels, physiological classification 103
– –, structural changes, hypertension 71
– –, volume-pressure relationship 104
– volume, abdominal 143
– – change 155–158
– – –, catecholamine induced 149
– – –, pulmonary 144, 145
– – – and veins 124, 125
– –, intrathoracic 35, 36
– –, reflexes in hypertension 68
bradycardia 34, 62
bradykinin 63
– -forming enzyme 49
breakdown, electrical, experimental
   arrangement 190
–, –, formation of pores 248
–, –, intercellular exchange 201
–, –, mechanism of 243ff.
–, –, molecular mechanism 177
–, –, – rearrangements 247
–, –, organelles sequestration 241
–, –, permeability in 249
–, –, reversible and irreversible 178
–, –, –, kinetics of 245
– voltage 178, 179, 182, 217, 243, 244
– –, asymmetric 297
– – and cell radius 185, 186, 188ff.
– –, critical 246

– –, natural occurrence 239
– –, pulse length dependence 245, 247
bulbospinal pathways, inhibitory 11
bypass, cardiac 145, 147
–, cardiopulmonary 37

capacitance, vascular, abdominal and
   sympathetic activity 142, 143
–, –, –, methods of study 128, 129
–, –, assessment of 106, 116
–, –, change in 105, 106
–, –, –, extent and magnitude of 154, 155
–, –, concept of 104
–, –, control 155
–, –, hepatic, techniques 136
–, –, importance of 154ff.
–, –, perspectives 158
–, –, pulmonary 144–147
–, –, –, importance of 146
–, –, responses 117, 119, 120ff.
–, –, –, active 136, 141, 159
–, –, –, importance of 126, 127
–, –, –, lack of 126, 127
–, –, –, methods used 107–109, 116
–, –, –, misconceptions 126
–, –, total body, methods of study 147
–, –, – and catecholamines 149–151
–, –, –, reflex responses 149, 152–155
–, –, – systemic 147
– vs. resistance responses 158
– vessels 10, 103
– –, abdominal and sympathetic stimulation
   129–133
– –, active responses 116
– – and chemoreceptors 131–133
– –, function of 157
– –, functional effect 154
– –, human 6ff.
– –, sympathetic activity 159
capacity 105
cardiac catheter 3
– denervation 11
– –, lack of arrhythmias 64
– failure, reflex control 58, 59
– output 44, 46, 47, 50, 146, 147, 157
– – and baroreflex 22
– –, decrease 31, 32
– –, in hypertension 65
– –, measurement of 3
– pain 63
– receptors 153
– reflexes upon chemoreceptor activity 131
cardiocytes, atrial 41
cardiopulmonary baroreflexes and muscle
   contractions 43ff., 49ff.
– mechanoreceptors, techniques 16
– receptors 9, 125
– –, deactivation 43
– –, in hypertension 69
– –, effects on resistance vessels 24, 25, 27,
   28, 32, 33, 35ff.
– –, in volume regulation 40

cardiopulmonary reflexes 35–37
– –, study of 16, 17
– –, venous responses 114, 115
cardiovascular control, veins in 154
– disease, reflex control in 56ff.
– events, measurement of 3ff.
– reflexes, human, techniques of study
   13–17
– system, human 2ff.
carotid artery occlusion 15
– body tumors 73
– nerve stimulation 15
– sinus, smooth muscle in 15
– – syndrome 73
catecholamines and adrenoceptors 6
–, arteriovenous difference 8
–, clearance 8
– and hepatic circulation 137
–, plasma levels 70
–, splenic circulation 138
– and total vascular capacitance 149–151
– as transmitters 12
cell alignment 203, 204, 214, 231
– formation, in evolution 242
– fusion, filter technique 216
– –, high suspension densities 216
– –, methods 215, 216
– – – without electrode contact 206
– isolation, magnetical 217
– rotation, technique 235
– suspension density 211, 213, 214
cells, giant, production of 231
Chagas' disease 58
chemical breakdown 249
chemoreceptors in man, venous response 123
–, pulmonary 37
–, stimulation of, in man 123
chemoreflexes 45
–, abdominal circulation 131
– and age 34
–, cardiogenic hypertensive 63
–, intestinal vascular 135
– in man 34
–, venomotor response 113
chloroplasts, field-induced, release of 180
cholinergic vasodilator system 52, 53
circulation, abdominal, blood reservoir 142,
   143
–, –, importance of capacitance 140–143
–, –, reflexes from cardiac receptors 134
–, central, redistribution 40, 41
–, cutaneous, reflex control 27
–, hepatic 135
–, –, reflex responses 137, 141
–, human splanchnic, methods of study 139
–, – –, responses in 139
–, muscular, reflex 23–26, 52, 53
–, musculo-cutaneous 107ff.
–, pulmonary, capacitance response 154
–, –, methods of study 144, 145
–, –, responses of 146
–, renal 47, 50

–, –, reflex control 26, 28
–, –, – effects 36
–, splanchnic 47, 50, 55, 150, 152
–, –, change in capacitance in 155
–, –, reflex regulation 28
–, splenic, reflex responses 138, 139, 141
cocaine 112
cold pressure test 38
– showers 53
compliance 104, 105
–, abdominal 143
–, hepatic vascular 136
–, pulmonary vascular 145
–, total body 149, 156, 159
compression forces, mechanical 244ff.
conductivity, external specific 183–186
–, hydraulic 238
–, internal 183–185
– term 183
constant flow perfusion 145
coronary artery disease, sudden death 64
– –, injection of contrast media 62
– –, reflex dilatation, vagal 34
– –, spasm 64
– bypass surgery 63
cuffs, congesting 17
Cushing response 153
cyanide 36
cyclic AMP 61

delay coils 113
denervation, sinoaortic 18
depressor drugs 15
diabetes 56, 57
dielectric levitation 206–209
dielectrophoresis 180, 201, 227
–, cell fusion with 201ff.
–, – – without 215
–, definition 201–203
– of liposomes 222
dielectrophoretic force 207, 214
digitalis and cardiovascular reflexes 59, 60
dipole-dipole interaction 203
discharge technique 227
discharges, electrical, atmospheric 240
dispase 188, 196, 197, 199, 232
distensibility curves of veins 104
diving reflex 34, 35
DNA transfection 187
– –, electropermeabilization 199
– –, various methods 199, 200
– transfer 226
– –, electrically induced 196
Doppler signal 3
dysreflexia, autonomic 72

eggs, electrofusion 236, 237
electrical breakdown see breakdown
– discharges see discharges
– field see field
electro-acoustic fusion 218–220
electrocompression, membrane 244

electrofusion, animal cells 231ff.
–, bacteria 22, 223
–, blastomeres and eggs 236, 237
–, chambers used 210–212, 226
–, critical field strength 183, 185
–, electromechanical model 243–245
–, enzyme pretreatment 232, 233
–, freezing and thawing 241
–, fungi 224
–, hybridoma 234–236
–, kinetics of 238
–, lightning-induced 240, 241
–, lipid bilayer membranes 220, 221
– in membrane research 237–239
–, plant protoplasts 226–230
–, role of, in evolution 239–242
– standard technique 201, 209, 210, 224,
    227
– techniques 177
–, universality of method 219
–, yeast 224–226
electroinjection 182
–, molecules into cells 190ff.
electron microscopy, freeze-fracture 238, 241
electropermeabilization, free suspended cells
    187
–, pulse properties 183
–, reversibility 177
electrophoresis 202
electroporation 181, 182
electrorotation experiments 196
electrotransfection 181
emotion, sympathetic response 12, 14
emotional fainting, blood pressure 55
– stimuli, venoconstriction 126
– stress, cardiovascular responses 51–54
endocytosis 181, 184, 185
eosin, electroinjection 191–193
ergoreceptors and blood pressure control
    43ff., 49ff.
ethidium bromide, uptake 198
exercise 143
– and arterial baroreflexes 16, 30–33, 43ff.
–, circulating epinephrine during 33
– and plasma catecholamines 10, 11
–, resistance changes during 58, 59
–, sympathetic activity 12
–, – response, hypertension 70
–, upright vs. supine 50
–, venoconstriction 125
extracellular fluid, reflex regulation 40

fainting with blood loss 54
–, blood pressure in 54
–, orthostatic 55
fibroblasts 234
Fick principle 3
field, electric, cell fusion 202, 207, 209, 213–
    215, 217, 224
–, –, cusped 207
–, –, natural production 239
–, –, external pulse 177

–, –, pulses, DNA transfection 182, 183
–, –, – and temperature 182
–, –, rotating 235
–, –, strength, critical 183, 185
–, –, – and frequency 189
–, –, –, supracritical 178, 179
filling pressure, cardiac 17
filtration rate, glomerular 3
Friend cell 180, 181, 187, 206, 229, 232
fused cells, production of 211, 212
fusion chamber, rotational 212

gene transfer, electrically induced 199
giant cells, production of 237
gravitation and abdominal circulation 156
gravity 49
Guillain-Barré syndrome 27

habituation 124
heart failure, congestive, disturbed reflex
    control 58, 59
– rate 30, 39
– –, aortic baroreceptors 29
– –, carotid baroreflexes 18
– –, reflex changes 31, 32
– –, response to standing 56
– transplantation 46
– –, supersensitivity after 11
heat stress 47, 48, 50
hemorrhage 38, 40, 124, 139, 143, 153,
    157
– in anesthetized patients 55
–, blood pressure in 54, 55
hepatic circulation, see circulation 135
– sphincter mechanisms 135
Hering-Breuer reflex 37
high suspension density technique 216
– voltage discharge technique 190
histamine receptors 10
human limbs, capacitance responses 120ff.
hybrid formation 225
hybridoma cells 188, 206, 211, 212, 216,
    226
– –, production 235
5-hydroxytryptamine see serotonin
hypercapnia 34, 113, 115, 123, 133
hypertension 22, 30, 34, 38, 43, 70, 71
–, blood pressure change 13
–, cardiovascular reflex events 65ff.
– and catecholamines 8, 9
–, norepinephrine release 69, 70
–, rat 42
–, reflectory 63
–, secondary 65
–, vessel structure 71
hyperventilation 55
hypocapnia 133
hypotension, orthostatic 56, 57
–, sympathotonic 57
hypothalamus, temperature regulation 12
hypoxia 113–115, 123, 139, 153
–, pressor responses 34

indicator-dilution method 3
– –, pulmonary circulation 145
infarction, myocardial 58
–, –, cardiovascular changes 61
–, –, reflex events 64
inspiration, constriction of veins 124
intermediolateral horn cells, inputs to 11, 12
intestinal veins, reflex responses 135
– –, sympathetic stimulation 134, 135
– vessels, methods of study 134
ischemia, cerebral 155
–, myocardial and cardiovascular reflexes 61
isoprenaline 110, 111, 150, 151

kanamycin 223
karyogamy 224, 225
kidney, arterial stenosis 65
–, autoregulation 28

lactic acid 35
laser beams, cell accumulation by 219
– microbeams, gene transfection 200
lateral diffusion coefficient 239
latex 194
leukocytes, beta-adrenoceptors 6
levitation, dielectric 206–209
–, electrostatic 207
–, magnetohydrostatic 207
Leydig cells 233
lightning discharges, simulated 240, 241
lipid bilayer membranes 182, 220, 221
liposomes, electrofusion 222
liver, see hapatic 135
lobeline 37
local anaesthetics 246
locus coeruleus 56
lymphoblasts 233
lymphocyte, cell uptake 180
lymphocytes, electrofusion 186, 187, 190
–, electroinjection 192

magnetic blood cell 216, 217
magneto-electro-fusion 217
mammalian cells, electrofusion 186
membrane asymmetry 181, 184, 185
– capacitance 183, 188
– capacitor, electrical breakdown of 177
– compression 244–246
– contact, chemically induced 215
–, lipid distribution 249, 250
– permeability increase 179
–, pore formation 178, 179
– potential, changes by electric field 177, 179
– – in electrofusion 184, 186, 187
– –, frequency dependence 188
– –, intrinsic 187
membrane potential, radius dependence of 188ff.
membrane proteins, conformational changes 248
–, relaxation time 183–186

– resistance, in changes 248
–, induced structural defects 249
– surface potentials 187
– thickness 178, 245
mental activity and veins 124
– stress, blood pressure 52, 53
mesophyll protoplasts, rounding-off 238
microneurography in man 5, 45
mitochondria, incorporation into cells 241
mitral stenosis 60
mobility, lipid molecules 181
muscarine receptors 10
muscle contraction, cardiovascular effects 44, 45, 50
–, venous reflex responses 120
– tissue, capacitance responses 120
– veins, sympathetic innervation 126
musculocutanous circulation 107ff.
– –, size of 127
myeloma cells 220
– –, conductivity 185
– –, electrofusion 179, 186, 187, 190, 234
– –, fusion with lymphocytes 234–236
myocardium, refractory period 64

Na^+-K^+-ATPase, endogeneous inhibition 69
natriuretic factor, atrial 41, 42
natriuresis 40, 42
nausea 61
neck suction 18, 19, 21, 22, 53, 65, 66
negative pressure application 124
– –, lower body 69, 139
nitroglycerine 15, 60
nitroprusside sodium 8, 15, 122
noradrenaline receptor affinity 112
noradrenergic activity to muscle 25
norepinephrine clearance 7, 8
– in CSF, hypertension 70
– outflow from organs 8, 10
–, plasma 41, 52, 56–58
–, –, aging 30
–, – concentration 7, 8, 42, 43
– release 10

occluded limb technique 116, 122
oocytes, electrofusion 236
organelles, incorporation into cells 200
orthostatic response 36, 124, 127
– –, cardiovascular 17, 20
– stress 55
– –, ventricular mechanoreceptors 21
osmoreceptors 40
osmotic processes 181
oxygenator 147

pain, cardiac 63
parasympathetic overactivity 61
parkinsonism 56
pear chain formation 203, 218
peptidoglycan 223
phenyl diguanide 36

phenylephrine 8, 15, 23, 31, 34, 53, 60, 150
–, baroreceptor stimulation 18
pheochromocytoma 6, 65
phospholipase 233
plant protoplasts 186, 187
– –, alignment 227
– –, fusion 226–230, 240
– –, hydraulic conductivity 238
– –, organelles, release 194
– –, regeneration of 229, 230
plasma flow, renal 3
– potassium 30
– volume, reflex regulation 39ff.
plasmid DNA 196, 199, 200
plasmogamy 225
platelets, alpha-adrenoceptors 6
plethysmography 3, 7, 116, 121, 134, 136,
138
Poiseuille relationship 110
postural reflexes 37, 38, 41
– stress 156
pressor response 43, 45, 46, 71
pressure approach, disjoining 246
–, cardiac filling 51, 58
–, carotid sinus 36
– changes, pulmonary 145
– – in vascularly isolated limb 122
– curve, carotid sinus 30
– -diameter relationship 121
– difference, hydrostatic 244
– drugs 15
– –, reflex changes 149
–, intracerebral 115, 153
–, mean circulatory filling 148, 149, 157
– to neck 13, 14
–, neck chamber 30, 31
–, negative to lower-body 16, 26, 28, 36, 41
–, right atrial 157
– ventilation, positive 156
– -volume curves 7
pronase 232, 233
propranolol 153
prostacyclin 63
prostaglandins, chemosensitive fibers 63
psychogenic stimuli and veins 124, 126
pulmonary receptors 115, 124
purine receptors 10

receptors, cardiac, and aortic stenosis 61
–, metabolic in muscle 45, 46
–, muscular, multimodal 44, 45
– for pain 45
–, ventricular, distribution 61
–, –, sensitivity and activation of 61, 62
red blood cells, chain formation 204, 205, 231
– – –, drug loading 194
– – –, electrical breakdown 194
– – –, electrofusion of 232
– – –, electropermeabilization 181, 184,
185
– – –, fusion of 231
– – –, giant 231

– – –, ghosts, electroinjection 190, 191
– – –, leaks induced 249
– – –, incorporation of magnetic particles
194, 195
– – –, magnetically coated 216
– –. –, osmotic injection method 200
– – –, turkey 205
reflex 37
– apnea 34
– bradycardia 29
– capacitance responses, lack of 126
–, cardioinhibitory 64
– responses, abdominal veins 130–134, 141
– –, total vascular capacitance 149, 152–
154, 155
– stimulation, pulmonary circulation 146
– –, vascular capacity 155
reflexes, cardiovascular 158
–, hepatic vessels 137
relaxation time 188, 202
renin activity, plasma 39
– -angiotensin system 41
–, reflex regulation 37–39, 68
– release 37, 38
– –, inhibition of 42
– secretion, nonneural mechanisms 37
– –, sympathetic stimulation 38
resealing process 193, 194, 235
– –, temperature dependence 197
resistance, hepatic outflow 142
–, vascular 108, 110
–, –, and catecholamines 9
–, –, cerebral 52
–, –, changes 117
–, –, in pulmonary circulation 144, 145
–, –, sympathetic responses, abdominal
circulation 129
–, –, total systemic 23, 27, 31, 54, 65, 109
– vs. capacitance curves 152
– vessels 10, 46, 103, 126
– – and chemoreceptors 131–133
– –, functional role 154
– –, renal 28
– –, splanchnic 28, 34–36
– –, sympathetic control 3
respiration, changes in baroreflex 19, 20
–, reflex inhibition 37
respiratory center, central command 45, 46
– oscillator, central 20
reticular nucleus, paramedian 11

salt-water-balance, reflexes 39ff.
sequestration, electrical 179
serotonin 63, 111
– receptors 10
– as transmitter 12
Shy-Drager syndrome 56
sickle cells 205
sinus nerves, electrical stimulation 21, 23
– node, sensitivity 68
sleep 124
– and baroreflex 29

smooth muscle, vascular, sensitivity 68
sodium content, intralymphocytic 43
— intake 43, 58
— and renal cardiovascular control 42, 43
—, total body, regulation 40
— transport in hypertension 69, 70
solitary tract nucleus 35, 56
— — —, connexions and cardiovascular
    control 11, 12
spaceflight 41
spheroplasts 223
spinal anesthesia 55
— cord, intermediolateral lesions 56
— — lesions, autonomic system 72, 73
spinothalamic tract, neuronal activity 63
splanchnic circulation, capacitance in 128
spleen, contraction of 155, 158
— vasculature, methods of study 137
stability analysis, linear 243, 245
standing, cardiovascular responses to 20, 21
stress 10, 12
stretch receptors, pulmonary 19
substance P 12
sucrose, permeability 191
sudomotor fibers 49
— impulses 5
surface charges, accumulation of 243
sweat glands, enzyme release 49
sympathectomy 111, 112
sympathetic activity 34
— — see also s. discharges, etc.
— —, abdominal vessels 142, 143
— — and blood pressure changes 23, 25
— —, capacitance vs. resistance vessels 133,
    134
— —, cutaneous nerves 5, 27, 35
— —, multiunit recordings 5, 71, 72
— — in muscle nerves 24, 25, 27, 35, 36
— — and plasma norepinephrine 8–10
— —, renal reflex 38, 40
— —, venomotor response 111–113, 117ff.,
    155
— afferents 35, 61
— cholinergic fibers 62
— control, descending 11
— discharges, differentiated pattern 9–12
— —, multiunit recordings 24, 26, 27, 45, 49
— dissociation 126
— function and sodium 43
— nerves, juxtaglomerular 37
— neuronal populations 49
— noradrenergic activity 48
— outflow during exercise 46, 49, 50
— —, fainting 54
— — in hypertension 71
— —, orthostasis 49
— —, regulation of 10–13
— —, renal 58
— — to sinus node 18
— — after spinal lesions 72, 73
— — in stress 52, 53
— overactivity 61

— reflexes, combined with parasympathetic
    53, 64
— stimulation and hepatic circulation 136
— — intestinal veins 134
— —, splenic vessels 138
— —, capacitance response 155, 156, 159
— vasodilator fibers 54
syncope during exercise 60, 61

Takayasu's disease 73
temperature changes and venoconstriction 125
tensor, Maxwell stress 209
thermodilution method 3
thermoreceptors, peripheral and central 47
thermoregulation, reflex events 47ff.
thermosensitivity, central 111
thin film models 243, 245
thrombolysis, intracoronary 64
thymocytes, resealing 193
tilting 17
trigeminal reflexes 34, 35
trinitroglycerine 23
trypsin 196, 232, 233

ultrasound, cell concentrating 218, 219
uptake, field induced 180, 181
urea 191

vacuoles and electrofusion 230
vagal activity, cardiac 18, 19
— —, recording of 5
— afferents 35, 61
— —, electrical stimulation 37
— C-fibers 64
— control, cardiac 14
— nuclei, cardiac 11
— neuropathy 56
Valsalva maneuver 53
vascular bed, abdominal, studies in 128ff.
— capacitance see capacitance, vascular
vasoconstriction, neurogenic 59
— and plasma norepinephrine 9
—, splanchnic 28, 35, 36
vasoconstrictor fibers, sympathetic 4, 5, 48,
    49
vasodilation in hypertension 71
—, local 4, 10
—, reflectory 60–62
vasodilator fibers, cholinergic 52
— —, sympathetic 4
— impulses 48
vasomotor center, central command 45, 46
— —, disinhibition 27, 50, 54, 58
— —, inhibition 35
— tone 12
vasopressin 35, 40, 41, 57, 58
vasovagal syndrome 54
— syncope 55
veins, blood volume 103, 106, 110
—, cutaneous effects of temperature 110–112
—, —, studies in 107ff.
—, diameter measurement 7

veins, diameter, radiographic method   107
–, hepatic, and sympathetic action   136
–, human limbs   120ff.
–, intestinal, nervous stimuli   155
–, measurement of volume changes   107, 108
–, pressure-flow relations   108
–, – gradients   108
–, regional responses, methods   116ff.
–, resistance-volume relation   110
–, role in cardiovascular control   154
–, smooth muscle in   102, 126
–, stress relaxation   149
–, sympathetic effects upon   108, 110–113,
    117ff., 155
–, see also venous etc.   112
–, volume changes   6, 7, 117–119
venoconstriction, functional importance   127
–, stimulus for   123
venomotor responses, studies of   104, 108, 109
– tone measurement   7
venous pressure   7
– reflex responses   112ff., 120
– – –, in humans   122ff.
– responses, CNS stimulation   120
– return   157
– segments in humans   121

– –, studies of responses   108, 109
– smooth muscle, effect of temperature
    111
– system, reflex regulation   28
– tone, history   102, 103
ventricular receptors   114
– –, activation of   64
veratridine   114
–, intracoronary   134
vesicles, drug incorporation   200
–, formation   238
vessels, muscular, baroreflexes on   23–26
volume see blood volume   145
– changes, active versus passive   155, 156,
    159
– – in human limb veins   121
–, unstressed   104, 105
– -pressure curves   104–106, 123
vomiting   61

weight changes   117–119
Workmann-Reynold effect   239

Xenopus eggs, electrical breakdown   185

yeast protoplasts, fusion of   224–226